Construction Economics

Construction Economics provides students with the principles and concepts underlying the relationship between economic theory and the construction industry. The *new approach* adopts an argument that economics is central to government initiatives concerning sustainable construction.

This edition has been revised to explain the effects of the current economic crisis on the construction industry. In addition, sections relating to less developed countries, the economics of sustainable development and theories relating to a firm's bid strategy have all been rewritten. With new data, examples, initiatives, readings, glossary items and references, the third edition of this established core text builds on the strengths of the previous edition:

- a clear and user-friendly style
- use of a second colour to highlight important definitions and formulae
- regular summaries of key points
- a glossary of key terms
- extensive use of tables and figures
- extracts from the academic journal *Construction Management and Economics* to consolidate and prompt discussion
- reviews of useful websites.

This invaluable textbook is essential reading across a wide range of disciplines from construction management and civil engineering to architecture, property and surveying.

Danny Myers is a lecturer and researcher based in the Department of Construction and Property at the University of the West of England, UK, and visiting lecturer at the University of Bath, UK.

Construction Economics
A new approach
Third Edition

Danny Myers

Routledge
Taylor & Francis Group

LONDON AND NEW YORK

First edition published 2004

Second edition published 2008

This edition published 2013
by Routledge
2 Park Square, Milton Park, Abingdon, Oxon, OX14 4RN

Simultaneously published in the USA and Canada
by Routledge
711 Third Avenue, New York, NY 10017

Routledge is an imprint of the Taylor & Francis Group, an informa business

British Library Cataloguing in Publication Data
A catalogue record for this book is available from the British Library

Library of Congress Cataloging-in-Publication Data
Myers, Danny.
Construction economics : a new approach / Danny Myers. — 3rd ed.
    p. cm.
Includes bibliographical references and index.
1. Construction industry. 2. Construction industry—Management. I. Title.
HD9715.A2M94 2013
338.4'7624—dc23   2012037199

ISBN13: 978-0-415-52778-1 (hbk)
ISBN13: 978-0-415-52779-8 (pbk)
ISBN13: 978-0-203-38443-5 (ebk)

Typeset in Sabon by DSM Partnership

Printed and bound in Great Britain by
TJ International Ltd, Padstow, Cornwall

Contents

LIST OF TABLES AND FIGURES

Tables

Figures

ACKNOWLEDGEMENTS

Although the book cover implies this is all my own work, it was not achieved alone. The staff and students that I have worked with have subtly, and not so subtly, made their mark. In particular, research undertaken with Kevin Burnside, a supportive colleague over a number of years, has helped to improve the clarity of the process of bidding for work in a construction industry dragged down by financial crisis and subsequent recession. The fruits of our discussion should be evident in Chapters 7 and 8.

Next, to bring life to a manuscript is no easy task and this would have been impossible without the graphic design skills, and patience, of Chris Wade, who has added some new artwork to this edition.

A major concern from the outset was to create a text that was easy to read and use. This is some challenge in the subject area, and again it has been a pleasure to be supported by the editing skills of Paul Stirner. His perceptive, informed and detailed line of enquiry has added greater clarity and rigour to the text, making the completed manuscript as accessible as possible.

Finally I should acknowledge the support of the publishers, in particular Brian Guerin who co-ordinates the team involved in commissioning a new edition and taking it through to publication, and you would be surprised at the number of departments that involves.

I hope you find the finished product interesting and relatively easy to use. If any errors or omissions remain, I apologise for these in advance, and would be grateful for correspondence bringing them to my attention. Enjoy the book!

Danny Myers
September, 2012

1 An Introduction to the Basic Concepts

This book is written for students from many backgrounds: architecture, surveying, civil engineering, mechanical engineering, structural engineering; construction, project or estate management, property development, conservation and, even, economics. Economics students may find it possible to skip over some of the standard analysis, but should be forewarned that in many ways construction is quite distinct from other sectors of the economy. An important aim of this text is to draw out these distinctions and clarify the unique nature of the industry. In this first chapter we begin to outline the main characteristics of firms involved in construction markets, introducing the complexity of the construction process and diversity of activities. As the chapter develops you will sense that there are a number of possible ways to describe the construction industry. Table 1.1 identifies a range of activities that can be included in a broad definition of the industry. By contrast, Table 1.2 (see page 11) divides the construction process into a number of professional stages and Table 1.3 (see page 19) outlines a simple classification system that narrowly defines the industry as firms that just construct and maintain buildings and infrastructure.

Table 1.1 The construction industry – broadly defined

The key actors include:

✓	Suppliers of basic materials, e.g. cement and bricks
✓	Machinery manufacturers who provide equipment used on site, such as cranes and bulldozers
✓	Manufacturers of building components, e.g. windows and doors
✓	Site operatives who bring together components and materials
✓	Project managers and surveyors who co-ordinate the overall assembly
✓	Developers and architects who initiate and design new projects
✓	Facility managers who manage and maintain property
✓	Providers of complementary goods and services such as transportation, distribution, demolition, disposal and clean-up

Source: Adapted from Manseau and Seaden (2001: 3–4)

The aim of the text is to demonstrate that underlying the construction process, from conception to demolition, is a lot of useful economics. As a discipline, economics should not be regarded as solely related to the appraisal of costs. The subject matter is far broader, and this text introduces a number of branches of economic theory. These have been selected to provide fresh insights into the performance of construction firms and a greater understanding of the need for a more holistic approach if the industry is to contribute to an efficient and sustainable economy in the future. These economic ideas should inform the work of all professionals concerned with the construction and maintenance of buildings and infrastructure – and, in particular, the way that they think.

The next section explains some of the key concepts used by economists. Further clarification is provided in the glossary at the back of the book, where all the economic terms highlighted in the text and other concepts and ideas relevant to construction economics are defined.

INTRODUCING CONSTRUCTION ECONOMICS

Construction economics – like pure economics, its mainstream equivalent – is concerned with the allocation of scarce resources. This is far more complex than it at first appears. Many of the world's resources (factors of production such as land, labour, capital and enterprise) are finite, yet people have infinite wants. We are, therefore, faced with a two-pronged problem: at any point in time there is a fixed stock of resources, set against many wants. This problem is formally referred to as **scarcity**. In an attempt to reconcile this problem, economists argue that people must make careful choices – choices about what is made, how it is made and for whom it is made; or in terms of construction, choices about what investments are made, how these are constructed and on whose behalf. Indeed, at its very simplest level, **economics** is 'the science of choice'.

When a choice is made, therefore, some other thing that is also desired has to be forgone. In other words, in a world of scarcity, for every want that is satisfied, some other want, or wants, remain unsatisfied. Choosing one thing inevitably requires giving up something else. An opportunity has been missed or forgone. To highlight this dilemma, economists refer to the concept of **opportunity cost**. One definition of opportunity cost is:

> the value of the alternative forgone by choosing a particular activity.

Once you have grasped this basic economic concept, you will begin to understand how economists think – how they think about children allocating their time between different games; governments determining what their budgets will be spent on; and construction firms deciding which projects to proceed with. In short, opportunity costs enable relative values to be placed on all employed resources.

This way of thinking emphasises that whenever an economic decision is made there is a **trade-off** between the use of one resource for one or more alternative uses. From an economic viewpoint the value of a trade-off is the 'real cost' – or opportunity cost – of the decision. This can be demonstrated by examining the opportunity cost of reading this book. Let us assume that you have a maximum of four hours each week to spend studying just two topics – construction economics and construction technology. The more you study construction economics, the higher will be your expected grade; the more you study construction technology, the higher will be your expected grade in that subject. There is a trade-off, between spending one more hour reading this book and spending that hour studying technology. In this example there is a fixed trade-off ratio. In practice, however, some people are better suited to some subjects than others and the same thing can be applied to resources. As a general rule, therefore, resources are rarely equally adaptable to alternative projects.

In construction, or any other economic sector, it is rare to experience a constant opportunity-cost ratio, in which each unit of production can be directly adapted to an alternative use. It is far more usual in business trade-off decisions to see each additional unit of production cost more in forgone alternatives than the previously produced unit. This rule is formally referred to as the **law of increasing opportunity costs**. This can be illustrated with the 'guns or butter' argument – this states that, at any point in time, a nation can have either more military goods (guns) or civilian goods (butter) – but not in equal proportions. For example, consider the hypothetical position in which all resources in the first instance are devoted to making civilian goods, and the production of military goods is zero. If we begin production of military goods, at first production will increase relatively quickly, as we might find some engineers who could easily produce military goods and their productivity might be roughly the same in either sector. Eventually, however, as we run out of talent, it may become necessary to transfer manual agricultural labour used to harvesting potatoes to produce military goods – and their talents will be relatively ill-suited to these new tasks. We may find it necessary to use fifty manual labourers to obtain the same increment in military goods output that we achieved when we hired one sophisticated engineer for the first units of military goods. Thus the opportunity cost of an additional unit of military goods will be higher when we use resources that are inappropriate to the task. By using poorly suited resources, the cost increases as we attempt to produce more and more military goods and fewer and fewer civilian goods.

The law of increasing opportunity costs is easier to explain using a **production possibility curve**. Using these curves, it is possible to show the maximum amount of output that can be produced from a fixed amount of resources. In Figure 1.1 (see page 4) we show a hypothetical trade-off between units of military goods and civilian goods produced per year. If no civilian goods are produced, all resources would be used in the production of military goods and, at the other extreme, if no military goods are produced, all resources would be used to produce civilian goods. Points A and F in Figure 1.1 represent these two extreme positions. Points B, C, D and E represent various other combinations that are possible. If these points are connected with a smooth curve, society's production possibilities curve is obtained, and it demonstrates the trade-off between the production of military and civilian goods. These trade-offs occur on the production possibility curve. The curve is bowed outwards to reflect the law of increasing opportunity cost. If the trade-off is equal, unit for unit, the curve would not bow out, it would simply be a straight line. Other interesting observations arising from the production possibility curve are shown by points G and H. Point G lies outside the production possibility curve and is unattainable at the present point in time, but it does represent a target for the future. Point H, on the other hand, lies inside the production possibility curve and is, therefore, achievable, but it represents an inefficient use of available resources.

Figure 1.1 The trade-off between military goods and civilian goods

Points A to F represent the various combinations of military and civilian goods
that can be achieved. Connecting the points with a smooth line creates the production
possibility curve. Point G lies outside the production possibility curve and is unattainable
at the present time; point H represents an inefficient use of resources at the present
time.

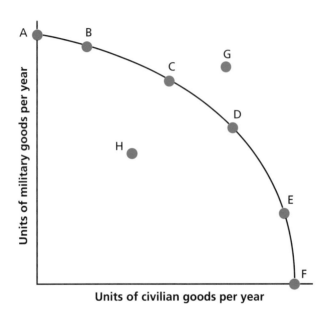

There are a number of assumptions underlying the production possibility curve.
The first relates to the fact that we are referring to the output possible on a yearly
basis. In other words, we have specified a time period during which production
takes place. Second, we are assuming that resources are fixed throughout this time
period. To understand fully what is meant by a fixed amount of resources, consider
the two lists that follow, showing (a) factors that influence labour hours available
for work and (b) factors that influence productivity, or the output per unit of input.

FACTORS INFLUENCING LABOUR HOURS AVAILABLE FOR WORK

The number of labour hours available for work depends on the nature of human
resources in society. This is determined by three factors:

- the number of economically active people that make up the labour force – this
  depends on the size of the population and its age structure, as children and
  retired persons will be economically inactive
- the percentage of the labour force who then choose to work
- prevailing customs and traditions (such as typical length of the working week,
  number of bank holidays, etc.).

FACTORS INFLUENCING PRODUCTIVITY

There are a number of factors influencing the productivity of an economy or sector of the economy:

- the quantity and quality of natural and man-made resources
- the quality and extent of the education and training of the labour force
- the levels of expectation, motivation and wellbeing
- the commitment to research and development.

The third and final assumption that is made when we draw the production possibility curve is that efficient use is being made of all available resources. In other words, society cannot for the moment be more productive with the present quantity and quality of its resources. (The concept of efficiency is examined more closely in Chapters 2, 5, 6, 7 and 8.)

According to several government reports (Egan 1998; NAO 2001, 2005 and 2007; Cabinet Office 2011), given the existing level of resources in construction it should be possible to increase productivity by at least 10 per cent. In other words, a production possibility curve representing all construction activities could be pushed out to the right, as shown in Figure 1.2 (see page 6). Several common sets of problem are identified as the root cause of this inefficiency. First, the industry demonstrates a poor safety record and an inability to recruit good staff. Second, there appears to be no real culture of learning from previous projects, and no organised career structure to develop supervisory and management grades. Third, concern is expressed about the poor level of investment into research and development that restricts the industry's ability to innovate and learn from best practice. The fourth, and possibly most worrying, problem is the fact that technology (in the sense of IT, innovation, prefabrication and off-site assembly) is not used widely enough across the construction sector.

Another plausible scenario suggested by the production possibility curve approach is that the construction industry may at present be working within the boundary of its production curve (say, point A in Figure 1.2). In which case, an increase in output could be simply achieved by greater efficiency. Supply constraints need to be reduced, the problems identified by the government reports resolved, and the factors generally acknowledged to increase productivity (listed above) must be addressed to achieve the full potential of the industry. Both these scenarios are shown in Figure 1.2 and they support the idea that the level of productivity in the construction industry needs to improve.

In very general terms, therefore, the study of economics (and construction economics) is concerned with making efficient use of limited resources to maximise output and satisfy the greatest possible number of wants. In short, the basis of the subject rotates around the concepts of choice, scarcity and opportunity cost.

Figure 1.2 Increasing output and the production possibility curve

In this diagram we show two scenarios: (a) improved productivity shifts the entire production possibility curve outwards over time; (b) output can be achieved more efficiently by moving to a position of full potential on the actual production possibility curve.

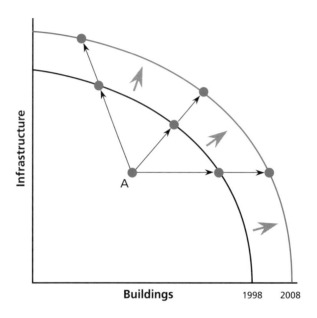

In modern society, economics is involved in all activities leading to the production of goods and services. Consequently a range of specialisms have evolved out of mainstream economics, such as transport economics, health economics, business economics, financial economics, agricultural economics, labour economics, international economics and, even, ecological economics. Hence it is not particularly surprising that many students in the twenty-first century are expected to read something called construction economics as part of their degree course. What is surprising, however, is that other vocationally oriented degrees do not have a similarly developed economics specialism. For example, students reading for degrees in catering, sports and leisure, publishing, retailing or computing do not benefit from a range of specialised literature in economics.

The reasons usually stated for construction warranting its own specialised economics is accounted for by the sheer size of the industry, its profound contribution to a nation's standard of living and its products' unique characteristics. Put very simply, the industry has five distinct qualities.

- The physical nature of the product is large, heavy and expensive.
- The construction industry is dominated by a large number of relatively small firms, spread over a vast geographical area.
- Demand for activity within the industry is directly determined by the general state of the economy as a whole.

- The method of price determination is unusually complex due to the tendering process used at various stages.
- Most projects can be considered as a 'one-off', as there is usually some defining quality that makes them in some ways unique.

These qualities alone have justified a number of dedicated academic publications. In 1974 the first edition of Patricia Hillebrandt's *Economic Theory and the Construction Industry* was published. Subsequently several other titles have appeared – for details see the reference section at the back of this book – in particular the two-part text co-authored by Ive and Gruneberg (2000) and the edited volume by Gerard de Valence (2011). In 1982, *Construction Management and Economics* a specialist refereed journal began to report on research contributing to the new subject specialism. This journal is published monthly and more than 1800 papers have appeared; several of them are drawn upon to support this text. Some extracts have been selected from these papers as case study readings to consolidate the three sections of this textbook. Another relevant academic journal is *Building Research and Information*. This has an interdisciplinary focus, with linkages made between the built, natural, social and economic environments. Consequently many of the papers in this journal contribute to our understanding of how buildings and infrastructure impact on ecology, resources, climate change and sustainable development; appropriately several examples are cited as references.

Alongside these academic developments, there has also been a succession of government reports investigating the problems of the construction industry (for example, see Latham 1994; Egan 1998; National Audit Office 2001; Fairclough 2002; HM Government 2008; IGT 2010; Cabinet Office 2011). These reports have highlighted the inefficiency caused by the sheer scale and complexity of the construction industry. A recurring recommendation is the need for the construction process to be viewed in a holistic way by a multidisciplinary team. This reflects the fact that construction draws knowledge from many areas, and an important but undervalued area is economics. Indeed, it is commonly observed that far too many projects run over budget and are delivered late, with a general disrespect for the client. Clearly it should not be acceptable for construction projects to fail cost-wise, time-wise or client-wise. An authoritative study by Professor Flyvbjerg (2003: 16–26) of 258 major public transport infrastructure projects constructed across Europe, USA, Japan and developing countries between 1927 and 1998 suggests that on average costs overrun by approximately 30 per cent, deadlines are missed by as much as ten years and the expected level of demand fails to meet targets by around 40 per cent. Worryingly the cost, time and quality dimensions continue to be problematic right up to the current day.

Each of the construction economics texts that have been published to date conveys a slightly different emphasis. For example, Hillebrandt (1974, 2000) defines construction economics as the application of economics to the study of the construction firm, the construction process and the construction industry. Whereas the preference of Ive and Gruneberg (2000: xxiii) is for a slightly less orthodox approach, adapting traditional economic models to capture local circumstances even if that means losing the ability to generalise about the economy at large. As a result, there is no coherent conceptual consensus about what constitutes the precise nature of construction economics. As George Ofori (1994: 304) bluntly concluded in his seminal review of the subject: 'Construction economics cannot be regarded as a bona-fide academic discipline. It lacks a clear indication of its main concerns and content.' A situation that de Valence (2011: 1) suggests still exists today.

The purpose of this text is to address this lack of consensus and make the case for a coherent economic vocabulary. The crux of the argument for this new approach is the increasing importance of strategies aimed at achieving **sustainable construction**. In other words, there is an increasing recognition that the industry makes an important contribution to a country's economic, social and environmental wellbeing.

INTRODUCING SUSTAINABLE CONSTRUCTION

The UK government published its first strategy for sustainable construction, *Building a Better Quality of Life*, in April 2000. This document aimed to provide a catalyst for change in the approach to construction processes. Subsequently it has been revised and extended, and the *Strategy for Sustainable Construction* published in June 2008 states the current UK position. Similar agendas have emerged in Europe, North America and some developing countries (see Chapter 15 for further discussion). Sustainable construction can be described in simple terms as comprising:

- efficient use of resources
- effective protection of the environment
- economic growth
- social progress that meets the needs of everyone.

Each of these strands is underpinned by economic concepts, which provide the rationale for this book.

Part A   Effective use of resources
This deals with microeconomics, and outlines the various ways of efficiently allocating resources between competing ends. In this section the prime focus is concerned with the determinants of demand and supply for infrastructure, housing, industrial buildings, commercial property, and repair and maintenance.

Part B   Protection and enhancement of the environment
This section considers failures of the market system, drawing upon various environmental economic concepts and tools to encourage future members of the construction industry to evaluate projects by more than just financial criteria.

Part C   Economic growth that meets the needs of everyone
This section incorporates coverage of the broader macroeconomic scene. It outlines the various government objectives that need to be achieved alongside sustainable construction. It highlights the difficulty of managing an economy and the need for professionals working in the construction industry to acquire an economic vocabulary.

Key Points 1.1

○ The construction industry can be described in a number of ways – for example, review the broad range of activities listed in Table 1.1 (page 1).

○ Construction has five distinguishing characteristics: (a) each project is regarded as a unique one-off product; (b) the industry is dominated by a large number of relatively small firms; (c) the general state of the economy influences demand; (d) prices are determined by tendering; and (e) projects are characterised by their 'lumpiness' in terms of their scale and expense.

○ The basis of economics rotates around the concepts of choice, scarcity and opportunity cost. Hence, economics is the study of how we make choices.

○ Any use of a resource involves an opportunity cost because an alternative use is sacrificed.

○ The graphic representation of the trade-offs that must be made can be displayed in a production possibility curve.

○ Sustainable construction is a strategy aimed to encourage the industry to (a) use resources more efficiently, (b) limit the environmental impact of its activities, and (c) produce buildings and infrastructure that benefit everyone.

INTRODUCING ECONOMIC VOCABULARY

The discipline of economics employs its own particular methodology and language. Consequently for the complete beginner it is necessary at the outset to clarify a few meanings.

Resources

Resources can be defined as the inputs used in the production of those things that we desire. Economists tend to refer to these resources as **factors of production** to highlight the fact that only by combining various factors can goods and services be produced. The factors of production are usually categorised into three general groups; namely, land, capital and labour – and sometimes the entrepreneur is specifically identified as a fourth entity. The point is that quantities of each factor are needed to make any good or service. To construct buildings or infrastructure, for example, labour is required to develop a plot of land, and plant and equipment, which may be hired or bought, is required to facilitate the process. To put it another way, land and labour are always combined with manufactured resources in order to produce the things that we desire. The manufactured resources are called **capital**, or more precisely physical capital, and consist of machines and tools.

The contribution of labour to the production process can be increased. Whenever potential labourers undergo training and learn new skills, their contribution to productive output will increase. When there is this improvement in human resources, we say that **human capital** has been improved. A relevant example is the effect that good trained management can have on the efficiency of a whole project. Indeed, according to Hillebrandt (2000: 104) management expertise is one of the scarcest resources of the construction industry throughout the world.

With each new construction project there is a choice to be made about the materials that will be used and the proportion of labour, plant and equipment required. In most instances, construction tends to be dominated by input costs relating to materials, components and labour. The importance, however, of the entrepreneur should not be overlooked, as without a dedicated resource managing and co-ordinating the other factors of production, virtually no business organisation could operate. In other words, an entrepreneur is sometimes regarded as a special type of human resource associated with the ability to make business decisions, take risks and foster innovation. In a small construction firm the manager-proprietor would be the entrepreneur; in a joint stock company the shareholders would take on that responsibility. (For further discussion about the role of the entrepreneur see Chapter 7.)

Each factor of production can be regarded as receiving a specific form of income. A landlord providing the use of land receives rent. Owners of physical (and monetary) capital are rewarded, directly or indirectly, and earn some form of interest payment to cover credit arrangements. Workers receive wages (salaries), and entrepreneurs gain profit. The distribution of these factor rewards (factor incomes) formed an important point of focus for the classical economists. Ricardo's work (1817) suggested that the rewards paid to the agricultural landlord determined all other payments and this inevitably led to a tension between the interests of the landlord and those of the consumer and manufacturer. Equally Marx (1844) was concerned about the inequalities that were rewarded to labour as he claimed that they were exploited by the owners of capital and land, as he observed: a worker cannot supplement his income with ground rent or interest on capital. For general introductory purposes, however, the significance of dividing income payments into four sets of factor rewards will become evident when we consider the measurement of **national income** in Chapter 13.

Market Systems

The concept of the **market** is rather abstract in the sense that it encompasses the exchange arrangements of both buyers and sellers for a particular good or service. Consequently, we can envisage many markets for specific building materials, housing, professional services, etc. The recurrent feature of any market is the exchange of information about factors such as price, quality and quantity. The difference, however, between one market and the next is the degree of formality in which it functions. The stock market in any Western economy, for example, provides instant information worldwide about the prices and quantities of shares being bought and sold during the current trading period. By contrast, construction markets are less structured and more informal, and they are usually determined by geographical location.

The construction industry is concerned with producing and maintaining a wide variety of durable buildings and structures, and as a consequence, there are many construction markets. As Drew and Skitmore (1997: 470) concluded in their analysis of the competitive markets for construction: 'The construction industry is highly fragmented, with the dominant firm being the small contractor.' The type of construction – particularly in terms of its size and complexity, its geographical location, and the nature of the client – will define the market in each case.

Let us consider in a little more detail what traditionally happens when a new project begins. Usually a contractor undertakes to organise, move and assemble the various inputs, and as such provides a service – a service of preparing the site

...bling and managing the process thereafter. ...rs add their services – such as plumbing, ...or whatever the specific job requires. As a ...asily become a series of 'separate' operations ...t in Table 1.2.

...pplying a construction project

...esponsibilities

...rovide specialist advice concerning structural, ...lectrical, mechanical and landscape details. ...lentify key specifications.

...lanages project in detail. ...aises between the client and the ...onstruction team.

...epares bills of quantities, cost plans, etc.

...anages work on site.

...pply specialist skills.

...ovide building materials and related ...mponents.

...his work depends upon the complexity of ...nt will be reflected in the cost per square ...icularly important in construction markets ...sses interested in competing for the work. ...ork beyond their local district as the costs ...d labour are relatively high. Travelling ...of work is available in the firm's own neighbourhood or catchment area. If, however, the construction project is very complex and/or very large, the costs per square metre are likely to increase and the relative costs of transport in relation to the total costs will decrease. The market catchment area for this highly specialised work will broaden. The following formula may make this clearer:

complexity + large size = competing firms from a wider geographical area

The converse of this rule explains why construction markets are so often dominated by small local firms subcontracting for work in or near their home towns. Indeed, it is only the biggest firms that can manage to compete on a national or international basis. Markets in the construction industry should, therefore, be defined as comprising those firms that are willing and able to compete for a contract in a specific geographical area. In other words, the total number of firms interested in work of a particular type can be referred to as comprising the local market.

In construction the services of one firm are often easy to substitute by contracting another firm with the same type of expertise. To the extent that prices in construction markets often find their own level, the theory behind this behaviour is examined in Chapters 3 to 8. For the moment, it will suffice to understand that the market for construction refers to a diverse and broad range of activities made

Figure 1.3 A complex set of markets for one building project

In the flow diagram the building project is represented as a sequence of stages.
Each stage is completed by a number of firms supplying their services.

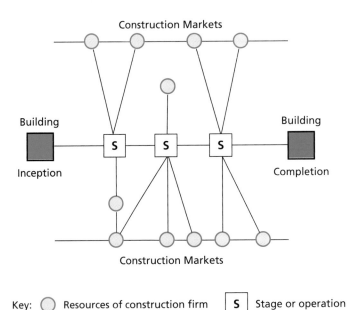

Source: Adapted from Turin (1975: 70)

up of many markets. To emphasise this point, consider the flow diagram set out in Figure 1.3 (page 12).

THE EXAMPLE OF ONE BUILDING PROJECT

In Figure 1.3 we represent a set of markets that could be involved in the construction of a small commercial building. The construction process from inception to completion is shown to comprise a number of separate markets. The completion of each operation, or stage, is the concern of several construction firms competing to supply materials, components, labour, etc. Figure 1.3 highlights the number of fragmented activities involved in completing just one small building. Each independent firm, in effect, is more concerned with its specific contribution than the project as a whole. In many ways, the next project in a firm's market sector may well be competing for its attention while it is still finishing the present project.

We discuss the characteristics of fragmentation and the resulting poor flow of information in Chapters 6 and 9 respectively. We also identify these issues as a problem to resolve in Chapter 15, where we analyse loosely connected activities as a barrier to achieving sustainable construction.

METHODOLOGY

This introductory chapter aims to explain what construction economics is all about. Therefore, apart from identifying the central concepts, we need to consider the methods employed by economists, as the approach taken to a discipline also helps to specify the nature of the subject. In general terms, economics is a social science and it attempts to make use of the same kinds of methods as other sciences, such as biology, physics and chemistry. Like these other sciences, economics uses models or theories.

> Economic models are simplified representations of the real world that we use to understand, explain and predict economic phenomena.

These models may take on various forms such as verbal statements, numerical tables and graphs – and, at the more advanced level, mathematical equations. For the most part, the models presented in this text consist of verbal statements and graphs.

A particular challenge faced by students of construction economics is that many of the processes in the industry do not lend themselves easily to generalisations and models. First, the construction industry involves a large variety of interests and parties that makes the process rather complex and plagued with unwarranted assumptions about what is possible. Second, economic analysis is only one of the disciplines contributing to the process as a whole. And, third, there is a distinct lack of vision about the role of construction in society and how it could better serve its clients. As Professor Duccio Turin poetically observed:

> [T]he building process is a world of 'as if'. It is 'as if' the client knew what he wanted when he commissioned the building from a designer; it is 'as if' the designer was in a position to advise the client on the best value-for-money he could obtain in the market; it is 'as if' contractual procedures

‒ were devised to ensure that the client would get the best possible deal from the profession and from the market place; it is 'as if' the manufacturer of building materials and components knew in advance what is expected of him and geared his production to such expectation; it is 'as if' the contractor knew how his resources were used, was in a position to control them, and was able to use this experience on his next job.

(Turin 1975: xi).

Although this summary of the industry was expressed nearly 30 years ago, as the text unfolds you will realise some striking similarity between then and now. It is this complex, fragmented and conservative nature that gives the subject matter of construction economics its appeal – as economists seek to unravel these seemingly unconnected threads of random behaviour. Economic models seek to identify the interrelationship between the key variables and simplify what is happening in the sector. So although some economic models may at first appear abstract, they do have practical applications. The important point we are trying to clarify is that an economic model cannot be criticised as unrealistic merely because it does not represent every last detail of the real world that it is seeking to analyse. If the model elucidates the central issues being studied, then it is worthwhile. For example, students may be expected to commence their course by completing an assignment based on a theoretical economic model of competition in the marketplace. This provides a simple introduction to the economic framework and the opportunity to demonstrate how construction deviates from or reflects this reference point. In short, the model provides a starting point – it enables us to proceed.

Following the recommendations of the Fairclough report (2002: 34) the construction industry should favour models that prioritise strategies aimed to improve sustainability, competitiveness, productivity and value to clients. In Part A, we present models of market behaviour that encourage a far better grasp of the meaning and purpose of efficiency, competition and profit. In Part C, we introduce a model of aggregation to study the operation of the whole economy that brings a fresh dimension to productivity by reviewing the total output of construction and reflecting on its contribution to the total output of an economy. In Part B, we bring the environment into the traditional model as a key variable for construction and the economy to consider. When the whole book has been studied, we identify a significant number of concepts that underpin an understanding of sustainability.

This leaves the precise nature and details of the models to emerge as the book unfolds and their purpose should become self-evident. Once we have determined that a model does predict real-world phenomena, then the scientific approach requires that we consider evidence to test the usefulness of a model. This is why economics is referred to as an empirical science – empirical meaning that real data is gathered to confirm that our assumptions are right.

An Example of an Economic Model

Before closing this section on models, we review one specific example to analyse and explain how income flows around an economy. Economists begin their explanation by ignoring the government sector, the financial sector and the overseas sector – that is, the **circular flow model** represents a simplified, scaled-down economy in which relationships are assumed to exist only between households and businesses.

To make the model effective, it is assumed that households sell factors of production to businesses and in return receive income in the form of wages, interest, rents and profits. This is shown in the bottom loop in Figure 1.4. The businesses sell finished goods and services to households in exchange for household expenditure. This is shown in the top loop in Figure 1.4. These assumptions are reasonably realistic. Businesses will only make what they can sell. Production will necessitate buying in land, labour, capital and enterprise, and the monies paid for these factors of production will generate respective income payments.

Already, without building in any of the complications of the real world, we begin to sense several insights or starting points. Clearly there is a close relationship between the income of a nation, its output and the level of expenditure, and we shall investigate this further in Chapter 13. Also, we can see how money enables households to 'vote' for the goods and services desired, and this will be developed further in Chapters 2 to 5. The perceptive reader will note that the model fails to include any reference to the environment and, as we explained above, this represents the contents of Part B. The model also explicitly excludes reference to the role of governments, overseas economies and financial institutions and these aspects are included in Part C. Again we can see how the model enables us to progress into the subject.

Figure 1.4 The circular flow model: a two sector economy

In this simplified model there are only households and businesses. Goods and services flow in one direction in return for money. This exchange can be thought of as a circular flow.

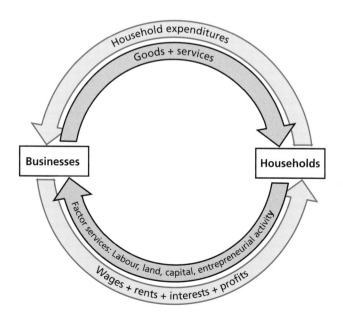

THE ROLE OF GOVERNMENTS AND FINANCIAL INSTITUTIONS

The importance of the construction industry to the overall wellbeing of the economy means that most governments are concerned that it becomes a highly efficient sector. As a consequence, the government's role as a client, regulator, policy-maker and a sponsor of change is raised at several points throughout the text. Equally the role of the financial sector makes a significant contribution to the effective management of the economy and the funding of construction projects. Any self-respecting text on economics must make some reference to the financial crisis sparked by the 2007 credit crunch that has troubled economies around the world. This crisis is still not fully resolved at the time of writing. In October 2010, Mervyn King, the Governor of the Bank of England, predicted that the aftermath of this crisis would hang over markets for many years to come. These themes are discussed throughout Part C of the text.

ENVIRONMENTAL ECONOMICS

Effective protection of the environment forms a key part of any text concerning sustainability. **Environmental economics** is important for several reasons: first, because the environment has an intrinsic value that must not be overlooked; second, because the sustainability agenda extends the time horizon of any analysis to assure equity between generations; and, third, demands must be viewed on a whole-life basis and this is particularly important in the context of products that last for more than 30 years. Any model of analysis that seeks to identify general principles of sustainable development must include, at the very least, these three dimensions. We explore these issues and related concepts in Part B and bring them all together in Chapter 15 where we review the possibility of achieving the government's sustainable construction agenda.

Microeconomics and Macroeconomics

Economics is typically divided into two types of analysis: **microeconomics** and **macroeconomics**. Consider the definitions of the two terms.

> Microeconomics is the study of individual decision-making by both individuals and firms.

> Macroeconomics is the study of economy-wide phenomena resulting from group decision-making in entire markets. As such, it deals with the economy as a whole.

One way to understand the distinction between these two approaches is to consider some generalised examples. Microeconomics is concerned with determining how prices emerge and change, and how firms respond. It involves the examination of the effects of new taxes, the determination of a firm's profit-maximising level of production, and so on. In other words, it concerns the economic behaviour of individuals – such as clients, contractors, surveyors and engineers – in various markets. We study this type of analysis in Part A. In contrast, questions relating to the rate of inflation, the amount of national unemployment, the growth rate of the whole economy and numerous other economy-wide subjects all fall in the realm of macroeconomic analysis. In other words, macroeconomics deals with aggregates or totals, and this forms the basis of the three chapters that comprise Part C.

You should be aware, however, of the blending together of microeconomics and macroeconomics in modern economic theory. Modern economists are increasingly using microeconomic analysis – the study of decision-making by individuals and by firms – as the basis for macroeconomic analysis. They do this because, even though aggregates are being examined in macroeconomic analysis, those aggregates are made up of the actions of individuals and firms. The study of any specific industry involves both microeconomic and macroeconomic approaches; particularly when the industry is multi-product, and has national and international significance.

Throughout this text the interaction between the construction sector and the other sectors of the economy is a constant reference point. In some texts a sectoral approach is referred to as **mesoeconomics**, derived from the Greek word *mesos* meaning intermediate. This is done to make it clear that the study of any specific sector or industry inevitably falls between the conventional microeconomic and macroeconomic categories. (These two terms are also of Greek derivation: *macros* meaning large and *micros* small.)

Consequently, to gain a comprehensive understanding of construction activity, it is advisable to embrace three perspectives – a broad macro overview of the economy, a specific sectoral study of the industry, and a detailed microanalysis of the individual markets in which construction firms operate. Studying the complete text, therefore, should provide a greater understanding of winning and completing projects in an efficient and sustainable manner. Figure 1.5 summarises how these three elements contribute to a fuller understanding of a project. In many ways, it models the overall approach of the text.

Figure 1.5 A model for construction economics: a new approach

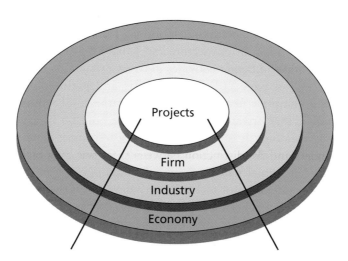

Sustainable construction issues

Key Points 1.2

○ We need to use scarce resources, such as land, labour, capital and entrepreneurship, to produce any economic good or service.

○ The exchange of information between buyers and sellers about factors such as price, quality and quantity happens in a market. Construction is made up of a diverse range of markets, as the industry comprises a large number of relatively small firms.

○ Every economic model, or theory, is based on a set of assumptions. How realistic these assumptions are is not as important as how effective they make the model or theory.

○ Microeconomics involves the study of individual decision-making. Macroeconomics involves the study of aggregates. Mesoeconomics combines the territory shared by microeconomics and macroeconomics to study a specific sector such as construction.

INTRODUCING CONSTRUCTION INDUSTRY ACTIVITY

The system of industrial classification used for statistical and government purposes favours a narrow definition of the construction industry that includes only firms that are involved with building and civil engineering. This categorisation is derived from the United Nations International Standard of Industrial Classification (ISIC). There are also American and European equivalents: the North American Industry Classification System (NAICS) and the General Industrial Classification of Economic Activities – otherwise known as NACE. In other words, firms generally recognised as officially comprising the construction sector tend to embrace a range of 'on-site' activities including those relating to infrastructure, new construction, repair, maintenance and (eventually) demolition. Table 1.3 shows the type of work that is classified into these various sectors and Table 1.4 gives some indication of the monetary value of these different activities in Great Britain.

As Table 1.4 (on page 20) shows, repair and maintenance are of major importance and comprise nearly 40 per cent of the total annual activity – this includes all public and private sector work carried out on houses, infrastructure and commercial buildings. It is also evident that government departments and their agencies are significant clients of the construction industry. As both Tables 1.3 and 1.4 suggest, official statistics often draw a distinction between public and private sector activity. The public sector includes everything that is owned and/or funded by national or local governments such as roads, schools, the National Health Service, and local council leisure centres. In fact, a close examination of output tables reveals that approximately 30 per cent of construction industry turnover relates to public sector clients. Obviously this includes a vast range of contracts, varying in size from £10,000 for a small flood defence scheme to £9.4 billion for the new venues and infrastructure for the Olympic Park in east London.

Table 1.3 The construction industry – narrowly defined

Areas of Construction	Examples of type of work
Infrastructure	Water and sewerage Energy Gas and electricity Roads Airports, harbours, railways
Housing	Public sector/housing associations Private sector (new estates)
Public non-residential	Schools, colleges, universities Health facilities Sports and leisure facilities Services (police, fire, prisons)
Private industrial	Factories Warehouses Oil refineries
Private commercial	PFI (and similar public private partnerships) Schools/hospitals (where privately funded) Restaurants, hotels, bars Shops Garages Offices
Repair and maintenance	Extensions and conversions Renovations and refurbishment Planned maintenance

The percentage of public sector construction work in the UK has fallen considerably since 1980, as many of the activities traditionally in the public domain have been privatised. Utilities and services such as gas, electricity, water supply, telecommunications and railways were previously 'pure' state owned activities; today they are quasi-public – privately owned but 'regulated' and controlled by specific government agencies. This means that there is a new expanding sector of 'regulated' private sector work that is reliant on public sector decisions before it can be executed. More recently, the private sector has also been given a greater role in the funding, building and maintenance of public facilities such as hospitals, schools, prisons and roads. In these **public private partnerships**, the private sector organises the funds and manages the risks, while the public sector specifies the level of service required and ultimately owns the assets – as they are commonly returned to public ownership after 10, 15 or 25 years. These 'regulated' and 'partnership' arrangements are explained further in Chapters 2 and 6 respectively. The important point for our purposes is that expenditure on the construction of public facilities is increasingly classified as private sector expenditure in the official data. (Figure 2.4 on page 38 shows the distribution of work across the public and private sector divide over the last 50 years.)

Table 1.4 Value of construction output in Great Britain

Types of work	Value of Output (£million) (Current prices)		
	2000	2005	2010
Infrastructure	7,421	7,702	12,660
Housing – public	1,054	2,207	4,770
private	9,200	20,112	14,281
Public non-residential	3,826	8,279	14,204
Private industrial	4,449	5,293	3,573
Private commercial	18,424	26,064	23,312
Repair and maintenance	26,655	39,427	41,630
Total (of all work)	**71,029**	**109,084**	**114,430**

Source: Construction Statistics Annual (ONS 2011a: Table 2.1).

Sources of International Data

A *narrow definition* of the construction industry confines official statistical data to the 'site-based' activities of firms involved with buildings and infrastructure. As Table 1.3 shows, this data is typically disaggregated into house building, private commercial and industrial building (that is, non-residential), infrastructure (civil engineering), repair and maintenance, and so on. Across Europe it is possible to see some common trends. Aggregating figures across 27 European countries, 26 per cent of construction output is repair and maintenance, 19 per cent is house building, 22 per cent is infrastructure and 33 per cent is non-residential (FIEC 2012).

A modern alternative, however, is to widen the statistical definition and go beyond the narrow boundaries created by the international classification to include the whole life cycle of construction: design, production, use, facility management, demolition, etc. In fact, the Pearce Report (2003) argued that to fully understand the extent of what is meant by a sustainable industry required data relating to the broad scope of construction productivity including its environmental and social impacts. (The new approach adopted in this text will help to clarify the contributions that the sector makes to these wider concerns.) As Pearce (2003: 24) defined it, a *broad definition* should include the mining and quarrying of raw materials, the manufacture and sale of construction products, and the related professional services such as those of architects, engineers and facilities managers. For example, a detailed analysis of current available data in Great Britain indicates that in addition to the 1,200,000 workers employed in the traditional construction sector, there are approximately 450,000 engineers, architects, facilities managers and chartered surveyors supplying professional services relating to construction and property, about 650,000 employed in the manufacture of building products and equipment,

25,000 involved in mining and quarrying construction materials and a further 100,000 workers selling building materials. In effect this broad approach increases the significance of construction as the size of the industry virtually doubles in terms of employment and output. The implications of the contrast between the narrow and broad definitions will be reviewed further in Chapter 13, where we explain the measurement of economic industry activity in more detail, and in the final reading where we review Squicciarini and Asikainen's (2011) paper that addresses the inability of statistical analyses to capture the true scope and impact of construction.

The International Council for Research and Innovation in Building set up a project to analyse the entirety of the broad construction sector and published its findings in 2004. The new broad system approach was tested in nine countries and the project's conclusions confirmed the fact that the construction system as a whole seems to be roughly twice the size of the conventional construction sector (Carassus 2004: 190). For example, in Canada it was assessed that the broad construction cluster employed nearly 1,800,000 people, of which only 900,000 worked in the traditionally defined construction sector. Similarly in France the whole sector on a broad definition was assessed to be responsible for the employment of 2,358,000 people, of which 1,600,000 were employed on site as contractors (Carassus 2004: 45 and 78).

Table 1.5 Sources of international data

The European Construction Industry Federation (FIEC) www.fiec.org

FIEC was created in 1905 to specifically promote the interests of the construction industry across Europe. The current membership is made up of 29 countries and 34 national federations. They represent 3.1 million construction enterprises employing a total of more than 15 million people. Publications include *Construction Activity in Europe* which gives a detailed statistical breakdown per member country.

United Nations Statistical Division http://unstats.un.org/unsd/default.htm

This organisation is committed to the advancement of a global statistical system. Consequently it develops standards and norms for statistical activities, and supports countries' efforts to strengthen their national statistical systems. Comprehensive sets of data can be accessed and individual macroeconomic variables can be examined by selecting countries and years of interest.

Organisation for Economic Co-operation and Development (OECD) www.oecd.org

The OECD encompasses 30 member countries sharing a commitment to the market economy (in some quarters the group is referred to as the rich countries club). Its work covers economic and social issues. Publications include *The OECD Observer*. a journal covering a number of themes including statistics and sustainable development, and a series of *Main Economic Indicators*, divided by subject (including construction) and by country.

Eurostat www.ec.europa.eu/eurostat

Eurostat provides the European Union with a high quality statistical information service, and co-operates closely with other international data organisations (such as the UN and OECD). *The Eurostat Yearbook* is published annually and presents a comprehensive selection of statistical data covering areas such as labour markets, economy, international trade, industry, services and the environment (it is freely available online as a pdf file).

The problem faced by academic researchers, however, is the different institutional arrangements that exist in each country and the different processes, agencies and systems used for data collection. In other words, government sources of data usually conform to the established narrow definition of construction, as collated and presented by the European Construction Industry Federation (FIEC). But access and symmetry is lacking when attempts are made to measure the size of the industry using the broader construction system approach. Some of the inconsistencies will be uncovered if you follow the links detailed in Table 1.5, as in many instances one series of data cannot be precisely compared to another across national boundaries.

Sources of UK Data

In the UK, economic and construction data is collected and published by the **Office for National Statistics** (commonly referred to as **ONS**). Its website (www.ons.gov.uk) provides a comprehensive range of statistical indicators and it is the starting point to search for most data relating to UK society, demography and the economy. The ONS is free from political influence and its role is to provide an evidence base

Table 1.6 A brief guide to official sources of UK statistics

UK National Accounts

This publication is normally referred to simply by the colour of its cover as the Blue Book. It is published annually in the autumn by the ONS, and is considered to be a most important source of data for the UK macroeconomy, since it provides a comprehensive breakdown of GDP. As with other ONS publications, recent editions have become more user friendly. For example, there are now useful notes explaining how to interpret the accounts, a subject index and a glossary of terms.

Economic & Labour Market Reviews

A monthly publication covering issues broadly relating to the economy, the labour market and price inflation. Each issue contains a current economic review, independent forecasts and a mix of articles on various aspects of data. Its primary objective, however, is to provide a gateway to access the current statistical releases. The content always includes a few pages of key data plus a directory of online tables, to highlight web pages on the ONS site that have been recently updated. Much of the data in Part C of this text is drawn from this source.

Monthly Digest of Statistics

As the name implies, this is an amalgam of statistics; it has been published monthly since January 1946. It covers a wide range of topics, including economic, social and demographic issues. It presents data on prices, the balance of payments and employment.

Financial Statistics

Monthly publication of the ONS relating to financial indices such as interest rates, exchange rates and the money supply. Editions since 1997 also show data relating to inflation.

Bank of England Inflation Report

This is published quarterly with the Bank of England's *Quarterly Bulletin*. The *Inflation Report* serves a dual purpose. First, it provides a comprehensive review of specialised indices and a commentary on their forecasts. Second, it is the official publication in which the minutes of the monetary policy committee (described in Chapter 12) are made available to the public.

for policy and decision making. The Bank of England and the Treasury are also responsible for a significant number of statistical releases relating to finance and the current economic climate, and this output is often supported by commentary from government economists.

For specific information relating to construction, the key source in the UK is the *Construction Statistics Annual*. A new edition is published each year around August, and it can be freely viewed in its entirety on the ONS website. Most of the tables are available in spreadsheet form, allowing the quarterly analyses to be annualised quite easily. The publication amalgamates all construction statistics produced by central and local government, together with data from a quarterly survey of private sector construction firms. It also carries an appendix that provides detailed notes on methodology and definitions to clarify the tables and figures. Overall it provides a comprehensive picture of the UK construction industry through the last decade, together with some international comparisons. (For example, the references to industry sectors in Table 1.3 and data on the value of output shown in Table 1.4 are derived from *Construction Statistics Annual*.) As with most economic data there is always a time lag, so the 2011 edition only presents data up until 2010. Consequently, data quoted in this text can appear out of date before the book even goes to print. It is important, therefore, that you have the confidence to research data for yourself, and this is one of the reasons that website sources are provided.

RESEARCHING DATA

When using official national statistics, in hard copy or from the Internet, it is useful to be aware of several conventions regarding their presentation. First, the symbols shown in Table 1.7 represent a summary of the main footnotes.

Table 1.7 Symbols used to annotate official statistics

●●	Not available
-	Nil or less than half the final digit shown
P	Provisional
R	Revised

These qualifying notes make it clear that published official statistics can be no more than an estimate. This observation is not made to discredit official statistics but to emphasise that any errors or omissions are corrected as soon as possible – the ultimate goal is to produce data sets that are as reliable and robust as possible.

Second, some statistical series do not have sufficiently consistent data to refer to the United Kingdom as a whole, and only refer to Great Britain or are restricted simply to England, Wales, Scotland or Northern Ireland. To give just one example, construction output figures tend to be specific to Great Britain. There is also a possibility that the data set may only relate to a quarter – in which case, it may be necessary to multiply the figures by four to get an approximation for the whole year. So take great care when reading the headings and footnotes that are associated with each table.

Figures relating to money can be expressed in three ways.

- The most straightforward is to represent economic activity in 'face value' terms; or put more simply, in prices used for everyday transactions. Such measurements are referred to as **current prices**; an example is shown in Table 1.4.
- A more sophisticated option is to adjust figures to allow for inflation as this makes comparisons across time more meaningful, such adjustments are referred to as **constant prices**, as they are expressed in terms of a specific base year. As an example scan forward to Figure 2.4 (on page 38) where levels of construction activity over a period of 50 years can be confidently compared as the distortions caused by changes in value across time have been removed by referencing all prices to the year 2005.
- Time series data may also be expressed in terms of **index numbers** – in this type of data set the starting point (or base year) is given the value of 100, which allows subsequent percentage changes to be quickly identified. Index numbers and examples of their use are the subject of Chapter 14.

Finally, when reviewing any construction data it must always be borne in mind that the construction industry (right across Europe) comprises a very large number of small, geographically dispersed firms, and this makes it difficult for government agencies monitoring the industry to compile comprehensive data sets. Furthermore, alongside any official activities 'put through the books' and recorded in national statistics, there may be unofficial work carried out for 'cash in hand'. As a consequence, the data agencies in some countries include an estimated value for so-called 'unrecorded output'. The relationship between the size of the official (formal) economy and its unofficial (informal) equivalent is discussed further in Chapter 13.

Key Points 1.3

○ The construction industry may be defined in many ways, but its impacts always cut right across the economy.

○ A narrow definition of the construction industry includes the erection of buildings, infrastructure work, repair, maintenance and demolition. Examples of the activities are presented in Table 1.3.

○ A broad definition of the construction industry adds to the narrow definition to include mining and quarrying of raw materials, the manufacture and sale of construction products, and related professional services. Examples of the broad range of activities are shown in Table 1.1.

○ The annual value of construction output (expressed in current prices), in Great Britain, for the decade 2000 to 2010 ranged from £71 billion to £114 billion.

Reading 1

George Ofori's review of construction economics has become a standard reference point for those commencing a study of the subject. In fact, Andrew Cooke (1996: 13) suggested 'that all students of economics in a construction related field should read it at least twice – once before they embark on their studies and once when they are about to complete them'. The following extract provides an opportunity to partly meet his recommendation.

Unfortunately, as you will see, one of Ofori's main concerns is that there is no real consensus about the exact definition of construction economics. Yet since his paper was written, the aim of sustainable construction has gained momentum and the importance of improving economic efficiency in the industry has been broadly discussed in various government reports (for example, Egan 1998; HM Government 2008; IGT 2010). This raises two related questions that you might consider now and as the text unfolds. First, what are the main hurdles in defining economics and, second, does the development of a sustainability agenda provide the necessary focus to create a more coherent approach?

The questions that relate to each of the readings are set to stimulate debate and further exploration. Those interested in pursuing the first two questions should see De Valence's (2006) paper where, across five pages, he manages to discuss the problems of classifying the boundaries of the subject area, or the chapter by Brochner (2011) that considers the current forces of change within the discipline and where construction economics appears to be heading.

George Ofori (1994) 'Establishing Construction Economics as an Academic Discipline', *Construction Management and Economics* 12: 295–306

What is construction economics?

Perhaps the most basic feature of a discipline is a clear idea among its practitioners and researchers about what it entails, its aims and its boundaries. In this section, adopting a chronological approach, definitions of construction economics offered by various writers are considered.[1]

Drewer (1978) suggests that Turin was perhaps the first to attempt to impose scientific order on a 'pre-Newtonian' situation. Turin (1975) remarked that '...if economics is concerned with the allocation of scarce resources, it follows that building economics should be concerned with scarce *building* resources' (p. ix). Stone (1976) observes that 'building economics' embraces 'those aspects of design and production, and the related problems of organisation which affect the costs of a building' (p. xi), including forms of construction, methods of production, organisation of the industry and the impact of new methods, materials and forms of organisation and contractual relationships.

Rakhra and Wilson (1982) distinguish between 'building economics' and 'economics of building'. They suggest: 'Building economics takes an aggregate view of the building sub-section of the construction sector' (p. 51), embracing levels of building activity, the industry's contribution to the economy, impact of changes in government's policies and the nature, structure and organisation of the industry. 'Economics of building' was an 'examination at the specific project level of the resource transformation that is known as building' (p. 51), embracing cost-benefit consequences of design alternatives and choice of building components, life-cycle costing, effect of various combinations of labour and plant on site productivity and analysis of project resource requirements. Few writers refer to or adopt Rakhra and Wilson's (1982) distinction, notable exceptions being Bowen and Edwards (1985), Ofori (1990) and Bowen (1993).

Seeley (1983) opted for a very narrow definition, remarking that '...building economics has been widely used ... to describe the investigation of factors influencing building cost, with particular reference to the interaction of building design variables' (p. v). Ahuja and Walsh (1983) define cost engineering, which may be considered a form of construction economics, as '...an active approach in the design, construction and commissioning phases of a project, aimed at extracting the best possible value for money throughout each activity that has cost implications' (p. ix).

Hillebrandt (1985) adopted a broad perspective, defining construction economics as 'the application of the techniques and expertise of economics to the study of the construction firm, the construction process and the construction industry' (p. 1). Similarly, this journal defines construction economics as including design economics, cost planning, estimating and cost control, the economic functioning of firms within the construction sector and the relationship of the sector to national and international economics. Ashworth (1988) considers construction economics as embracing clients' requirements, impact of a development on its surrounding areas, relationship between space and shape, assessment of capital costs, cost control, life-cycle costing and economics of the industry in general.

Bon (1989), whose book was 'to offer a first step toward a theoretical framework for building economics' (p. xiii), suggests that 'building economics is about economising the use of scarce resources throughout the life cycle of a building...' (p. xiii) and concerns the 'application of standard investment decision criteria to buildings as a special class of capital assets' (p. xiii). Johnson (1990) adopts a similar definition, suggesting that '...knowledge of economics can provide a basis for making difficult trade-offs associated with both design and long-term management of buildings' (p. 9).

Ruegg and Marshall (1990) promise to show readers '...how to apply the concepts and methods of economics to decisions about the location, design, engineering, construction, management, operation, rehabilitation and disposition of buildings' (p. xi). Drake and Hartman's (1991) perspective is similarly project oriented, considering construction economics as being concerned with ends and scarce means in the Construction Industry' (p. 1057) and listing the, mainly surveying, techniques it embraces. Raftery (1991) suggests that 'building economics' could be said to be primarily about a combination of technical skills, informal optimisations, cost accounting, cost control, price forecasting and resource allocation. Finally, Bowen (1993) describes 'economics of building' as focusing 'on the application of quantitative techniques using financial criteria for the provision of financial advice to the design team' (p. 4).

From the above discussion, a common definition of construction economics does not exist. The chronological approach adopted helps to show that the issue has not become any clearer over time. For construction economics to develop into a discipline, a common definition is required to set the framework for issues to be considered and methodological approaches to be adopted. The definition should relate to the economic principles of scarcity and choice, refer to what is being studied (projects, practices, organisations and enterprises and industry) and state the overall aim of the discipline.

Segments

Two distinct segments of construction economics emerge from the discussion in the previous section. The first relates to construction projects, whereas the second concerns the industry. Ofori (1990) terms these 'construction project economics' and 'construction industry economics', respectively (these terms are used in the rest of the paper). However, again from the above definitions, some writers consider one or the other of the segments to be the entire field of construction economics or building economics. For example, Bon's (1989), Johnson's (1990) and Ruegg and Marshall's (1990) 'building economics' and Drake and

Hartman's (1991) 'construction economics' relate only to projects and are similar to Rakhra and Wilson's (1982) 'economics of building' and Ofori's (1990) 'construction project economics', respectively. However, Stone's (1976) 'building economics' and Hillebrandt's (1985) 'construction economics' incorporate both segments. The present author prefers and adopts the title construction economics as a perspective encompassing both the project and the industry, as this enables all aspects of the field to be studied.

The project-related segment is basically about techniques (such as cost planning, life-cycle costing and value engineering) – Raftery (1991) likens it to 'cost accounting and management'. It is better known, as it has received greater attention from researchers and in course syllabi (Ofori, 1990). However, even here, there is some confusion. Seeley (1983, p. 1) appears to make 'cost control' synonymous with construction project economics and Kelly (1983), as well as Ferry and Brandon (1991), seem to equate 'cost planning' with construction project economics. Male and Kelly (1991) refer to cost management and define it as 'a synthesis of traditional quantity surveying skills ... with structured cost reduction/cost substitution procedures using the generation of ideas by brainstorming ... in a multidisciplinary team' (p. 25). Finally, Kelly and Male (1993) have combined cost management, which emphasises cost reduction at the design stage, with value management, which focuses on clients' needs prior to design, to obtain the 'comprehensive service' of project economics, which '...seeks to control time, cost, and quality during design and construction within the context of project functionality' (Marshall, 1993, p. 170).

It is necessary to delineate, agree upon and continuously research into and improve its segments and construction economics should be developed as an integrated whole.

Conceptual structure

Cole (1983) distinguishes between the core of a discipline, the 'fully evaluated and universally accepted ideas' (p. 111) found in all undergraduate textbooks and the research frontier which includes all on-going studies, most of which eventually turn out to be of little or no significance. Does construction economics have a core of confirmed and accepted concepts?

Some key terms

Precise and common definitions are indispensable building blocks in any discipline, a key base of its conceptual structure. Authors in construction economics often find it necessary to define their main terms (e.g. Batten, 1990; Ive, 1990). Bowen and Edwards (1985) and Bowen (1993) define such basic terms as estimating, forecasting, cost and price. Some construction economics terms, each of which has a clear definition in general economics, are considered in this section.

The industry

There is, as yet, no accepted definition of the construction industry (Ofori, 1990). Some writers consider it as involving only site activity, others include the planning and design functions and yet others extend it to cover the manufacturing and supply of materials and components, finance of projects or management of existing construction items (Turin, 1975; Hillebrandt, 1985). This leads to difficulties. For example, writers' basic data and inferences often differ, simply because they adopted different definitions of 'construction'.

Notes

[1] As you will see in Ofori's search for consensus he was prepared to include other authors' references to building economics. He regards this as simply a narrower version of construction economics. As Ofori (1994: 296) explained at the start of the paper, 'of the two commonly used titles, building economics and construction economics, the latter embraces the former, covering also civil engineering and other forms of construction'.

References

Ahuja, H.N. and Walsh, M.A. (1983) *Successful Cost Engineering*, Wiley, New York

Ashworth, A. (1988) *Cost Studies of Building*, Longman, Harlow

Batten, D.F. (1990) Built capital, networks of infrastructure and economic development. In *Proceedings CIB 90 Joint Symposium on Building Economics and Construction Management*, Sydney, 14–21 March, 1, pp. 1–15

Bon, R. (1989) *Building as an Economic Process: An Introduction to Building Economics*, Prentice Hall, Englewood Cliffs, NJ

Bowen, P.A. (1993) A communication-based approach to price modelling and price forecasting in the design phase of the traditional building construction process in South Africa, PhD thesis, University of Port Elizabeth

Bowen, P.A. and Edwards, P.J. (1985) Cost modelling and price forecasting: practice and theory in perspective. *Construction Management and Economics*, 3, 199–215

Cole, S. (1983) The hierarchy of the sciences? *American Journal of Sociology*, 89, 111–39

Construction Industry Development Board (1988) *Construction Sector in a Developed Singapore*, CIDB, Singapore

Drake, B.E. and Hartman, M. (1991) CEEC: The organisation and activities of European construction economists. In *Management, Quality and Economics in Building*, Bezeiga, A. and Brandon, P.S. (eds), Spon, London, pp. 1056–69

Drewer, S. (1978) In search of a paradigm: notes on the work of D.A. Turin. In *Essays in Memory of Duccio Turin*, Koenigsberger, O.H. and Groak, S. (eds), Pergamon, Oxford, pp. 47–51

Ferry, D. and Brandon, P.S. (1991) *Cost Planning of Buildings*, Blackwell Scientific Publications, Oxford.

Hillebrandt, P.M. (1985) *Economic Theory and the Construction industry*, 2nd edn, Macmillan, London

Ive, G. (1990) Structures and strategies: an approach towards international comparison of industrial structures and corporate strategies in the construction industries of advanced capitalist countries. *Habitat International*, 14, 45–58

Johnson, R. (1990) *The Economics of Building: A Practical Guide for the Design Professional*, Wiley, New York.

Kelly, J. (1983) Value analysis in early building design. In *Building Cost Techniques. New Directions*, Brandon, P.S. (ed.), Spon, London, pp. 116–25

Kelly, J. and Male, S. (1993) *Value Management in Design and Construction: The Economic Management of Projects*, Spon, London

Male, S. and Kelly, J. (1991) Value management and the economic management of projects. *The Building Economist*, 29(4), 25–30

Marshall, H.E. (1993) Value management in design and construction: The economic management of projects. *Construction Management and Economics*, 11, 169–71

Ofori, G. (1990) *The Construction Industry: Aspects of its Economics and Management*, Singapore University Press, Singapore

Raftery, J. (1991) *Principles of Building Economics*, Blackwell Scientific Publications, Oxford

Rakhra, A.S. and Wilson A.J. (1982) Building economics and the economics of building. *The Building Economist*, 21, 51–3

Ruegg, R.T. and Marshall, H.E. (1990) *Building Economics: Theory and Practice*, Van Nostrand Reinhold, New York

Seeley, I.H. (1983) *Building Economics: Appraisal and Control of Building Design Cost and Efficiency*, 3rd edn, Macmillan, London

Stone, P.A. (1976) *Building Economy: Design, Production and Organisation – A Synoptic View*, 2nd edn, Pergamon, Oxford

Turin, D.A. (1975) Introduction. In *Aspects of the Economics of Construction*, Turin, D.A. (ed.), Godwin, London, pp. viii-xvi

....

Extract information: Pages 298–9 of original plus relevant references from pages 304–6.

Part A

Effective Use of Resources

WEB REVIEWS: Effective Use of Resources

On working through Part A, the following websites should prove useful.

www.bis.gov.uk

Following the general election of May 2010, sponsorship of the construction industry was transferred from the Department for Business Enterprise and Regulatory Reform (which was abolished) to the Department for Business Innovation and Skills. At the department's home page it is possible to select a 'business sector' and navigate to construction, and from here it is easy to download related statistics, up-to-date information on sustainable construction and the low carbon agenda; these aspects were introduced in Chapter 1. Many of the other issues and reports referred to in the text are also supported. According to its website one of the department's objectives 'is to secure an efficient market in the construction industry, with innovative and successful firms that meet the needs of clients and society while being competitive at home and abroad'.

www.competition-commission.org.uk
www.oft.gov.uk

The Enterprise Act 2002 made major changes to UK competition law and sets the legal framework for two agencies responsible for monitoring market behaviour. The *Competition Commission* tends to focus on large firms concentrating on mergers and maintaining a level of competitive tension between companies, and the *Office of Fair Trading* (OFT) seeks to protect consumers against unfair market behaviour such as collusion. Their websites include information on what is currently under enquiry, summaries of reports, press releases, annual reviews and fuller information on their specific roles.

www.carol.co.uk
www.ft.com

Public limited companies must make their annual reports available to the public. These two sites facilitate the process. *Carol* is the acronym for Company Annual Reports Online, and an interesting development in recent years is that many annual reports are now designed specifically for the Internet. Hard copies of company reports can also be requested by phoning the *Financial Times* – and www.ft.com provides the same opportunity online. This newspaper's website also offers unrestricted access to every story in the current edition and its vast archive. The site provides its own business search engine and the stated aim is to be the leading Internet resource for business people everywhere. It is certainly of use to students studying economics, especially as it is updated on a daily basis. It may be of particular use if you follow the advice in Chapter 8 and choose to study a specific construction firm.

www.ons.gov.uk

The Office for National Statistics is the UK government agency responsible for compiling, analysing and disseminating official statistics on Britain's economy, society and demography at a national and local level. Detailed data are available from this site and it will enable the various statistical series used in this book to be updated.

Decisions about resource allocation are necessary because we live in a world of scarcity. A review of the ideas listed at Key Points 1.1 and 1.2 should remind you of how central this basic premise is to the study of any branch of economics. To take a surreal example, when you open your front door in the early morning there are not millions of bottles of milk covering the neighbour's lawn; nor is there no milk. There is just enough bottled milk to meet the demand: say, the two pints your neighbour ordered. What this chapter seeks to explain is how this finely tuned allocation of resources can occur, given the multitude of construction, manufacturing and service resources that simultaneously need to be allocated.

The problems of **resource allocation** are solved by the **economic system** at work in a nation. In the case of construction, resource allocation has been strongly influenced by the public sector. *What* is produced, *how* it is produced and *for whom* can be determined by central government – and it was frequently – but governments across Europe now prefer to use a market system to answer the *what, how* and *for whom* questions. In general terms, therefore, it is possible to envisage two model systems. Each economic model brings together producers and consumers in different ways and each needs to be appreciated in order to understand how the universal questions about resource allocation are resolved.

Economic Systems: Two Extremes

The problem of resource allocation is universal as every nation has to tackle the issue of determining what, how and for whom goods and services will be produced. In Figure 2.1 we begin our presentation of the economic systems of the world by introducing two extremes: the **free market model** and the **centrally planned model** (along with two exemplar nations).

Figure 2.1 A spectrum of economic systems

On the extreme right-hand side of the diagram is the free market model, and on the extreme left-hand side, the centrally planned model. Cuba is a country whose system closely resembles the centrally planned model. At the other extreme is the USA, which comes close to the free market model. In between are the mixed economies of the remaining nations of the world.

FREE MARKET MODEL

The free market system is typified by limited government involvement in the economy, coupled with private ownership of the means of production. Individuals pursue their own self-interest without government constraints: the system is decentralised.

An important feature of this system is **free enterprise**. This exists when private individuals are allowed to obtain resources, to organise those resources and to sell the resulting product in any way they choose. Neither the government nor other producers can put up obstacles or restrictions to block those in business from seeking profit by purchasing inputs and selling outputs.

Additionally, all members of the economy are free to choose what to do. Workers may enter any line of work for which they are qualified and consumers may buy the goods and services that they feel are best for them. The ultimate voter in a free market, capitalist system is the consumer, who votes with pounds and decides which product 'candidates' will survive. Economists refer to this as **consumer sovereignty** as the final purchaser of products and services determines what is produced – and, therefore, 'rules' the market.

Another central feature of the free market economy is the **price mechanism**. Prices are used to signal the value of individual resources, acting as a kind of guidepost which resource owners (producers and consumers) refer to when they

Figure 2.2 The price mechanism at work

Price changes co-ordinate the decision-making processes of consumers and producers. When supply exceeds demand, the price of a product will need to fall for the market to clear. Conversely if demand exceeds supply, the price of the product will rise. For the price mechanism to function in all markets, it is important that resources are owned privately and can move freely between competing uses.

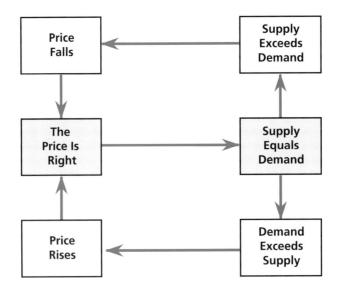

make choices. The flow chart in Figure 2.2 suggests how the price mechanism works. For example, when supply exceeds demand a price change occurs which brings the producers and consumers into harmony. This is precisely what happens during the January sales: the price of stock that has not been previously sold is reduced, to the point where demand is sufficient to clear the market. Conversely, when demand exceeds supply, the price of the good in question will rise until the market is in balance. This may be seen at a property auction where, to begin with, several buyers compete for a specific property: together they bid the price up, until finally there is only one interested party prepared to pay the final purchase price to the **vendor**.

Prices can thus be seen to generate signals in all markets (including factor markets): they provide information, they affect incentives and they enable buyers and sellers to express opinions. And, providing that prices are allowed to change freely, markets will always tend towards equilibrium, where there is neither excess demand nor excess supply. To put it another way, in a truly free market the price determination of goods and services has nothing to do with the government. By allowing the forces of supply and demand to operate freely, the economy finds its own natural balance.

It may be self-evident by now that other terms used to describe the free market economy are 'market' or 'capitalist' economy.

Summary: What? How? For Whom?

What   In a free market economy, consumers ultimately determine what will be produced by their pattern of spending (their voting in the marketplace). As far as producers are concerned, their decisions about what goods to produce are determined by the search for profits.

How   Since resources can substitute for one another in the production process, the free market system must decide how to produce a commodity once society votes for it. Producers will be guided (by the discipline of the marketplace) to combine resources in the cheapest possible way to achieve a particular standard or quality. Those firms that combine resources in the most efficient manner will earn the highest profits and force losses on their competitors. Competitors will be driven out of business or forced to combine resources in the same way as the profit-makers.

For whom The 'for whom' question is concerned with the distribution of goods after production. How is the pie divided? In a free market economy, production and distribution are closely linked, because incomes are generated as goods are produced. People get paid according to their productivity; that is, a person's income reflects the value that the market system places on that person's resources. Since income largely determines one's share of the output 'pie', what people get out of the free market economy is based on what they put into it.

Key Points 2.1

○ We can illustrate and simplify the different types of economic systems by looking at two extremes: the free market model at one end of the scale and the centrally planned model at the other.

○ The key attributes of a free market (model) economy are (a) limited government involvement, (b) individuals and producers express their desires through the price system, and (c) free enterprise – producers, consumers and resource owners all have complete freedom of choice over the range of products bought and sold.

○ A free market economy is also called a market or capitalist economy.

○ Any economic system must answer three questions. *What* will be produced? How will it be produced? For whom will it be produced?

THE CENTRALLY PLANNED MODEL

A centrally planned system (also referred to as a command economy) is typically characterised by a dominant government sector, coupled with the common ownership of resources. In other words, there is a central planning authority that takes the place of the price mechanism in allocating resources. The precise nature of a central planning authority depends upon the political system governing the economy. Indeed, it is worth noting that the terms 'socialist' and 'communist' properly refer to political systems and not economic systems. In fact, a right-wing dictatorship could operate a centrally planned economic system as effectively as a left-wing commune.

The common motivation for having a centrally planned system is the conviction that government commands are more likely to produce the 'right' mix of output, while the market mechanism may seem to operate in favour of the rich. Central planning creates the opportunity to direct resources to the society's most pressing needs. As the respected American economist J.K. Galbraith once observed, one of the few saving graces of the disintegrating communist economies is that nearly everyone has some kind of home, whereas many capitalist economies have not yet resolved the problem of providing affordable housing for the poor.

The flow diagram in Figure 2.3 outlines what a centrally planned system might involve. The three-stage process shown is a simplification of a bureaucratic reality. For instance, at stage one, various planning committees would exist to consider specific economic sectors and/or geographic areas. Similarly, at stage two, production targets and wages would be 'negotiated' with factory officials, workers, management and others involved in the chain of production. Finally, by stage three, the plans often become fraught with many difficulties, to the extent that there may be shortages and/or surpluses.

Figure 2.3 The general principles of a centrally planned economy

For such a system to work, the resources need to be centrally owned and controlled.

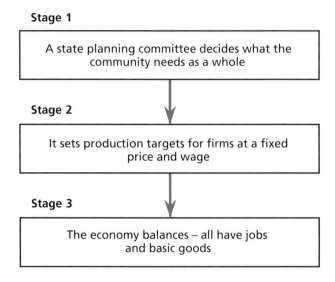

Stage 1

> A state planning committee decides what the community needs as a whole

Stage 2

> It sets production targets for firms at a fixed price and wage

Stage 3

> The economy balances – all have jobs and basic goods

In today's market driven, global culture, such a system may seem hard to imagine, but it was used for more than 60 years from 1928 onwards when Stalin introduced Russia's first Five-Year Plan. For example, the State Planning Committee (Gosplan) in the former Soviet Union used to make sure that all construction projects identified the required quantity of labour, plant and materials. Then, all the plant and material requirements from projects due to come on stream in each region within a five-year period were added together. This enabled the planning committee to place the necessary orders with state contractors, material suppliers and plant production factories. Understandably this system was typified by cost estimating problems and over-employment, because in a planned economy there is no competition – no tendering – no risk and little chance of being laid-off work.

A similar system of resource allocation had been adopted by the People's Republic of China after its revolution in 1949. Following the death of the revolutionary leader Mao Zedong, the economic reforms of 1978 began the process of opening up trade links and allowing competition to operate. China is now fast becoming the workshop of the world and the Chinese clearly embrace the market system – although they still regard themselves as socialist.

As a consequence, most of the large, previously centrally planned, nation state economies are now involved in a transition towards market-oriented systems – and this will be reviewed in the next section on mixed economies. This transition is not without problems, however, as much is still influenced by the old regime of the centrally controlled plan. There is a kind of culture shock as a withdrawal of the 'visible hand' – as Rod Sweet (2002) refers to the planners and tweakers of a command economy – leads to the need to interpret the signals of the 'invisible hand' of the free market. In terms of the construction sector, this involves acquiring new skills to manage time, cost and quality.

Summary: What? How? For Whom?

What   In a centrally planned economy, the collective preference and wisdom of the central planners ultimately determines what is produced.

How   The central planners decide on the methods of production. This means that they need to know how many resources to allocate to each industry, many of which interrelate.

For whom   The relative rewards that people get are set by the central planners rather than the market. Thus market forces are not all important in determining factor rewards. There may be more opportunity to achieve some kinds of equality.

Key Points 2.2

○ The centrally planned (model) economy relies on government commands rather than decentralised markets. It is also referred to as a command economy.

○ The key attributes of a centrally planned (model) economy are (a) the government owns and controls most of the resources, (b) a planning committee identifies production targets, (c) the rewards for producing are usually set by the state rather than the market, and (d) there is a desire to achieve the 'right' mix of output.

○ In any centrally planned system there are problems co-ordinating the different sectors.

○ The difficulties of formulating and implementing the central plans have caused many command economies to introduce the market mechanism.

THE MIXED ECONOMY

The two economic systems introduced in this chapter do not exist in a pure form. The economic models used are simplified representations of the real world. In practice, most economic systems are far more complex; countries are neither purely free market nor purely planned. In the complex setting of everyday life, all nations have a **mixed economy**. Economists do not, therefore, study systems in which the activities of consumers and producers interact freely through a market or simply according to government plans. Economists study systems that contain mixtures of private decision-making and central organisation; the private decisions (being made in response to market forces) exist alongside the centralised controls of state legislation and economic plans.

One way of comparing the range of economic systems that presently exist in the world is to imagine them located on a wide spectrum (similar to that in Figure 2.1). In theory, each nation could be positioned according to the proportion of resources owned by the public and private sectors. The nations that have a high proportion of government owned resources would be located close to the centrally planned model, and those dominated by privately owned resources would be located at the other extreme near to the free market model. Such an exercise illustrates that all economies of the real world are mixed economies; some may come close to one or other of the economic models, but no nation fits precisely into the pure planned or pure free market category. During the last two decades there has been considerable movement along the spectrum. Indeed, we do not actually show the precise position of any countries in Figure 2.1, because as soon as the economies are located, their position becomes out of date.

In general terms the **transition economies**, such as China and the states of the former Soviet Union, have shifted slowly away from the pure centrally planned model, while many European economies (particularly Britain) have moved closer towards the free market model under the influence of **privatisation** and **deregulation**. But transitions are rarely smooth, as it is difficult to extend entrepreneurship, liberalise prices formerly controlled by the government, encourage competition and generally challenge established cultural and economic norms. The *Transition Report* published annually since 1998 by the European Bank for Reconstruction and Development (EBRD) offers a reasonably comprehensive analysis of the progress of transition economies. The 2011 report noted that recovery from the late 2000s global recession was under way in virtually all countries of the EBRD region, although its pace varied significantly from one region to the next. The recovery was distinctively slow in central and eastern Europe (the former Soviet Union) and fastest in the countries of central Asia, such as Hong Kong, India and Indonesia, with seven-percentage points difference in GDP between the slowest and fastest transition economies (EBRD, 2011: 16).

The transition of command economies since the Berlin Wall fell in November 1989 has attracted considerable coverage in both the news media and academic literature, as economies shifting towards free market systems reflect changes in national aspirations and culture. Both Russia and China, for instance, have tried to harness some of the efficiency of the market mechanism in an attempt to raise the living standards of the Russian and Chinese people and, consequently, the government's role in these nations has been reduced in recent years. These changes have not gone unchallenged. For example, China rejected restructuring if it meant destroying the socialist principles upon which the country was built.

In the UK, and elsewhere in Europe, the mix of public sector and private sector activity has varied over time because of the differing philosophies that the main political parties have adopted towards state intervention. For much of the post-war period the desirability of less or more public ownership of industry lay at the heart of the political divide in the UK between the Labour and Conservative parties. In terms of construction activity, from 1955 to 1985, much of the output was commissioned on behalf of a public authority such as a government department, local authority, or nationalised industry. For example, up until the 1980s it was common for construction work relating to railways, airports, telecommunications, ports, gas, electricity, water and council housing to be commissioned by public sector authorities. In Figure 2.4 the outputs of construction (excluding repair and maintenance) across Britain, in constant (2005) prices, are plotted to show the historic divide between public and private sectors, and clearly the balance has swung towards private enterprise. In fact in recent decades, the way that public capital is managed has undergone profound change, with much of the ownership and operation transferred to the private sector – albeit within a regulatory framework. This has been the case in many countries in the world, particularly in Europe, and as a consequence the output of infrastructure has begun to be separately recorded (as in Britain since 1980). Today the distinction between the public and private sectors is blurred. Construction output in the water, sewage, energy and telecommunications sectors is typically funded by the private sector and regulated by government agencies. Health, education, housing and rail networks are funded by both the public and private sectors, while only funding relating to the building of roads, footpaths, bridges, tunnels, nuclear power plants and government facilities tends to remain a responsibility of the public sector.

In simple terms, therefore, economies are no longer planned or market led; as Samuelson and Nordhaus (2010: xvi) are keen to suggest the global trend today

Figure 2.4 New construction output 1955–2010 (Great Britain)

£Bn (2005 prices)

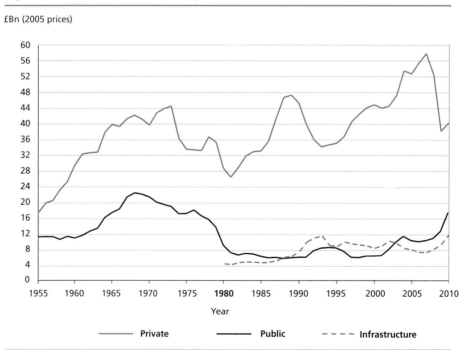

Source: Adapted from ONS (2011b)

is to operate in a *centrist* (mixed) mode, an approach that 'proclaims the value of the mixed economy – that combines the tough discipline of the market with fair-minded government oversight'. This approach recognises that neither unregulated markets nor over-regulated planned systems can organise and allocate resources efficiently. Indeed the global financial crisis of the late 2000s has highlighted how the innovative, unregulated, market products of modern finance can produce chaos, debts and economic collapse. As a result, during the next decade, societies will be redirected back to a fairer and more efficient centre – and the leitmotif for the foreseeable future seems to be an increasing reliance on the mixed economy to allocate resources fairly and efficiently.

Key Points 2.3

○ In reality all economies are mixed economies, since elements of private markets and state coexist in all nations.

○ It is the degree of market orientation, or of state intervention, that distinguishes one economic system from another.

○ Economies of the world are currently in a state of flux and political pressures for greater regulation of financial systems will bring change.

○ The current preference is for mixed systems of resource allocation, relying on the market wherever possible but within a regulated framework.

EQUITY, EFFICIENCY AND THE ENVIRONMENT

Much of the opposition that exists to market reforms relates to questions concerning the environment and the issue of social justice. There are certainly many environmental problems and social inequalities spawned by the market mechanism. For example, those with the greatest wealth have most 'votes' about what is produced, while those with no incomes, if left to fend for themselves, get nothing. On the other hand, the market mechanism does provide an efficient system of communication between producers and consumers that effectively signals 'what' to produce and 'how'. It does, however, fail to capture common goods such as environmental pollution.

Since the United Nations Conference on Environment and Development at the 1992 Earth Summit in Rio de Janeiro set the stage for sustainable development and related issues, the ideas and concepts lying beneath **equity, efficiency** and the **environment** have taken on an increasing significance. The basis of any sustainable development agenda, in any sector of the economy, must involve a mutual consideration of the business case, the wider community and the natural environment. Any sustainable outcome, satisfying the stakeholders representing these three interests simultaneously (whose needs often conflict) could genuinely be described as an 'elegant solution'. An important proposition of this text is that such a solution depends upon economics – as ultimately any policy that is fully sustainable boils down to questions of resource allocation.

Efficiency

In economics, efficiency is mainly concerned with resolving the questions of 'what' to produce and 'how'. The concept is accordingly often divided into two parts – **productive efficiency** and **allocative efficiency**, both of which are satisfied in a pure free market economy.

PRODUCTIVE EFFICIENCY

Productive efficiency means using production techniques that do not waste inputs. Expressed in the language of policy documents concerning sustainability, it means increasing growth rates while reducing the use of resources. In any free market economy businesses will never waste inputs. A business will not use 10 units of capital, 10 units of labour and 10 units of land when it could produce the same amount of output with only 8 units of capital, 7 units of labour and 9 units of land. Productive efficiency therefore refers to output that is produced at the lowest possible cost. During the last decade, for example, the increasing use of prefabricated components on site has enabled improvements in construction productivity – as this has encouraged greater levels of output with fewer workers. These developments depend upon managers responding 'correctly' to the various input prices facing them. The more expensive the inputs, the more incentive managers have to economise. The market signals, therefore, 'how' production should technically occur.

ALLOCATIVE EFFICIENCY

Allocative efficiency relates to maximising the total value (sometimes called utility) of the available resources. This means that resources are moved to their highest-valued uses, as evidenced by consumers' willingness to pay for the final products. The process of demand and supply guides resources to their most efficient uses. Individuals, as business people looking after their own self-interest, end up – consciously or unconsciously – generating maximum economic value for society. 'What' is produced, therefore, should involve no welfare losses; the utility of all groups in society should have been considered.

Equity

Equity does not, in its economic sense, simply mean equality. In this discipline, equity relates to fairness and social justice. From an economist's point of view, therefore, discussions of equity become closely related to considerations of sustainability such as respecting and treating stakeholders fairly – the 'for whom' question. Equity may also be broken down into two parts, **horizontal equity** and **vertical equity**, both of which depend upon government intervention.

HORIZONTAL EQUITY

This concept involves treating people identically. For example, a government policy of horizontal equity would support and promote equal opportunities between people of identical qualifications and experience, regardless of race or gender.

VERTICAL EQUITY

This concept is more contentious since it is concerned with being 'fair'. Vertical equity is about reducing the gap between the 'haves' and the 'have-nots'. It can involve governments providing targeted support to specific categories of people. For example, it may involve taxing the rich more heavily to provide services to support the poor.

It should be clear from these explanations that equity and efficiency often conflict. In fact, they can be imagined as polar extremes; as far apart as the two economic models shown in Figure 2.1. A centrally planned model economy, based on government intervention, is more able to support the concept of equity; whereas a free market model economy is able to foster a greater level of efficiency. It is the trade-off between these two qualities that, to some extent, accounts for the changes that nations are experiencing with their economic systems. The equity-efficiency trade-off certainly explains why all nations have become mixed economies. As *The Economist* (2002: 9) observed more than a decade ago: 'The vision is for a Britain which sits mid-way between America (free enterprise) and Europe (fairness), allowing business to grow fast while providing its people with good social services.' Ten years later governments continue to wrestle with the same types of challenge.

Environment

Compounding the two way trade-off between efficiency and equity is the environment. The agenda of sustainability has significantly raised its profile. As the media continually reminds us, seas are being polluted, forests devastated, species are becoming extinct, rain is acid and the ozone layer is being destroyed. These environmental issues are usually considered as **market failures** and will be dealt with fully in Chapter 10. They need to be introduced here, however, as they are central to understanding sustainability and implicated in all the resource allocation questions of 'what', 'how' and 'for whom'.

In economics, environmental goods and services are often discussed as a means of highlighting common problems of resource allocation regardless of the system. The explanation relies upon distinguishing between two categories of cost: **private costs** and **external costs**.

PRIVATE COSTS

These are the market costs of an individual's actions that are known and paid for directly. For example, when a construction business has to pay wages to site workers, it knows exactly what its labour costs are. When it has to buy materials or build a factory, it knows quite well what these will cost. Similarly, if individuals have to pay for car repairs, shoes or concert tickets, they know exactly what the cost (market price) will be. In short, private costs are those borne solely by the companies or individuals who incur them. They are 'internal' in the sense that a firm or household must explicitly take account of them.

EXTERNAL COSTS

These are not so straightforward, as they represent the costs of actions borne by people other than those who commit them. They are also referred to as third-party or neighbourhood costs. For example, consider the situation in which a construction business dumps waste products from its building site into a nearby river or in which an individual litters a public park or beach. Obviously, a cost is involved in these actions. When the construction firm pollutes the water, people downstream suffer the consequences. They may not want to swim in or drink the polluted water. The pollution may also ruin the prospects for fishing. In the case of littering, the people who come along after litter has cluttered the park or the beach are the ones who bear the costs. In other words, the creator of cost is not necessarily the sole bearer of that cost. The individual or firm does not internalise all costs – some external ones are overlooked.

Environmental problems may be described as situations in which the external costs exceed private costs. This is difficult for either market or command systems to control. The problem is that some collective resources are taken for granted – and the full costs of the economic activity are not accounted for, only the smaller private costs relating to some of the resources. This is neither efficient nor fair, and environmental problems can therefore be regarded as sitting midway on the spectrum of economic systems shown in Figure 2.5.

A society has to rely on public bodies such as the **National Audit Office**, or its equivalent, to assess the effectiveness of government spending and measure its impact on all resources. These activities assure that some kind of balance is achieved

Figure 2.5 The trade-off between equity, efficiency and the environment

Adding the environment to the spectrum showing the trade-off between equity and efficiency brings in a third dimension which does not particularly lend itself to the market or command system. In effect, the problem is triangulated, and meeting all three criteria simultaneously is a difficult challenge.

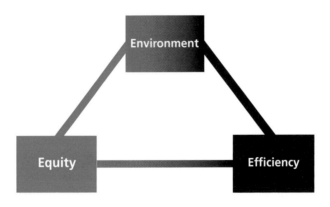

between economic efficiency, protection of the natural environment and the spending of taxpayers' money. Systems of **value management** can also be used to assess the efficiency, environmental impact and general effectiveness of projects. The use of value management is slowly increasing in both public and private sectors and will be discussed further in relation to construction projects in Chapter 6.

A recurring theme of this text involves studying how the construction industry can simultaneously embrace environmental, social and business concerns. Hence we shall, implicitly and explicitly, revisit the three strands of efficiency, equity and environment several times before reaching the concluding chapter of the book – which reviews the possibility of the construction industry becoming sustainable.

Key Points 2.4

○ Ideas relating to efficiency, equity and the environment have gained validity since the sustainable development agenda has increased in importance.

○ There are two concepts of efficiency: productive efficiency – when inputs are not wasted; and allocative efficiency – when resources are employed in their highest-valued uses.

○ There are two concepts of equity (or fairness): horizontal equity – equality of opportunity; and vertical equity – actions to achieve social justice or fairness.

○ A free market system encourages efficiency and a centrally planned system promotes equity.

○ There are two concepts of cost: private costs – the costs to the first party; and external costs – the costs to third parties. The prevalence of environmental problems can be explained by the fact that external costs can be ignored or overlooked by those considering just private costs.

○ Problems relating to the environment seem to pervade both free market and centrally planned systems of resource allocation.

○ A recurring theme of this text is sustainable construction, which depends upon the industry embracing the three concerns of efficiency, equity and the environment.

3 The Market Mechanism

In Chapter 2 we emphasised that a central tenet of almost any modern economic system is a strong belief in the market mechanism (see Key Points 2.1 and 2.3). We also began to recognise that the forces of supply and demand are closely associated with this mechanism. This was highlighted in Figure 2.2. To interpret market signals perceptively – whether in the construction industry, building materials sector, property market or whatever – it is necessary to consider the price mechanism in more detail. We now begin to elaborate our understanding of the price mechanism. We can define it more fully as:

> an economic system in which relative prices are constantly changing to reflect changes in supply and demand for different goods and services.

The **price mechanism** is synonymous with the term **market mechanism**.

The forces of supply and demand are all encompassing within the market economy. Even before any good or service can be produced, the various factors of production need to be employed and their prices are affected by supply and demand. The forces of supply and demand bring together producers and consumers in such a way that appropriate goods and services are produced and appropriate incomes are rewarded. To highlight this interrelationship between supply and demand in different markets, we first consider an overview of a market-based economy.

ALLOCATION IN A MARKET ECONOMY

Figure 3.1 (see page 46) displays how the wishes of producers and consumers are 'signalled' by the price mechanism. To begin with **factor markets**: households (consumers) supply economic resources (factors of production) to firms that demand them to undertake productive activity. This is shown in the top half of Figure 3.1. The supply and demand of these factors (resources) determine the prices paid for them in the particular sector in which they are put to economic use. Householders receive differing levels of wages for their labour input, interest for the capital services they provide, rents for the land that they own and profits for their entrepreneurial abilities, according to the sector supplied.

In **product markets**, a similar scenario exists: firms supply various goods and services according to consumer demand. Again, it is the market price that balances the two parties' interests. This is shown in the bottom half of Figure 3.1.

Price Signals and Self Interest

For a market economy to function effectively, it is important that every individual is free to pursue 'self interest'. Consumers express their choice of goods or service through the price they are prepared to pay for them – in their attempts to maximise

Figure 3.1 Product and factor allocation via the price mechanism

In this simplified model households supply their services – labour, capital, land and entrepreneurial skills – to firms that demand them for production. The prices paid in factor markets are the balancing items that determine resource allocation. In product markets, firms supply goods and services to households that demand them, and prices provide the balancing item to determine distribution.

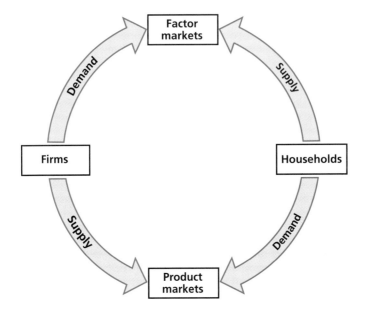

satisfaction. Producers, and owners of resources (and for most people this is their own labour power), seek to obtain as large a reward as possible in an attempt to maximise profit.

If consumers want more of a good than is being supplied at the current price, this is indicated by their willingness to pay more to acquire the good – the price is 'bid up'. This in turn increases the profits of those firms producing and supplying the good – and the incomes paid to the factors producing that good increase. As a result, resources are attracted into the industry, and supply expands.

On the other hand, if consumers do not want a particular product, its price will fall, producers will lose money and resources will leave the industry. This is precisely what happened in the 'new build' market during the recession that followed the credit crunch in 2007. The demand for new houses declined and prices fell; as a result, producers either concentrated on other construction work or went bankrupt. Consequently, between 2008 and 2010, the completion of new homes fell by more than 30 per cent, reaching the lowest level since 1924.

In simple terms, therefore, the price system indicates the wishes of consumers and allocates the productive resources accordingly. Or, in the terms we used in Chapter 2, the price mechanism determines *what* is produced, *how* it is produced and *for whom* is it produced. Now we shall examine how well these general principles apply to the UK construction industry.

Price Signals in the Construction Industry

According to the Interdepartmental Business Register (which is used for all ONS business surveys) there are more than 220,000 firms that make up the construction industry. This makes the market rather fragmented and very distinct from manufacturing. In the manufacturing sector there are usually a few firms producing a similar product that can be freely examined before purchase. In construction the opposite seems to apply – and a construction project usually involves many small firms combining their skills on site to produce a 'unique' specified product. This process means that the construction industry tends to be less efficient than manufacturing.

The reason individuals and businesses turn to markets to conduct economic activities is that markets generally reduce the costs of trading. These costs are called **transaction costs** because they are part of the process of making a sale or purchase. Economists define transaction costs as all of the costs enabling exchanges to take place. They include the cost of being informed about the qualities of a particular product, such as its availability, its durability record, its servicing facilities, its degree of safety, and so on. Consider, for example, the transaction costs in shopping for a new computer. These costs include phone calls or trips to sellers in order to learn about product specifications and prices, and subsequently there is the cost of negotiating the sale and, potentially, the cost of enforcing the contract should the machine fail to operate. In a purely theoretical or highly organised market these costs do not exist, as it is assumed that everybody has access to the knowledge they need for exchange to take place.

In the context of construction, however, these transaction costs are significant and go some way to explaining the adversarial nature of competition within the industry. A particular problem is the fact that often the product that is being created by a construction firm does not exist before the exchange is made. The product has to be precisely specified to assure the quality and quantity before an exchange can be agreed. And, complicating matters, firms will offer the product at differing prices. Consequently, contractual costs arise to clarify what is expected and to cover any contingencies in the proceedings, such as what happens if the work is handed over late, over price or fails to meet the anticipated quality. Gruneberg and Ive (2000) have identified six types of transaction costs that anyone demanding the services of a construction firm may have to meet. These transaction costs are summarised in Table 3.1 (see page 48).

Some economists argue that the increased use of electronic communication will reduce transaction costs both between businesses and between businesses and consumers. In fact, there is an ongoing academic debate about the significance of the **new economy** and its implications for B2B (business to business) and B2C (business to consumers) transactions. At present the construction industry does not effectively feature in this debate, as e-commerce has made little impact on the way construction firms communicate with one another or with their customers. High transaction costs, therefore, are destined to remain a hurdle to the efficiency of the modern construction industry working within a market context and we will revisit elements of this debate in Chapter 6.

Table 3.1 Transaction costs which affect construction

Search costs	The cost of having to find out what is on offer
Specification costs	To clarify precisely what is required to assure the desired quantity and quality
Contract costs	The cost of finding or creating conditions of contract to clarify what is expected and cover contingencies
Selection costs	The cost of choosing the best tender
Monitoring costs	To measure and control price, timing and quality
Enforcement costs	The legal bills relating to breached contracts

Source: Adapted from Gruneberg and Ive (2000: 123–4)

A belief in the importance of market signals dates back to the classical economists. They emphasised how prices and wages continually adjust to keep the general levels of supply and demand in balance. In fact, their belief in market forces was so strong that for a significant historical period economists recommended that the management of an economy could simply follow a laissez faire approach. Milton Friedman, a modern exponent of the market mechanism, strongly argued that the need for government intervention is minimal – as governments are only needed to provide a forum for determining the rules of the game and to act as an umpire to assure the rules are enforced. Interestingly, when this classical approach was first challenged, during the depression of the 1930s, construction was one of the first sectors that governments chose to manipulate.

Key Points 3.1

○ The price mechanism is synonymous with the market mechanism.

○ The forces of supply and demand in factor and product markets are reconciled by price (see Figure 3.1).

○ Price movements provide the signals that freely responding individuals interpret, determining what is produced, how it is produced and for whom it is produced.

○ Transaction costs are costs associated with exchange.

○ Transaction costs are significant within the construction industry, and this reflects the fragmented and adversarial nature of firms involved in the process.

○ Recognition of the importance of the market mechanism dates back to the classical economists.

GRAPHICAL ANALYSIS

Analysis of the price (market) mechanism has played a significant role within the history of economics. As far back as 1776, Adam Smith wrote in *The Wealth of Nations* about the 'hidden hands' of supply and demand determining market prices. Since then a standard part of any economics course has involved the study of supply and demand graphs. This is probably because it is easier to communicate an idea visually – as the saying goes 'a picture is worth a thousand words'. A supply and demand graph enables the relationship between price and quantity to be explored from the consumers' (demand) perspective and the producers' (supply) perspective. The standard layout of each axis is shown in Figure 3.2.

Figure 3.2 The axes of a supply and demand graph

On the vertical axis it is customary to plot the price per unit. On the horizontal axis we plot the quantity demanded and/or supplied per period of time.

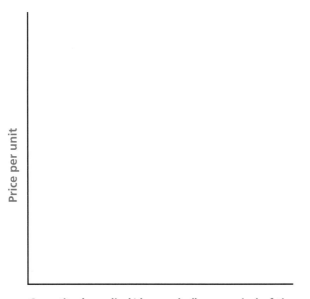

Using the labelled axes in Figure 3.2 can you determine what the pattern of demand in relation to the price would look like? To put the question in more formal terms, can you plot a **demand schedule**? Clearly, as the price of a commodity rises, the quantity demanded will decrease and as the price falls, the quantity demanded will increase. That is, from the demand side there is an **inverse relationship** between the price per unit and the quantity purchased: higher prices lower (that is, cause smaller quantities of) demand. This is because consumers seek to maximise their satisfaction and get best value for money.

Using the labelled axes in Figure 3.2 can you determine what the pattern of supply in relation to price would look like? To rephrase the question in more formal terms, can you plot the **supply schedule**? Clearly, as the price of a commodity rises the quantity supplied will increase and as the price falls the quantity supplied will decrease. That is, from the supply-side there is a **direct relationship** between the price per unit and quantity supplied: an increase in price usually leads to an increase in the quantity supplied. This is because suppliers seek to maximise their profit and get the biggest possible return for their efforts.

As suggested, these basic principles seem easier to appreciate when plotted on a graph. See if you agree by considering Figure 3.3.

Figure 3.3 A simple supply and demand diagram

The demand curve displays the fact that the quantity demanded falls as the price rises. The supply curve displays the converse relationship: that the quantity supplied rises as the price rises.

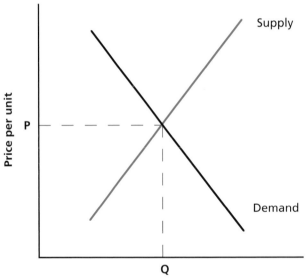

Three Qualifying Remarks

Economists have devised various methodological procedures to give their statements rigour and academic value. Three of these small but important techniques need to be highlighted especially as textbooks often fail to constantly enforce them.

PER PERIOD OF TIME

On the horizontal axis of Figures 3.2 and 3.3 we ended the statement with the qualification *per period of time*. This is to highlight that supply or demand is a flow that takes place during a certain time period. Ideally, the time period should be specified as a month, year, week, or whatever. Without a time dimension, the statements relating to quantity become meaningless.

CETERIS PARIBUS

The second qualifying remark relates to the Latin phrase **ceteris paribus**, which means other things being equal or constant. This is an important assumption to make when dealing with a graph showing two variables. For example, price is not the only factor that affects supply and demand. There are many other market conditions that also affect supply and demand and we cover these in Chapters 4 and 5. In the exercise above, constructing Figures 3.2 and 3.3, we assumed *ceteris paribus*. We did not complicate the analysis by allowing, for example, consumers' income to change when discussing changes to the price of a good. If we did, we would never know whether the change in the quantity demanded or supplied was due to a change in the price or due to a change in income. Therefore, we employed the *ceteris paribus* technique and assumed that all the other factors that might affect the market were held constant. This assumption enables economists to be more rigorous in their work, studying each significant variable in turn. The *ceteris paribus* assumption approximates to the scientific method of a controlled experiment.

SUPPLY AND DEMAND CURVES

When using supply and demand curves to illustrate our analysis, they will frequently be drawn as straight lines. Although this is irritating from a linguistic point of view, it is easier for the artist constructing the illustrations and acceptable to economists, since the 'curves' very rarely refer to the plotting of empirical data. It is worth noting, therefore, that so-called supply and demand 'curves' are usually illustrated as straight lines that highlight basic principles.

Key Points 3.2

- ○ On a supply and demand graph, the horizontal axis represents quantity and the vertical axis represents the price per unit.

- ○ Supply and demand curves illustrate how the quantity demanded or supplied changes in response to a change in price. If nothing else changes *(ceteris paribus)*, demand curves show an inverse relationship (slope downward) and supply curves show a direct relationship (slope upward) as shown in Figure 3.3.

- ○ To understand the premise upon which supply and demand diagrams are drawn, it is important to remember three criteria: (a) the time period involved, (b) the *ceteris paribus* assumption and (c) the shape of the 'curves'.

THE PRICE IS RIGHT

Look again at Figure 3.3: inevitably there is a point at which the two curves must cross. This point represents the market price. The market price in Figure 3.3 is P and this reflects the point where the quantity supplied and demanded is equal, namely point Q.

At price P the market clears. There is no excess supply; there is no excess demand. Consumers and producers are both happy. Price P is called the **equilibrium price**: the price at which the quantity demanded and the quantity supplied are equal.

Most markets tend towards an equilibrium price, including the labour market, housing market, foreign exchange market, drainpipe market, or whatever. All markets have an inherent balancing mechanism. When there is excess demand, price rises; and when there is excess supply, prices fall. Eventually a price is found at which there is no tendency for change. Consumers are able to get all they want at that price, and suppliers are able to sell the amount that they want at that price. This special market concept is illustrated in Figure 3.4.

Figure 3.4 The determination of equilibrium price

In this example, the equilibrium price for each new flat is £100,000 and the equilibrium quantity is 2,000 units. At higher prices there would be a surplus: flats would be in excess supply and they will remain empty. For example at £150,000 the market would not clear; there needs to be a movement along the demand curve from H to E and a movement along the supply curve from h to E. These movements necessitate the price to fall. At prices below the equilibrium, there would be a shortage: flats would be in excess demand and there would be a waiting list. For example, at £50,000 the price would rise, reducing the demand from F to E and increasing supply from f to E. This market will tend to settle at a price of £100,000.

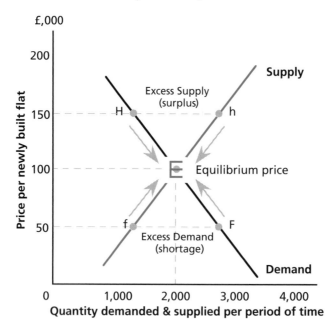

The Concept of Equilibrium

The concept of equilibrium is important in economics and we will be referring to it in different markets and in different contexts as we study the economy. **Equilibrium** in any market may be defined as:

> a situation in which the plans of buyers and the plans of sellers exactly mesh.

Equilibrium prevails when opposing forces are in balance. In any market, the intersection of a given supply curve and a given demand curve indicates the equilibrium price. If the price drifts away from this equilibrium point – for whatever reason – forces come into play to find a new equilibrium price. If these forces tend to re-establish prices at the original equilibrium point, we say the situation is one of **stable equilibrium**. An unstable equilibrium is one in which if there is a movement away from the equilibrium, there are forces that push price and/or quantity even further away from this equilibrium (or at least do not push price and quantity back towards the original equilibrium level).

The difference between a stable and an unstable equilibrium can be illustrated with two balls: one made of hard rubber, the other made of soft putty. If you squeeze the rubber ball out of shape, it bounces back to its original form. On the other hand, if you squeeze the ball made of putty, it remains out of shape. The former illustrates a stable equilibrium and the latter an unstable equilibrium.

Now consider a shock to the system. The shock can be shown either by a shift in the supply curve, or a shift in the demand curve, or a shift in both curves. Any shock to the system will produce a new set of supply and demand relationships and a new price-quantity equilibrium. Forces will come into play to move the system from the old price-quantity equilibrium to a new one. Now let us consider a specific example in the housing market.

A Change in the Conditions of the Market

To illustrate the dynamics of the market imagine what might happen if mortgage interest rates rise, while other things remain constant. This will reduce the demand for owner-occupied property at each and every price. This decrease in demand is shown in Figure 3.5 (see page 54) in the traditional economist way, by shifting the demand curve to the left from D_1 to D_2. If property prices now stay at P, consumers will only demand Q_a while suppliers (sellers) will continue providing Q. Consequently there will be an excess amount of supply in the market place equal to $Q - Q_a$. However, providing prices are allowed to move to make the amounts supplied and demanded equal again, suppliers will be able to off-load vacant properties by reducing their prices. As the price falls consumers will become interested in buying and demand will increase. Consequently a new equilibrium price will be arrived at. This new price is P_1 in Figure 3.5 and the new quantity being demanded and supplied will now be equal to Q_1.

The shifting of the demand curve (such as in Figure 3.5) only occurs when the *ceteris paribus* assumption is violated. In other words, the curves only shift to a new position when the market conditions change. We will explore these 'shifts' in more detail in the next two chapters.

Figure 3.5 Changing market conditions lead to a new equilibrium price

The leftward shift of the demand curve indicates that consumers are now willing and able to buy fewer properties in every price range due to the increase in mortgage rates. The excess supply of properties on the market at the old price P causes a new equilibrium to be found at the lower price P_1, at which the quantity demanded and quantity supplied are once again found to be equal.

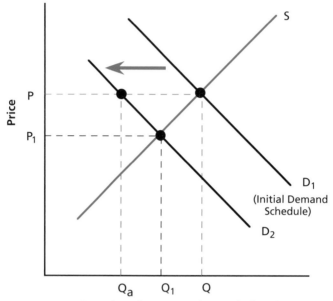

Quantity of property demanded and supplied for owner occupation per year

A CLOSING NOTE

Finally, to complete our introductory overview of market forces, we also need to recognise the existence of **market failure**. So far we have not assumed any market distortions. For example, we have assumed that labour will move freely to wherever work is most profitable and consumers will buy whatever they desire in freely determined markets. Yet, in reality, monopolies, oligopolies, subsidies, trade unions, externalities, high transaction costs and other market imperfections distort the situation. We shall be examining these issues in more detail in Chapter 10. But they are worth bearing in mind throughout the book, especially as our analysis is also concerned with how modern economies can manage to achieve construction that is sustainable, and these imperfections tend to create a gap between theoretical solutions and market-driven practice.

Key Points 3.3

○ When we combine demand and supply curves, we find the equilibrium price at the intersection of the two curves. At the equilibrium price there is no tendency to change, the market clears (see Figure 3.4).

○ Equilibrium exists whenever the separate plans of buyers mesh exactly with the separate plans of the sellers. Price points the buyers and sellers in the right direction.

○ If conditions in the market change, the relevant curve shifts and a new equilibrium position is established (see Figure 3.5).

○ Market failures exist, and market distortions form an important consideration when setting any agenda aimed at achieving sustainability.

4 The Theory of Demand

As suggested in the previous chapter and reviewed in Key Points 3.1 the concepts of supply and demand are the basic building blocks of economics, whether the outcomes are sustainable or not. In this chapter, we focus specifically on demand and in the next chapter we deal with supply.

When economists speak of demand they mean **effective demand**. Effective demand is money backed desire. It is not the demands of a crying baby or of a spoilt child wanting and grabbing at everything it sees. Demand from an economist's point of view is real, 'genuine' demand backed by the ability to make a purchase. It is distinct from need. For example, in 2012 the total number of households in England needing accommodation exceeded the total number of homes in the housing market. Only those who had sufficient means to 'demand' accommodation – that is, they could afford to buy or rent at market prices – were confident of securing somewhere to live. This anomaly explains the number of homeless people living in bed and breakfast hostels at a cost to the government.

Figure 4.1 A standard market demand curve

The demand curve for most goods and services slopes downward from left to right, as the higher the price, the lower the level of demand (other things being equal).

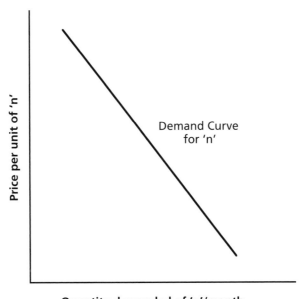

Quantity demanded of 'n'/month

THE BASIC LAW OF DEMAND

We already established in the last chapter (for example see Key Points 3.2) that a demand curve has a negative slope. It moves downward from left to right. This is illustrated in Figure 4.1.

The shape of the demand curve for most goods or service is not surprising when one considers the basic **law of demand**. This may be stated formally as:

> at higher prices, a lower quantity will be demanded than at lower prices (and vice versa), other things being equal.

The law of demand, therefore, tells us that the quantity demanded of a product is inversely related to that product's price, other things being equal.

To continue with this analysis, we must consider the 'other things being equal' phrase more carefully. Clearly, demand is not only affected by price. As already implied in Chapter 3, it is easy to see that other factors may also affect demand. The demand curve shown in Figure 4.1 may shift to another position. Conditions in the relevant market may change significantly enough to cause consumers to change the quantity demanded at each and every price. For instance, imagine that Figure 4.1 represents the demand curve for a product of the construction industry. What events may cause more or less demand for that product at every price?

DEMAND IN THE CONSTRUCTION INDUSTRY

The determination of demand for goods and services produced by the construction industry is a complicated process. This is partly due to the size, cost, longevity and investment nature of the products and partly due to the broad range of what constitutes construction activity. This will become clearer as we consider factors affecting demand for various sectors of the industry.

Demand for Housing

In 2010 there were in excess of 27 million households in the UK. A large majority of these 27 million households could afford to demand a home, but it is important to remember the introductory point about effective demand and avoid confusion between 'need' and 'demand'. Each household requires some kind of shelter – a flat, bungalow, terraced house, maisonette, semi-detached house, cardboard carton, or whatever. The related resources are allocated through the market mechanism and the public sector, or through some mix of the two. It is difficult to envisage just one housing market. Indeed, when statisticians discuss the UK housing market it is usual to distinguish four sectors.

1 The *owner occupied sector* – that is, accommodation owned outright or being bought with a mortgage.
2 The *private rented sector* – property that is let at a market rent deemed 'fair' to tenants and landlords.
3 The *local authority rented sector* – housing made available by the local authority (council) at a subsidised rate from public funds.

4 The *registered social landlord sector* – property managed by non-profit making organisations that combine public and private funds to provide housing for those in need. A large proportion of this sector is made up of properties managed by housing associations.

These four sectors have been listed in order of size according to the format used to officially measure tenure in the UK. In general terms, however, it is sufficient to understand that housing demand may be either for owner occupation or rent, or for some combination via shared equity. We now consider the main factors determining demand for housing within three main markets.

DEMAND FOR OWNER-OCCUPIED HOUSING

Most households in the UK, Ireland, Scandinavia, Australasia and much of Europe demand homes to own and occupy. Some 65 per cent of the households in the UK live in this form of tenure, and the relative size of the owner-occupied sector in Spain (80 per cent), Ireland (76 per cent) and Finland (78 per cent) is even larger. This form of ownership is generally supported by government initiatives that encourage demand by making the process of home buying as fast, transparent and as consumer friendly as possible. In the UK there have even been tax incentives for owner occupation, in the form of a subsidy on mortgage repayments offered regardless of income, though these have now been phased out. The logic is that if people own the property that they occupy, they will maintain it better. The feel-good factor derived from ownership makes the transaction costs of choosing and funding worthwhile, especially as a house provides an investment as well as a shelter. As you can imagine there are several factors that determine the demand for privately owned housing, and in Table 4.1 we identify the main ones.

Table 4.1 Factors affecting demand for owner-occupied housing

✓	The current price of housing
✓	The price of other forms of housing
✓	Income and expectations of change
✓	Cost of borrowing money and expectations of change
✓	Government incentives such as tax benefits
✓	Demographic factors such as the number of households
✓	Price of associated goods and services, such as maintenance, furniture, council tax, insurance, etc.

DEMAND FOR PRIVATELY RENTED HOUSING

Since the Second World War there has only been a small market demand in the UK for privately rented accommodation. At present approximately 16 per cent of UK households demand this type of accommodation. This is in direct contrast to some European economies where as many as 40 per cent of households are living in private rented accommodation. There has been a strong change in the pattern of housing demand in the UK. At the beginning of the twentieth century, people from all income levels routinely rented from private landlords. In 1915, for example, 90 per cent of UK families lived in the private rented sector. This change in the pattern of demand is closely associated with the supply drying up. This happened following the introduction of **rent controls** by government; these imposed a ceiling on rents – and created a big disincentive for landlords. As a result, current demand is mainly met by small-scale individual landlords who maintain and manage properties in their spare time. In an attempt to reverse this trend, the private rented sector in the UK has been largely deregulated in recent years.

In general, the market in private housing to rent varies greatly from country to country for a number of cultural and economic reasons. The main economic factors affecting the demand in this sector are listed in Table 4.2.

Table 4.2 Factors affecting demand for privately rented housing

✓	Current rent levels and expectations of change
✓	Income distribution – which determines affordability
✓	The cost of borrowing and expectations of change
✓	The law on rents and security of tenure
✓	Demographic factors, such as household formation
✓	The price of owner occupation

DEMAND FOR SOCIAL HOUSING

Housing provided by local authorities and registered social landlords, such as housing associations, is generally referred to as social housing. The origins of social housing lie in the idea that governments should pay a subsidy towards housing to make up for the shortage of accommodation available to low-income families. In the UK during the 1980s and 1990s – with both Conservative and Labour governments favouring free market policies – much of the local authority housing stock has either been transferred to housing associations to allocate, manage and maintain or sold to tenants thereby transferring stock to the owner-occupied sector. A similar process has been evident in the former Soviet Union, China, Czechoslovakia and Poland where privatisation of the social housing stock has been a key feature of the transition process.

Combined, the local authority and registered social landlords sectors still represent 'home' for approximately five million (18 per cent) of UK households. The factors that determine the demand for social housing are quite different from those driving demand for owner-occupied and privately rented housing. The main factors of demand for social housing are listed in Table 4.3.

Table 4.3 Factors affecting demand for social housing

✓	The current price (rent) of social housing
✓	The price level of other forms of tenure
✓	Assessment of need
✓	Availability of finance, such as income support and mortgages
✓	Levels of government subsidy

Demand for Industrial and Commercial Buildings

We now consider the demand for a whole range of buildings, including offices, factories, warehouses, hotels, garages, shops – in short, nearly all buildings except houses. Industrial and commercial buildings are not required for their own sake, but for the services they can provide. Consequently, demand for an industrial or commercial building is based on factors related to the specific sector in which the building will be used. Demand of this type is known as **derived demand**.

Derived demand implies that buildings are rented or purchased not because they give satisfaction, but because they can be used to produce goods or services that can be sold at a profit. This is different from the factors affecting the demand to buy a house. Following the emergence of problems in the sub-prime mortgage element of the US residential market towards the end of 2007, global business confidence was significantly dented and the demand for industrial and commercial property in the UK declined as a consequence during 2008 and 2009.

Investments in industrial and commercial building, therefore, depend on the expectation that the buyers or renters – that is, businesses – will make profits in the future. If business confidence is low, investment will not take place – even if there is current demand for an increase in production or sales. The factors affecting demand for industrial and commercial buildings are largely dependent on the state of the economy, and business expectations concerning output and profit. In other words, because demand is derived, it is dependent on many things other than price. Some of the main factors of demand for this broad category of industrial and commercial buildings are shown in Table 4.4.

Table 4.4 Factors affecting demand for industrial and commercial buildings

✓	Technological developments
✓	Changes in taste or fashion
✓	Expected levels of cost, including interest rates
✓	The state of the economy and government policy
✓	Business confidence
✓	The age and condition of the existing premises

Demand for Infrastructure and Other Public Sector Construction

We now consider demands for major infrastructure projects such as hospitals, roads, schools, tunnels, prisons, museums, bridges, and police and fire stations. These demands are created by large numbers of individuals who, left to their own devices, are not able or willing to pay the market price for the desired facility. In these cases, the government decides the level of service that should be available. This does not mean that the public sector necessarily finances these major infrastructure projects. In many countries public policy, or a lack of public funds, means that the provision of these facilities has been transferred to the private sector. Examples of private sector provision include toll motorways on the Continent. In the UK, the private finance initiative (discussed in Chapter 6) ensures that many facilities previously provided by the public sector are now financed by the private sector.

The demand for these products is again largely derived. An excellent example comes from China where recent highway construction has taken place alongside the development of new cities, changing lifestyles and rapid economic growth. In other words, China does not want highways per se, but it demands the modern ways of transacting economic activity rather than rickety, congested local roads.

The demand for community oriented facilities can be judged on much the same basis as industrial and commercial buildings. However, assessing the demand for these products is even more complex as it also depends on the assessment of need and the funds available. Some of the main factors of demand for this broad area of construction activity are shown in Table 4.5. In considering these factors, one can understand the political difficulties of choosing which of the various public needs should be transferred into effective demand.

Table 4.5 Factors affecting demand for infrastructure and public sector construction

✓	Assessment of need – present and future
✓	Availability of finance and levels of government subsidy
✓	Government policy
✓	The age and condition of the existing stock

Demand for Repair and Maintenance

The demand for the repair and maintenance of buildings and infrastructure cannot be ignored, since it accounts for the biggest area of activity in the construction industry. In the UK, for instance, it represents around 40 per cent of all construction activity, and that does not account for the repair and maintenance activities carried out by the informal, black market economy which probably contributes significantly to meeting this type of demand.

Generally it seems that the poorer the country, in terms of GDP per capita, the lower the proportion of construction work is repair and maintenance. This can be explained by the fact that developed economies have greater stocks of construction products – so there is less need to add to it by new building, but there will be a greater need for repair and maintenance. Furthermore, as countries develop they often become concerned with maintaining and protecting their cultural heritage. In some cases, buildings are conserved even though it would be cheaper to knock them down and build anew. This trend may become more pronounced as the campaign for sustainability encourages the preservation of non-renewable materials.

As a result, there are very many factors determining the level of demand for repair and maintenance work, which embraces a wide range of activity. In general terms the key determinants are listed in Table 4.6.

Table 4.6 Factors affecting demand for repair and maintenance

✓	The current cost of repair and maintenance
✓	The cost of new building
✓	Level of current income
✓	Government policy associated with heritage and conservation, etc.
✓	The age and condition of existing stock (and its state of disrepair)
✓	The ownership pattern, since owner-occupied property tends to be better maintained

A GENERALISED DEMAND EQUATION

If we carefully study the factors affecting demand in the various sectors of the construction industry set out in Tables 4.1 to 4.6, we can quickly recognise some general themes. For example, it seems that recurring determinants of demand are the price of the good being considered, the price of related goods, the level of income and government policy. Therefore, we can quickly move towards stating a generalised tool for analysis.

Economic theory does not normally refer to just one sector of the economy. The analysis is undertaken in such a general way that it can equally apply to any sector. For example, the demand for any good is sometimes presented in the form of a general equation as follows:

$$Q_{nd} = f (P_n, P_{n-1}, Y, G, ...)$$

This is formally referred to as a **demand function**. It may look complicated but it is only a form of shorthand notation. The demand function represents, in symbols, everything we have discussed above. It states that Q_{nd} the quantity demanded of good 'n' is f a function of all the things listed inside the bracket: P_n the price of the good itself, P_{n-1} the price of other goods, Y income, G government policy, and ... a host of other things. (The notation relating to the basic demand function is set out in Table 4.7, see page 64.) These equations may be adapted and extended as necessary to support the analysis of a specific sector. For example, in our analysis earlier in this chapter, we were able to specify some of the other determinants – the other things – because we identified what 'n' specifically represented. For example, we often referred to the age and condition of the existing stock, and an assessment of need.

Table 4.7 Factors affecting the demand for any product

P_n	The price of the product
P_{n-1}	The price of other products
Y	Income
G	Government policy
...	A host of other things

Key Points 4.1

○ The basic law of demand is that as price rises, lower quantities are demanded; and as price falls, higher quantities are demanded. There is an inverse relationship between the price and the quantity demanded, other things being equal (see Figure 4.1).

○ In the construction industry there appear to be many determinants that affect demand, including the price (rent) of the building or infrastructure, the price of other goods (such as substitute goods), current level of income and government policy.

○ Much of the demand for construction activity is of a derived nature, in as much as the goods are not necessarily demanded in their own right but for what they can add to the final good or service being produced.

○ The relationship between the quantity demanded and the various determinants of demand can be expressed as an equation. The general terms used in the equation are shown in Table 4.7.

CHANGING MARKET CONDITIONS

Clearly there are many non-price determinants of demand, such as the cost of financing (interest rates), technological developments, demographic make-up, the season of the year, fashion, and so on. For illustrative purposes we will consider just four generalised categories – income, price of other goods, expectations and government – taking each in turn and assuming *ceteris paribus* in each case.

Income

For most goods, an increased income will lead to an increase in demand. The phrase increase in demand correctly implies that more is being demanded at each and every price. For most goods, therefore, an increase in income will lead to a rightward shift in the position of the demand curve.

Goods for which the demand increases when income increases are called **normal goods**. Most goods are 'normal' in this sense. There are a small number of goods for which demand decreases as incomes increase: these are called **inferior goods**. For example, the demand for private rented accommodation falls as more people become able to buy their own homes. (It is important to recognise that the terms normal and inferior in this context are part of an economist's formal language, and no value judgements should be inferred when the terms are used.)

Price of Other Goods

Demand curves are always plotted on the assumption that the prices of all other commodities are held constant. For example, when we draw the demand curve for lead guttering, we assume the price of plastic guttering is held constant; when we draw the demand curve for carpets, we assume the price of housing is held constant. However, the prices of the other goods that are assumed constant may affect the pattern of demand for the specific good under analysis. This is particularly the case if the other good is a **substitute good** (as in the example of guttering) or a **complementary good** (as in the carpet and housing example). Economists consider how a change in the price of an interdependent good, such as a substitute or complementary good, affects the demand for the related commodity.

Let us consider the guttering example a little more fully. Assume that both plastic and lead guttering originally cost £10 per metre. If the price of lead guttering remains at £10 per metre but the price of plastic guttering falls by 50 per cent to £5 per metre, builders will use more plastic and less lead guttering. The demand curve for lead guttering, at each and every price, will shift leftwards. If, on the other hand the price of plastic guttering rises, the demand curve for lead guttering will shift to the right, reflecting the fact that builders will buy more of this product at its present price. Therefore, a price change in the substitute good will cause an inverse change in the pattern of demand for the other alternative.

The same type of analysis also applies for complementary goods. However, here the situation is reversed: a fall in the price of one product may cause an increase in the demand for both products, and a rise in the price of one product may cause a fall in the demand for both.

Expectations

Consumers' views on the future trends of incomes, interest rates and product availability may affect demand – and prompt them to buy more or less of a particular good even if its current price does not change. This is particularly evident when we consider the demand for construction based activities. Consider the demand for housing (see Tables 4.1, 4.2 and 4.3). For example, potential house purchasers who believe that mortgage rates are likely to rise may buy less property at current prices. The demand curve for houses will shift to the left reflecting the fact that the quantity of properties demanded for purchase at each and every price has reduced due to consumer expectations that mortgage rates will rise.

Government

Legislation can affect the demand for a commodity in a variety of ways. For example, changes in building regulations such as the Code for Sustainable Homes, introduced in 2007, have placed a greater focus on the standards expected for water and energy usage. This, in turn, has influenced the design of, and demand for, the standard fittings and appliances used in kitchen, bathrooms and general heating. The demand for these products has increased, regardless of their present price. Or to put it another way, the demand curve for all code compliant products has shifted to the right, reflecting the fact that greater quantities of these units are being demanded at each and every price. (It will take some time for suppliers to adjust to these new standards, so prices might at first rise before technology catches up.) The government can also influence the level of demand by changing taxes or creating a subsidy. Tables 4.1 to 4.6 allude to the influence of government on construction demand, and this will be considered more fully in Chapters 9, 10 and 11 which review government policy relating to effective protection of the environment.

Revisiting Ceteris Paribus

When we first introduced the idea of holding other things constant, it may have appeared that these 'other things' were unimportant. The previous section, however, should have highlighted how wrong this interpretation would be. Indeed the *ceteris paribus* assumption enables economists to emphasise the fact that price and a host of other factors determine demand. Whenever you analyse the level of demand for any construction product there will always be a need to consider both the price and many other related factors. To clarify this important distinction between the price determinant and the non-price determinants, economists are careful to distinguish between them when they discuss changes in demand.

UNDERSTANDING CHANGES IN DEMAND

We have already explained that changes of non-price determinants cause the demand schedule to shift to the right or to the left, demonstrating the fact that more or less is being demanded at each and every price. These changes are often referred to as *increases* or *decreases* of demand.

Let us consider one example in detail. How would we represent an increase in the quantity demanded of naturally ventilated commercial buildings (at all prices) due to a respected piece of research concluding that air-conditioned buildings caused **sick building syndrome**? The demand curve for naturally ventilated buildings would shift to the right, representing an increase in the demand at each and every price. This is shown in Figure 4.2.

We could use a similar analysis when discussing decreases in demand due to a change in non-price determinants. The only difference would be that the demand curve would shift to the left, demonstrating that the quantity demanded is less at each and every price.

Figure 4.2 Change in a non-price determinant causing a shift in demand

If a non-price determinant of demand changes, we can show its effect by moving the entire curve from D to D_1. We assumed in our example that the move was prompted by some research in favour of naturally ventilated buildings. Therefore, at each and every price, a larger quantity would be demanded than before. For example, at price P the quantity of naturally ventilated buildings demanded increases from Q to Q_1.

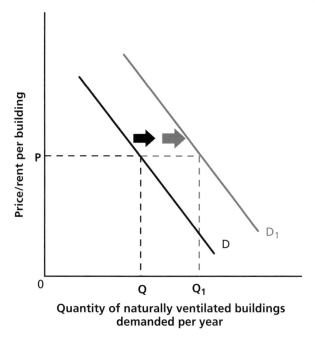

By contrast, the price determinant causes a movement along the demand curve. This is obvious when one remembers that the demand curve represents price and quantity relationships. Changes to the quantity of demand due to price alone are often referred to as an *extension* or *contraction* of demand. This involves a move along the demand curve. When more is demanded at a lower price, this may be regarded as an extension from one coordinate on the demand curve to another. When less is demanded due to a rise in price, demand contracts. Such movements along the demand curve are described further in Figure 4.3 (see page 68).

Before we can begin to apply our theoretical knowledge of demand, it is particularly important to remember the distinction between a movement along, and a shift in, a demand curve. These rules will not only help us to understand the graphical analysis, but they will also enable us to acknowledge the numerous factors that come into play when interpreting demand in the construction industry.

**Figure 4.3 Change in price causing a movement along a given
demand curve**

We show the demand curve for a hypothetical good X. If the price is P_1, then the
quantity demanded would be Q_1; we will be at coordinate A. If the price falls to P_2,
and all other factors in this market remain constant, then there will be an extension of
demand to Q_2 – from coordinate A to coordinate B

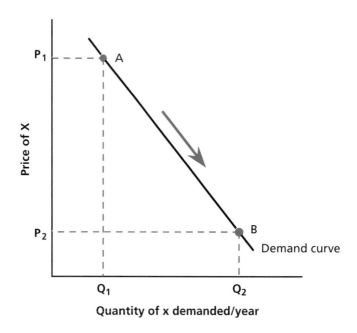

Quantity of x demanded/year

Key Points 4.2

○ Four major non-price determinants are (a) income, (b) price of other goods,
(c) expectations and (d) government policy.

○ If any of the non-price determinants changes, the demand schedule shifts
to the right or left and we refer to an increase or decrease of demand
(see Figure 4.2).

○ Movements along a given demand curve are caused by price changes and
these are described as contractions or extensions of demand (see Figure 4.3).

ELASTICITY OF DEMAND

To complete this introduction to the theory of demand we need to briefly add that professional economists are also interested in measuring the degree of response of demand to a change in market conditions. The measurement of response is termed **elasticity**. It may be defined as:

> a measurement of the degree of responsiveness of demand to a change in an external variable.

A numerical value for the elasticity of demand may be calculated in relation to price, income or a related good using the formula:

$$\text{elasticity of demand} = \frac{\text{percentage change in demand}}{\text{percentage change in the explanatory variable}}$$

In this text we focus on price as the key variable. For instance, we may wish to know the extent to which a change in the price of petrol will cause the quantity of demand for petrol to change, other things held constant. Let's say that petrol prices rise by 10 per cent and this leads to a reduction in demand by 1 per cent. The calculation would be as follows:

$$\text{price elasticity of demand} = \frac{-1\%}{+10\%} = -0.1$$

This is a very small response and, in technical terms, whenever this measure is less than one demand is referred to as inelastic. The theory of demand states that quantity demanded is inversely related to the relative price, consequently price elasticity is always a negative number – if the price rises, which is a positive percentage change, the quantity demanded falls, which is a negative percentage change – but it is a convention that economists ignore the negative sign when discussing the price elasticity of demand. Therefore, if the price elasticity is stated as 0.3 it can be inferred that if price goes up by 1 per cent demand will go down by 0.3 per cent – so a 10 per cent increase will cause a decrease in demand of about 3 per cent. (By contrast, calculations relating to price elasticity of supply will nearly always be positive, as there is usually a direct relationship between the two – as price increases or decreases so does the quantity supplied.)

Another qualifying comment is the importance of percentage changes, as they allow us to measure price changes in terms of pence, pounds or thousands of pounds, and quantity changes in some other unit of measurement. The important thing is to be able to express the changes as relative values. A technical problem, however, arises from the way that percentage variables can be calculated. The standard way to compute a percentage change is to divide the change by the initial, or starting, level; but the same overall change will generate different answers depending on whether the starting point is the higher price and smaller quantity or the lower price and greater quantity. The way out of this difficulty is to take the average of the two prices and the two quantities over the range in question and divide the change with the average. This is known as the **midpoint method** of calculation, as it uses the average of the initial and final values to compare the percentage changes in price and quantity.

Even if you understand the mathematical rigour necessary to calculate the midpoint elasticity it is still not easy to be 100 per cent accurate. This is because changes in price are not the only thing to affect changes in quantity demanded: other factors, such as changes in income, population and technology also affect demand. So estimating the price elasticity of demand is problematic, but the basic model is useful, and it will be examined in greater detail in Chapter 5, when we discuss supply, as it will be easier at this point to introduce elasticity applications relevant to construction economics.

Key Points 4.3

○ Elasticity of demand is given by the percentage change in quantity demanded divided by the percentage change in the explanatory variable (such as: price, income, or a related good).

○ Price elasticity of demand is given by the percentage change in quantity demanded divided by the percentage change in price.

○ The standard way to compute a percentage change is to divide the change by the starting point. The midpoint method calculates a percentage change by dividing a change by the average of the initial and final levels.

5 The Theory of Supply

In considering the theory of supply, we adopt a similar approach to that of the previous chapter on the theory of demand and, to some extent, this gives us an intellectual start. We will not, however, fully appreciate the suppliers' side of the market until we have worked through the next three chapters. Chapters 6, 7 and 8 are particularly significant. Chapter 6 explores the day-to-day relationship of contractors and clients and Chapters 7 and 8 deal with the theory of the firm.

Before formally commencing with supply it is useful to distinguish between cost and price. The basic difference arises from the perspective you are considering. For example, from a supplier's point of view, when a producer sells a good to a consumer the cost and price should not be the same. Normally, the producer seeks to make a profit – it is important, therefore, that the cost of the good is less than the selling price. Consequently, it is quite usual in construction for the cost of a project to be estimated and a mark-up for profits (risks) and overheads added before arriving at a price for the job. The contractor's mark-up is the difference between price and cost. In the present environment in the UK, however, many clients have become more knowledgeable and powerful; the client (or consultants acting on behalf of the client) predetermines an acceptable price and the contractor has to try to meet this figure.

The process is complicated further by the fact that for most construction work a price needs to be stated before the activity commences – when all the costs are not yet known. This contrasts with manufacturing: here, the producer does not have to determine the price until the activity is complete and all the costs have been revealed. Furthermore, it is important to understand that the most usual form of price determination in the construction industry is through some form of **competitive tendering**. This, in turn, makes it difficult for potential contractors supplying their services to take advantage of the market, as the lowest price bid is often seen as the most acceptable. To avoid the obvious problems that this may lead to, in some countries the practice is to adopt the average or second lowest bid as clients are aware that competition based solely on the lowest possible price does not always represent best value. Similarly on large-scale public sector contracts in the UK it is becoming common practice to add a small percentage to the price of the winning bid to allow for the unrealistic cost estimates that may have occurred as ambitious contractors try to submit the lowest possible price at the bidding stage – this adjustment is known as **optimism bias**. Competitive tendering seeks to achieve fair comparisons on a like-for-like basis. What the subsequent chapters will highlight is the need for the efficient contractor to submit realistic tenders in relation to their costs and capacity.

THE BASIC LAW OF SUPPLY

We have already encountered the basic idea of supply in Chapter 3 and it might be useful to review Key Points 3.2 and 3.3. The supply curve slopes upwards from left to right, demonstrating that as price rises the quantity supplied rises and, conversely, as price falls, the quantity supplied falls. This is the opposite of the relationship that we saw for demand. The basic law of supply can be stated formally as:

> the higher the price, the greater the quantity offered for sale, the lower the price, the smaller the quantity offered for sale, all other things being held constant.

The law of supply, therefore, tells us that the quantity supplied of a product is positively (directly) related to that product's price, other things being equal. Or in terms of the discussion above, the number of potential contractors interested in bidding to supply a project will increase as the profit margin the client is prepared to meet rises. This is displayed in Figure 5.1.

Figure 5.1 The supply curve for an individual firm

The standard supply curve for most goods and services slopes upward from left to right. The higher the price, the higher the quantity supplied (other things being equal).

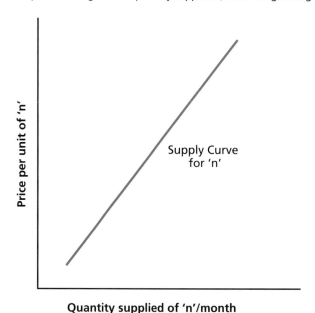

The Market Supply Schedule

The incentives within a specific market – and the constraints faced – are roughly the same for all suppliers. Each individual firm seeks to maximise its profits, and each firm is subject to the law of increasing costs. As we noted in Chapter 1 (see page 3), as a firm (or society) uses more and more of its resources to produce a specific item, the cost for each additional unit produced increases disproportionately. We referred to this as the law of increasing opportunity costs, and it highlights that resources are generally suited to some activities better than others. It is not possible to increase continually the quantity supplied of a specific item without cost increasing at a disproportionately high rate. In other words, when we utilise less well-suited resources to a particular production activity, more and more units of these resources have to be used to achieve an increase in output.

A firm's costs will also be affected by its fixed overheads. These vary according to the size of the firm and the nature of its activities. As we will explain in Chapter 7, the typical construction firm has relatively low fixed costs. The construction contractor's output is not based in a permanent factory, with all its related fixed costs. Each new construction site represents the firm's factory, and much of the fixed capital is hired as and when required.

We are now in a position to begin to appreciate the concept of a **market supply schedule**. The market supply of a product is given by the sum of the amounts that individual firms will supply at various prices. For example, at a price of £6 per unit, we might find that three firms are willing to supply 400, 300 and 200 units per day respectively. If these three firms make up the whole industry, we could conclude that at a price of £6 the market supply in this hypothetical industry would be 900 units per day. Let us consider this example in more detail. The relevant data is presented in Table 5.1.

Table 5.1 The individual and market supply schedules for a hypothetical three-firm industry

We see from the data that as price increases suppliers are willing to produce greater quantities. At the other extreme, low prices may actually discourage some firms from operating in the market. By combining the supply from each firm within the industry, we can identify the total market supply at each price; we do this in the final column.

Price	Quantities Supplied			
£/Unit	Firm A Units/day	Firm B Units/day	Firm C Units/day	Total Market Supply
4	0	0	0	0
5	300	0	0	300
6	400	300	200	900
7	500	380	250	1,130
8	580	460	280	1,320
9	620	500	290	1,410
10	650	520	295	1,465

In Table 5.1 we see how three firms comprising an industry perform individually at various prices. At low prices, producers B and C offer nothing at all for sale; most probably because high production costs constrain them. At higher prices, the law of increasing opportunity costs imposes constraints. By adding up each individual firm's output, at each specific price, we can discover the total supply that firms would be willing and able to bring to the market. We have highlighted the combinations at £6 per unit. As a brief educational exercise you could try to plot the market supply schedule on a graph. If you do it correctly, the supply curve for the market should be similar to that for an individual firm – a curve sloping upwards from left to right, as represented in Figure 5.1.

Key Points 5.1

○ Before a final price to supply work can be determined a construction contractor needs to carefully estimate costs and assess an acceptable level of profits.

○ The basic law of supply is that as price rises larger quantities are supplied, and as price falls smaller quantities are supplied. There is a direct, or positive, relationship between price and the quantity supplied, other things being constant (see Figure 5.1).

○ The size of each individual firm will determine how much it can produce at various prices. Fixed overheads and the law of increasing opportunity costs affect each firm differently.

○ The market supply of a product is derived by the summation of the amounts that individual firms will supply at various prices. Plotting these total amounts against their related prices enables one to construct a market supply curve.

SUPPLY IN THE CONSTRUCTION INDUSTRY

Many firms contribute to the supply of construction products, including large national contractors, material manufacturers, plant hirers and local site labourers. So while it may be theoretically possible to estimate construction supply by summing what the firms in the market are willing to supply at various prices, the huge range of private contractors involved in construction complicates the process of simply aggregating individual supply curves. There are approximately 220,000 firms in the UK supplying construction-based activity and, as we have discussed in earlier chapters, the industry is clearly not one simple market. There can be little competition in the supply of products between the local builder undertaking repair and maintenance in a small town and a large national civil engineering firm. They supply separate markets. The important point to note is that we need to consider factors affecting supply in specific sectors of the industry. This mirrors the approach taken in the previous chapter on demand. We need to envisage the construction industry as several different markets, each with distinct factors that affect supply and demand.

The total supply of construction output in Great Britain is broken down into specific activities according to monetary value in Table 5.2. House builders supply approximately 17 per cent of the annual output. Contractors supplying industrial and commercial buildings account for 24 per cent of the total output in value terms. The large building and civil engineering firms undertaking complex infrastructure projects such as motorways, power stations, airports and other public sector activity relating to education, health and the police account for a further 24 per cent of the industry's supply. This leaves the large number of small general builders dealing with repair and maintenance contracts to supply the largest share of the activity with 36 per cent.

Table 5.2 Construction industry supply in Great Britain, 2010 (at current prices)

Type of work	Value (£million)	Percentage of total
New public housing	4,770	
New private housing	14,281	
Total	19,051	16.6
Private industrial	3,573	
Private commercial	23,312	
Total	26,885	23.5
Infrastructure	12,660	
Public (non residential)	14,204	
Total	26,864	23.5
Total repair and maintenance	41,630	36.4
Grand total	**114,430**	**100.0**

Source: Construction Statistics Annual (ONS 2011a: Table 2.1).

It should be pointed out that the labour, capital and management resources employed on any one construction project could transfer to another. In fact, it is quite common for a specialist trade firm to move from site to site as contracts are fulfilled, and few firms remain in place for the whole duration of a project. It is this overlapping nature of the sectors comprising construction that gives rise to some common reference points for factor rewards across the industry. In other words, rates of profit, wages and material prices tend towards some kind of equilibrium. Indeed, as competition intensifies or diminishes in particular sectors across the industry, suppliers could decide to shift the use of their resources to gain higher rewards. This will become clearer as we discuss in subsequent chapters patterns of cost and contracting characterised by differing levels of competition.

SUPPLY AND THE PRICE DETERMINANT

The law of supply states that more goods are supplied at higher prices, other things being held constant. This is because at higher prices there is greater scope for firms to earn a profit. Firms already in the market have an incentive to expand output, while higher prices may also enable those firms on the fringes of the market to enter the industry. At higher prices, therefore, the increased quantity supplied is made up by existing firms expanding output and a number of new firms entering the market. For example, in Table 5.1 we showed that in our hypothetical industry at a price of £5 per unit market supply was 300 units per day, but higher prices enticed other firms into the market and total supply increased.

SUPPLY AND NON-PRICE DETERMINANTS

Up until now, we have discussed supply and its related curve on the assumption that only price changes. We have not effectively considered any other determinants that influence producers' behaviour. We have constantly reiterated the ceteris paribus qualification, that other things are held constant. Some of these 'other things' assumed constant are the costs of production, technology, government policy, weather, the price of related goods, expectations, the goals of producers (do they wish, for example, to maximise profits or sales), and so on. Now, we shall broadly consider four of these non-price determinants.

Cost of Production

We have implied that producers are seeking to maximise their profits. Therefore, any change in production costs will, *ceteris paribus*, affect the quantity supplied. To illustrate this principle, return to Table 5.1. If unit production costs increase by £1, and this additional cost cannot be passed on by suppliers, then they will supply less to the market at each price. These changed conditions will cause the market to shrink so that, for example, only 300 units per day would be supplied at a price of £6 per unit.

In technical terms, what is happening is that the supply curve has shifted to the left: less is now supplied at each and every price. The opposite would occur if one or more of the inputs became cheaper. This might be the case if, say, technology improves, but such opportunities seem slow to emerge in a construction industry that is both labour intensive and culturally inclined to invest little in research, development and training.

Government

In a similar way, taxes and subsidies also affect costs and thus supply. For example, the landfill tax has increased construction costs and reduced supply at each price. A subsidy would do the opposite, and increase supply at each price, since every producer would be 'paid' a proportion of the cost of each unit produced by the government. A more complicated issue is the impact of general taxation, as much of the construction industry's demand is derived and depends on how others forecast their requirements. The most direct impact that the government has on construction markets is through legislation. Obviously the industry is affected by changes in statutory regulations that apply to building, planning, and health and safety.

Supply Chain Management

It would be a very rare for a contracting firm to be able to complete any construction activity entirely alone – just consider, for example, the variety of materials that need to be supplied for a typical project. Most construction activity normally involves integrating and managing a whole host of activities and processes to reach the final product, including subcontracting skilled work and purchasing materials. One medium-sized contractor with an annual turnover of £240 million (made up of contracts averaging £7 million each) claimed that 75 per cent of its total work was subcontracted, 8 per cent was spent on materials and 4 per cent was used to hire plant (Jessop 2002: 7).

The larger firms, especially the huge conglomerates, ensure their clients are provided with prompt and reliable services by diversifying into other businesses to extend their range of operations; for example, a construction firm may choose to merge with its material supplier to guarantee it meets completion targets on time. Such a merger would also eliminate many of the associated transaction costs. In some recent procurements (which we discuss in Chapter 6), the contractor has even agreed to take a stake in the completed project.

This discussion emphasises that construction firms not only produce different products but they also operate outside their immediate business and, to understand the supply implications, we find ourselves considering changes in many related markets as well as the conditions in the construction industry.

Expectations

A change in the expectations about future prices or prospects of the economy can also affect a producer's current willingness to supply. For example, builders may withhold from the market part of their recently built or refurbished stock if they anticipate higher prices in the future. In this case, the current quantity supplied at each and every price would decrease, the related supply curve would shift to the left.

Key Points 5.2

○ Supply within construction is made up of several interrelated markets. Output is determined by thousands of firms that can transfer to other sectors of the industry if they think it would be worthwhile.

○ The supply curve is plotted on the assumption that other things are held constant. Four important non-price determinants are (a) costs of production (including technological changes), (b) government, (c) supply chain issues and (d) expectations.

UNDERSTANDING CHANGES IN SUPPLY

Just as we were able to distinguish between shifts of, and movements along, the demand curve, so we can have the same discussion for the supply curve. A change in the price of a good itself will cause a movement along the supply curve, and be referred to as an *extension* or *contraction* of supply. A change in any non-price determinant, however, will shift the curve itself and be referred to as an *increase* or *decrease* in supply.

Let us consider one example in detail. If a new computer-assisted design and cost estimating package reduces fees relating to new builds, then design and build contractors will be able to supply more new buildings at all prices because their costs have fallen. Competition between contractors to design and build will ultimately shift the supply curve to the right, as shown in Figure 5.2. By following along the horizontal axis, we can see that this rightward movement represents an increase in the quantity supplied at each and every price. For example, at price P, the quantity supplied increases from Q to Q_1. Note that if, on the other hand, the costs of production rise, the quantity supplied would decrease at each and every price and the related supply curve would shift to the left.

For analytical purposes, it is helpful to distinguish the cause of changes in supply. In our example about computer-assisted design, it would have been wrong to conclude that price has simply fallen and quantity supplied expanded accordingly. The reason for the increase in supply, at all prices, is due to a change in technology.

Figure 5.2 A shift of the supply curve

If price changes, we move along a given curve. However if the costs of production fall, the supply curve shifts to the right from S to S_1 representing an increase in the quantity supplied at each and every price.

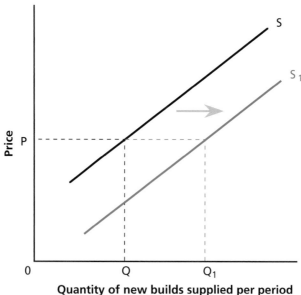

Elasticity

We introduced the concept of elasticity in Chapter 4 (see Key Point 4.3 for a summary). Economists are often interested in the degree to which supply, or demand, responds to changes in market conditions. For example, the measurement of price responsiveness is termed **price elasticity**. It may be defined as:

a measurement of the degree of responsiveness of supply to a change in price.

A numerical value for the **price elasticity of supply** (PES) is calculated using the same midpoint approach outlined in the elasticity of demand section on page 69. In this instance, the formula is:

$$PES = \frac{\text{percentage change in quantity supplied}}{\text{percentage change in price}}$$

What the formula tells us is the relative amount by which the quantity supplied will change in relation to price changes. For example, if a 10 per cent increase in price leads to a 1 per cent increase in the quantity supplied the price elasticity of supply is 0.1. That is a very small response. There are in effect three types of measure that economists use as a reference point to discuss price elasticity.

1 **Price-inelastic supply**

   When the numerical coefficient of the price elasticity of supply calculation is less than 1, supply is said to be 'inelastic'. This will always occur when the percentage figure for the change in supply is smaller than the percentage figure for the change in price. A PES coefficient of anything between 0 and 1 represents a situation of inelastic supply. The introductory example in which a 10 per cent increase in price led to a very small response in supply suggests a price-inelastic response: the measured coefficient was 0.1. In most cases where firms are supplying into the construction industry the price elasticity of supply, in the short term at least, will be inelastic.

2 **Price-elastic supply**

   When the numerical value of the price elasticity of supply calculation is greater than 1, supply is said to be 'elastic'. This will always be the case when the percentage change in supply is larger than the percentage change in price. For example, if a 5 per cent rise in price leads to a 50 per cent increase in quantity supplied, the PES coefficient will be 10. In other words, a small change in price elicits a large response in supply. This would be an unusual occurrence in the markets for construction or property – but not impossible.

3 **Unit-elastic supply**

   This is the most hypothetical case, as it describes a situation in which a percentage change in price leads to an identical percentage change in supply. This will always produce a coefficient value of 1, since the same figure appears on both the top and bottom lines of the price elasticity of supply formula.

Elasticity calculations played a significant part in the review of housing supply undertaken by Barker for the UK government in the early 2000s. For example, by drawing on several comprehensive academic estimates of the price elasticity of supply of housing in the UK carried out between 1974 and 2003, there appeared to be a considerable degree of consensus that UK housing supply is relatively unresponsive to price changes, with output increasing by proportionately less than price. The estimates taken from different time periods suggested that the price elasticity of supply of housing is always less than 1. Most estimates tended to be in a range between 0.3 and 0.8 (Barker 2003: 42).

Much of Barker's (2003) work examined the way households and house builders react to changing circumstances, such as rising incomes and price changes. For example, it was suggested that UK households have a high propensity to consume housing services as incomes rise over time. These are the comparative income elasticities estimates calculated by the Treasury for various European countries that were presented as evidence: Germany 0.0; Italy 0.4; Finland 0.5; France 0.6; Sweden 0.7; UK 1.0; Belgium 1.0; Ireland 1.1; Netherlands 1.2; and Spain 1.9. Compared to many other European countries, the UK has a high income elasticity of housing demand. But this position is complicated by the fact that the UK also has a low responsiveness of demand to house price changes. A low price elasticity of housing demand means that a one per cent rise in house prices results in less than a one per cent fall in spending on housing. The Barker study estimates that the UK's price elasticity of demand is 0.5. These factors, combined with low elasticity of supply in response to price changes, suggests that the UK's housing market is set on a path of long-term rising prices (Barker 2003: 39).

PRICE ELASTICITY OF SUPPLY AND TIME LAGS

For our purposes it is sufficient to understand that in the short run the price movement of construction goods tends not to affect supply. To increase the supply of any good or service takes time. If firms have some surplus capacity, they may be able to increase production fairly rapidly, but once they reach full capacity supply is fixed until extra capacity can be installed. For construction this is a particular issue – and it is common to talk about short-run and long-run supply. The **short run** is defined as the time period during which full adjustment – to, say, a change in price – *has not* yet taken place. The **long run** is the time period during which firms have been able to adjust fully to the change in price.

In the short run, rental values and house prices are demand determined because adjustments cannot quickly be made to the supply of property. The markets for construction in the short term are price inelastic in supply. In fact, it is the inelastic supply relative to demand that causes property markets to be unstable and characterised by fluctuating prices. In the extreme short run, the supply of buildings or infrastructure is fixed and the supply curve is a vertical straight line. Such a scenario is shown in Figure 5.3. It shows that the quantity supplied per year is 100,000 units regardless of price. For any percentage change in price, the quantity supplied remains constant. Look back at the formula for calculating elasticity. If a *change* in the quantity supplied is zero, then the numerator is zero, and anything

Figure 5.3 Perfectly inelastic supply

The supply curve is vertical at the quantity of 100,000 units per year, as the price elasticity of supply is zero. Producers supply 100,000 units no matter what the price.

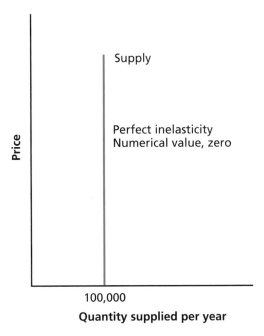

Quantity supplied per year

divided into zero results in an answer of zero. This is defined by economists as **perfect inelasticity**. (Note, exactly the same situation can be envisaged for demand: that is, a vertical demand curve represents zero elasticity at every price, too. For example, the price of electricity or gas may increase but in the short term consumers will continue to demand the same amount of energy until they have had time to switch over to more energy efficient options.)

Time tends to be the main determinant of the elasticity of supply. In the immediate time period supply is fixed, and inelastic to the value of zero; but given time for adjustments, supply increases can be organised and responses become elastic. This feature of supply inelasticity is particularly notable within property markets. Land is characterised by being perfectly inelastic; that is, as property prices increase the quantity supplied does not alter. Undeveloped areas can be developed and existing areas of land can change use, but both these possibilities take time.

So there will always be a time lag between price signals and changes to output in the construction industry. The duration of the delay is determined by a number of factors including site identification, site acquisition, negotiating planning permission, and the period of time needed for construction. These factors are examined in more depth in the Barker report (2003), where it is confirmed that the supply elasticity will be higher in the longer term as companies have time to change their factors of production and/or methods of construction. As Barker (2003: 41) specifically points out, in the very short run there is some scope to respond to changes in demand, as builders currently building a phased development could increase temporarily the build rate on that site (though this might be at the expense of less output in the next time period if the supply of land does not accelerate). Furthermore, in principle it should be possible to switch land from one use to another. However, as parcels of land in any region or area become scarcer, opportunities for substitution become extremely limited. This is particularly important given the localised nature of demand for housing and development generally, and Barker concludes that housing supply in areas of high demand is likely to exhibit far less responsiveness to price than other consumer durable products, such as cars, televisions and washing machines (Barker 2003: 42).

Key Points 5.3

○ If only price changes, we move along the supply curve and there is an extension or contraction of supply.

○ If any of the non-price determinants changes, the entire supply curve shifts to the left or right and we refer to a decrease or increase in supply (see Figure 5.2).

○ Price elasticity of supply is given by the percentage change in quantity supplied divided by the percentage change in price.

○ Long-run supply curves are more elastic than short-run supply curves because over a longer period of time more resources can flow into or out of an industry when price changes. This is especially the case in property markets.

○ It is the inelastic supply of buildings that causes the related markets to be unstable and characterised by periods of escalating prices.

COMBINING SUPPLY AND DEMAND

In these chapters on demand and supply, we have confined our discussion to isolated parts of the market relating to the consumer or producer. Obviously this separation is theoretical and only useful for educational purposes. In reality, there is a very close relationship between the forces of demand and supply. Indeed, we have already discussed in Chapter 3 how the interaction of supply and demand determines prices. We introduced the concept of an equilibrium (or market) price at which both consumers' and producers' wishes are met. We even extended our discussion to consider the effects of changes to market conditions. Knowing and understanding how supply and demand interact is an essential prerequisite for interpreting many markets, including construction-related markets. It may be worthwhile, therefore, to spend a couple of minutes reviewing the market mechanism by attempting the exercise set out in the next section. An outline answer is provided in Figure 5.4 (shown on page 84).

EXERCISE ON SUPPLY AND DEMAND

Draw a basic supply and demand diagram and consider the new equilibrium point in each of the three changing markets described in Table 5.3.

Table 5.3 Changing market conditions

Market	Product	Change in conditions	New equilibrium point
A	Energy efficient buildings	A successful advertising campaign by the energy efficiency office	See coordinate A in Figure 5.4
B	Computer assisted design	Improved microchip reduces cost	See coordinate B in Figure 5.4
C	New houses	A fall in the cost of a mortgage and an increase in the price of land	See coordinate C in Figure 5.4

Obviously this should be done before looking at the answers suggested overleaf (on page 84) in Figure 5.4.

Key Points 5.4

- By combining the forces of supply and demand we can begin to understand many markets.
- The concept of equilibrium demonstrates how the wishes of sellers and buyers are brought together via price.
- In reality, the determinants of supply and demand need to be considered simultaneously.

Figure 5.4 Changing market conditions across three markets

The three hypothetical markets and the changes described in Table 5.3 are plotted.
In market A we assume an increase in demand causing quantity and price to increase.
In market B we assume an increase in supply causing quantity to increase and price to
decrease. In market C we assume an increase in demand and a long-run decrease in
supply causing a higher equilibrium price but no significant short-run change to quantity.

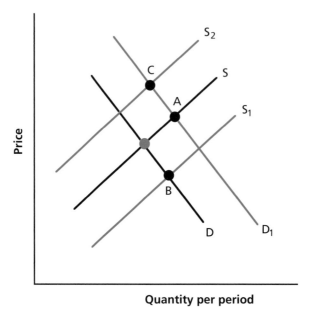

6 Clients and Contractors

In Chapters 4 and 5 we discussed two central ideas in explaining the market: demand and supply (see the Key Points in these chapters). In terms of construction economics, it is important to appreciate that those making the demands are referred to as clients and those who respond to their instructions by supplying the products are referred to as contractors. Both groups can take many forms, and it is common to emphasise that each project is a one-off. The analogy of a film set is sometimes employed to illustrate how the various contractors move in, do their work (complete a project) and then leave for another location to do 'similar' work for another client. Hence the industry is characterised by a fragmented approach, with each sub-group concentrating on its own specialism.

Extending an idea of Ive and Gruneberg (2000: 151), Table 6.1 tries to represent the innumerable projects that contractors could meet in response to their clients' demands. The location of a particular project, its design and production are some of the reasons why each project is unique. In the table, we also emphasise the fact that the construction stage is carried out by a vast combination of firms selected from the thousands comprising the industry. On each project there will be various teams of labour on site, working with different sets of people. This is in stark comparison to manufacturing and service industries where the work is usually repetitive and the workforce is normally permanent.

Table 6.1 Contractors involved in construction

	Project Types				
	BUILDING	**CIVIL ENGINEERING**	**REFURBISHMENT**	**REPAIRS AND MAINTENANCE**	
	A B C D	E F G H	I J K L	M N O P	etc.
DESIGN					
CONSTRUCTION	220,000 FIRMS				
OWNERSHIP					
USE					

Source: Adapted from Ive & Gruneberg (2000) and Construction Statistics Annual (ONS 2011a: Table 3.1)

The perceptive reader may question the idea that only registered construction firms are involved in the construction stage, as closer links are evolving between clients and contractors. These links will be discussed later in this chapter, under the heading of partnering. But first we need to describe the traditional arrangement of client and contractor(s) as two separate parties involved in the construction of one project.

CLIENTS

At the core of any construction process are the clients. Some are well informed and know precisely what they want and how it can be technically achieved, but the majority seem to know nothing. Within the discipline of economics, a whole literature has begun to emerge that discusses transactions in which some of the parties involved know more than others. (This area of study is referred to as the economics of **asymmetric information** and will be considered further in Chapter 10 when we discuss market failure.) It has been suggested that independent advisers may be required to help clients. With a similar role to financial advisers, they would assist inexperienced clients to decide what is specified and how it can be best achieved. The case for client support has become stronger in recent years. As suggested in Chapter 2, the government can no longer act as a monolithic client, as changes in public sector funding, management and accountability have added a new layer of complexity to the client base. For example, the National Health Service is far from being a single organisation as it comprises a network of GP practices, laboratories, hospitals, trusts and care homes. So the Department of Health, the government body with overall responsibility for the NHS, oversees a diverse range of funders managing a broad estate of buildings. The department has no means of ensuring that all managers that might commission building work on behalf of the NHS are aware of current best practice and changes in the construction industry.

To address these problems the Construction Task Force led by Sir John Egan made comparisons with other industries that have increased efficiency as market forces expanded. The task force recognised that, in the best companies, the customer drives everything – the customer is 'king'. It is clear that the modern client in markets such as cars, steel and engineering products expects value for money, products that are free from defect, goods delivered on time, worthwhile guarantees and reasonable running costs. Unfortunately, the picture painted of construction in the late twentieth century was of an industry that 'tends not to think about the customer (either the client or the consumer), but more about the next employer in the contractual chain' (Egan 1998: 16).

CONTRACTORS

One of the most striking features of the construction industry across the world is the large number of firms. In the United Kingdom alone, there are approximately 220,000 private construction contractors recorded in the official government statistics and 90 per cent of them employ fewer than eight people. Across the

European Union, there are more than 3 million contractors – employing 15 million people. Construction accounts for nearly 10 per cent of the total employment in Europe. Much of this labour works in small enterprises. Small firms dominate the industry for two main reasons:

- they can supply services which do not suit the nature of large firms, such as repair and maintenance
- they can supply labour on a subcontract basis to large firms.

Labour-only Subcontractors

It is common practice throughout Europe for the majority of work carried out on a construction site to be done by subcontractors. In addition, subcontractors often organise the materials and maintain the equipment. To the main contractor, labour-only subcontracting is the cheap and efficient option, as the self-employed worker is not entitled to holiday pay, redundancy money, sick leave, pension rights or any other benefits that accrue to permanent members of staff.

As a result of this tradition, a high level of fragmentation is often associated with the European construction industry. After being in post for six months a UK construction minister once remarked in an interview that in terms of meeting all the trade associations he was only half way down the alphabet. The interviewer sarcastically added that 'by the time he gets to Z he'll probably have moved on to his next job' (Broughton 2001). In fact it is estimated that there are more than 500 separate trade associations representing different groups of UK construction workers (Wolstenholme 2009: 22). This characteristic fragmentation leads to many of the industry's recognised strengths and weaknesses.

WEAKNESSES

There is a recognised lack of collaboration within construction teams. This, in turn, often leads to a lack of trust between the various parties and explains the adversarial nature that typifies the relationships. Another weakness of fragmentation is the lack of commitment to education, training, research and safety on site. It is often suggested that subcontractors do not seem to learn from one project to the next – a common myth is that each building is regarded as a 'prototype'. The general level of tension between contractors and subcontractors was interestingly captured at the start of the recession in 2008, as the first thing that the larger contractors chose to do to protect their business was streamline their supply chains and cut back on the number of specialist subcontractors they used. For example, Wates cut its suppliers list from 4,000 to 500 and Costains was equally drastic with a cut from 18,000 to 2,500. To justify their actions the directors of these large organisations used derogatory phrases such as 'we will drop "white van man" in favour of bigger companies', 'we will weed out the small "fly-by-night" organisations as we seek to work with fewer quality subcontractors' and 'our aim is to be responsible for between 10 and 25 per cent of a subcontractor's work – [we want] to assure that our position (as the main contractor) is important to them, but not too important!' (McMeeken 2008). Furthermore, to compound the nature of this problematic relationship, specialists fortunate enough to remain on the main contractor lists were increasingly expected to cut their margins as well as jump through all kinds of hoops relating to health and safety and sustainability, which often involves paying for associated accreditation schemes.

A generic outcome of these types of problems is the way it limits any attempts that the construction industry seeks to make in the long term to construct on time, on budget and to the quality expected in a world concerned with sustainability.

STRENGTHS

On the positive side, it is usual to identify one strength of fragmentation: it enables the industry to be sufficiently flexible to deal with the highly variable workloads that accompany the changing economic circumstances.

Clearly the weaknesses outweigh the strengths and this is why the various types of partnering agreements that are emerging are regarded as such promising developments. These are discussed in the next section.

Key Points 6.1

○ Clients create demand for the construction process. Increasingly, these clients are based in the private sector.

○ Contractors liaise with the client to supply the products, which are typically produced by subcontractors.

○ The construction team is typified by fragmentation, which leads to weaknesses such as a lack of collaboration, trust, training, etc.

PARTNERING

Since the publication of the Banwell Report in 1964, various government committees and academics have been advocating partnering as an important way of improving the performance of the construction industry. For example, a National Audit Office report (NAO 2001: 58) concluded that 'partnering is the way to deal with the inherent problems of construction which is still widely regarded as a design to order industry'. A comprehensive review of the academic literature suggests a consensus that partnering leads to better cost control, better quality product, closer relationships, enhanced communication, continuous improvement, potential for innovation, lower administrative cost, reduced litigation, increased satisfaction and improved culture (Chan et al. 2003: 524).

There are many definitions of **partnering** and different forms have been used in the UK, US and Japan. In general terms, however, partnering refers to some form of collaborative approach in which clients and contractors are increasingly open with one another in order to meet common objectives.

The idea of partnering began to gain momentum through the 1990s and there were a number of government reports and academic papers supporting its use. For example, Bennett and Jayes (1995) argued that partnering delivers cost savings ranging from as low as two per cent to as much as 30 per cent depending on the type of partnership. They found that **project partnering**, which is based on a client and contractor working openly on single project, produced cost savings of around 2–10 per cent. **Strategic partnering**, which involves the client and contractor working on a series of construction projects, was regarded as capable of delivering savings of up to 30 per cent. In addition, Bennett and Jayes (1995) specifically claim that partnering delivers better designs, makes construction safer, enables deadlines to be met with ease and provides all parties with increased profits. Some current examples of partnering will be dealt with in the following two sections to see how these possibilities can develop.

Prime Contracting

Prime contracting is a good example of partnering since it involves the integration of design, construction and maintenance under the leadership of one main contractor. The prime contractor takes on responsibility for everything from the selection of the subcontractors through to the delivery of the product. Consequently the client has only one point of contact, and the prime contractor strives to look after the client's interests and assure that quality, budgets and delivery targets improve.

Prime contractors are presently looking after estates run by the Ministry of Defence and new supermarkets being built on behalf of J. Sainsbury plc. Such contracts are usually continued for a number of years – on a repeat business basis. Accordingly, both parties seem to gain a greater opportunity to learn what is needed.

Private Finance Initiative (PFI)

This form of partnering has a far higher profile, since it involves the public and private sectors collaborating over some of the biggest projects in construction. These projects include highways, hospitals, prisons, schools and government offices. Since its launch in November 1992, more than 650 PFI contracts have come into force representing future commitments of around £260 billion (HM Treasury 2011). Similar arrangements, such as public private partnerships (sometimes referred to as 3P), are a growing phenomenon worldwide.

A project procured under the **private finance initiative** is based on a distinctive kind of relationship between a public sector client and the private sector. The general procedure is as follows: private firms operating in a consortium agree to design, build, *finance* and *manage* a facility traditionally provided by the public sector. (Finance and manage are italicised to emphasise that a distinguishing characteristic of these schemes is that under PFI arrangements the private sector is expected to raise the initial finance to fund the project and thereafter manage its operation.) In return, the public sector client agrees to pay annual charges (called a unitary charge) during the life of the contract and/or allows the private sector to reap any profits that can be made for a specified period. The contracts usually run for 25 to 30 years or more. In this way, both sectors can be seen to be specialising in what they do best: with the public sector client initiating the development, specifying the requirements and vetting the tenders, and the private sector taking the risks and determining the best way to deliver the service.

The group formally awarded with the contract is referred to as a consortium or special purpose vehicle (SPV). The SPV may comprise several companies such as a construction and facilities management contractor, a bank and/or an equity investor, a management company, and separate liaising companies. (Subsequently, the SPV may subcontract its responsibilities.) This complex set of relationships is implied in Figure 6.1, where the PFI relationships have been simplified into a flow diagram.

Figure 6.1 Private finance initiative

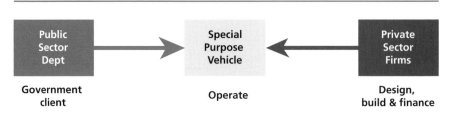

PFI arrangements have the obvious advantage that the contractor no longer 'builds and disappears' – as in the BAD old days – because the *finance* and *manage* elements of the contract legally tie the contractor to the project once it has been built. When the private sector has money at risk in this way, there are far greater incentives to get everything right, especially as the consortium contractors now have to consider the running and maintenance costs.

In effect, PFI procurement creates a far better level of communication between public sector clients and private sector contractors, as both sides are linked in a common objective. The collaborative agreement provides mutual benefits to both parties. In the terms introduced in Chapter 2, the completed project should be more economically efficient and more sustainable (see Key Points 2.4 and see Figure 6.1).

Using private capital and expertise in the provision of public sector services is not entirely new. For example, joint working between sectors has a reasonably long history in housing – local authorities have transferred the ownership and management of large estates to a government regulated private sector group of **registered social landlords**. The point is that apart from the much debated private finance initiative, there are other forms of public private partnerships, where the public, private and voluntary sector work together to achieve a wide variety of objectives.

CONCLUSION

In conclusion, it is easy to see why partnering in any form is an attractive idea. In purely economic terms it clearly helps to eliminate inefficiency in so far as costs per unit of output are reduced. Leibenstein's (1973) concept of **x-inefficiency** can be applied. He argued that when a government department or individual firm is left entirely to its own devices to choose the APQT – that is, the activity, pace, quality of work, and time spent – it is unlikely to choose a combination that will maximise efficiency. In other words, if a public sector authority or individual contractor can choose what to build, how to build it, how long to spend on the project and so on, it is less likely to maximise the value of output. The resulting loss of value is what Leibenstein referred to as x-inefficiency and his solution relies on securing a more competitive market structure. Partnering potentially improves the dynamics of a market by putting the client more firmly in the driving seat, improving the flow of information between the participants and providing greater incentives to complete the contract on time, to budget and to the expected quality. Some of the benefits of partnering listed by the National Audit Office in its report on how the construction industry in the UK could become more efficient are summarised in Table 6.2.

Table 6.2 The benefits of partnering

✓	Reduced need for costly design changes
✓	Increased opportunities to replicate good practice learned on previous projects
✓	Avoids adversarial relationship between client and contractors
✓	Contractors have good incentives to deliver on time, on budget and to a high standard
✓	The liaison of clients and contractors should improve the overall efficiency of the building – particularly in terms of its operation and maintenance
✓	It should be possible to drive out inefficiency and waste from the construction process

Source: Adapted from National Audit Office (2001: 6, 31)

There are, however, some reservations about whether the benefits of partnering are fully realised under PFI. For instance, changes are inevitably required during the life of a contract running over 25 to 30 years, either as result of changing client needs or because of new technological developments, and these can prove to be particularly expensive to implement, especially when one company has already been nominated as the main contractor. The National Audit Office (2008) has confirmed that changes to PFI contracts tend to be more expensive than equivalent changes in conventionally procured projects by a margin of 5 to 10 per cent of the cost of the change. When it is noted that public authorities spent over £180 million in 2006 on changes to operational PFI projects, the scale of the problem becomes evident.

A second issue is that the annual (unitary) costs have increased as the PFI programme has grown. For example, the annual PFI bill for the National Health Service alone increased by more than 500 per cent between 2002 and 2012, from £270 million to £1.5 billion, as the number of health-related schemes running in England increased from 32 to 114. Clearly such commitments could create financial difficulties in managing future health budgets, and some have argued that it will eventually lead to service cuts and closures (Hellowell and Pollock 2007).

Another significant drawback is that successful partnering is far easier for big firms than small firms – and construction is still dominated by the latter. In a truly competitive perfect market all information is freely available – everybody has access to the knowledge they need for exchange to take place – and transaction costs are zero. Governments often try to ensure these conditions prevail by standardising legal and financial procedures, in order that market participants know where they stand. However, in construction markets where partnering and, in particular, PFI is emerging these conditions do not prevail. Many firms lack the necessary resources to understand the complex legal information that is inevitably associated with these forms of procurement. Transaction costs are prohibitively high, with architects', lawyers' and accountants' fees to be met by all the participating parties. As a result, it is unusual for more than three or four consortium groups to find sufficient resources to engage in the tedious, lengthy and detailed bidding processes involved. Indeed, PFI bidding costs can commonly exceed £1 million per project. The firms that are able to take on such large-scale operations and risks are few and far between, and it is a common concern that partnering arrangements often exclude smaller contractors.

The credit crunch compounded these issues even further. It led to a shortage of finance across public and private sectors, bringing the questions raised above into sharper focus. In some ways PFI has stopped looking so attractive: it financially ties the government in for too long and does not offer enough flexibility to adapt to change. As a consequence during the public sector spending reviews of 2010 and 2011 the Treasury took the opportunity to reassess the future expansion of PFI and opened up a debate about the best way to attract private finance into public sector projects. At the time of writing this debate continues and new PFI schemes have been put on hold (The City UK 2012).

Key Points 6.2

- ○ Partnering is a broadly used term describing several different types of collaborative arrangement.

- ○ Partnering has made great strides in recent years, particularly in the public sector, as few governments can meet social and infrastructure needs from public funds alone. PFI and other forms of PPP contracts are used to unlock private sector capital.

- ○ Partnering assumes a win-win scenario for all parties. Several of the benefits are summarised in Table 6.2.

- ○ Discussions about the pros and cons of using PFI tend to focus on value for money but the debate needs to broaden to recognise the social capital that PFI delivers.

RETHINKING CLIENT AND CONTRACTOR RELATIONSHIPS

From the commentary in this chapter, it appears that the construction industry may be characterised by dissatisfaction and, compared to other industries, may seem inefficient. Indeed, Sir John Egan alluded to the severe problems facing the industry in the late 1990s when he commented on the low and unreliable rate of profits that prevent the necessary investment in training, research and development to sustain a healthy industry. In fact, he noted that the City's view of construction confirmed its position as a poor investment. As Egan (1998: 11) observed: 'The City regards construction as a business that is unpredictable, competitive only on price not quality, with too few barriers to entry for poor performers.' From the client's perspective investment was portrayed in a similar way, as unpredictable in terms of delivery, budget and the standards of quality expected. Again in Egan's (1998: 10) words: 'Investment in construction is seen as expensive, when compared both to other goods and services and to other countries. In short, construction too often fails to meet the needs of modern business that must be competitive in international markets, and rarely provides best value for clients and taxpayers.'

A subsequent review of government R&D policies for the construction industry, *Rethinking Construction Innovation and Research* (Fairclough 2002), further confirmed the impression of underachievement. Indeed, Fairclough portrayed an industry at the beginning of the twenty-first century that was 'dirty, dangerous and old fashioned' (2002: 30). He emphasised that for construction to evolve a strategic vision was needed. Interestingly, the vision that was articulated centred on the industry's contribution to the overall aims of economic, environmental and social sustainability.

The findings of Egan (1998) and Fairclough (2002) were drafted at the request of the UK government, and they captured the state of play at the beginning of the twenty-first century. The image they presented was in many ways more 'Victorian' than 'modern', and it applied with equal measure to most other nations. As evidence there is an international plethora of government-sponsored reports and in-depth reviews into the construction industry. For example, to paraphrase the opening paragraph of a review of the state of construction in Singapore, the construction industry has a poor image and it can easily be singled out from the rest of the economy by attitudes, technologies, processes and culture that are at least half a century old (Dulaimi et al. 2001: 1). Similar sentiments are expressed in the conclusion to Manseau and Seaden's (2001) international review of public policies in Europe, North and South America, South Africa and Japan, where they conclude that 'construction almost everywhere is perceived as being "in trouble", with low margins of profit, high costs of production and lack of concern for the end user'.

During the first decade of the new century the industry was sheltered by a healthy economy and cultural inertia prevailed. In other words the communication, sustainability and levels of efficiency achieved between clients and contractors still languished in the past. As Wolstenholme's (2009) review of the progress made since Egan's report confirmed, there had been little change; and in colourful terms once used by Elvis Presley the industry was called upon to have 'a little less conversation and a lot more action please' (Wolstenholme, 2009: 5). One particularly strong finding of the Wolstenholme review was that people tend to cherry pick the behaviours that they adopt, based on their own self-interest. So, many clients may say they want a solution that represents the best whole-life cost, yet they still start out by pursuing the lowest tender price. Similarly the contractor's commitment to partnering is often found to be no more than skin-deep, as so-called partners still seek to maximise their own profits rather than find ways to share risk and collaborate genuinely so that all can profit. So, at the beginning of the financial crisis, we witnessed another well-intended call for cultural change in the industry to 'integrate and embrace the complex picture of how clients and contractors can interact sustainably with the environment to maximise health, wealth and happiness' (Wolstenholme 2009: 5–8).

To some extent these difficulties explain the current emphasis that is being placed on **value management**. Although the process originated more than 60 years ago within the US manufacturing sector and was referred to as value engineering, it has been increasingly applied to large-scale construction projects since the 1980s. For our purposes the terms value management and value engineering are synonymous, but in the UK the former term seems to be used more frequently.

VALUE MANAGEMENT

In the most general sense, value management is a specific event organised to identify and eliminate unnecessary costs, although it should not be interpreted as solely a cost-cutting exercise. This might sound confusing but the aim is to optimise value for money by reviewing costs, quality and function. There is no guarantee, therefore, that the initial costs will be cut, but a more efficient project should almost certainly emerge.

The process developed in the industrial community of the United States and an early anecdote outlines its origins. The story goes that Lawrence Miles a purchasing engineer of the General Electric Company during the Second World War was assigned the task of procuring materials to expand the production of the B24 bomber from 50 to 1000 aircraft per week. At the time there was a shortage of materials and consequently he initially had problems sourcing supplies. However, Miles discovered many substitute components that were not only cheaper but better than the original products. Hence the idea of value management was born and was associated with cost cutting. Today, however, it has broadened in scope and is far more closely associated with securing value for money, in the sense of optimising the whole-life costs of a project to meet the client's long-term requirements (HM Treasury 2006: 7).

The usual way that value management is implemented is through structured workshops, led by an independent facilitator, to take account of issues relating to design, construction, operation and management. In short, value management workshops are multi-disciplined. The aim is to review the performance and costs of a project throughout its life cycle. The prototype value management workshops were organised to take place over several days (often up to one week), but recently the trend has been for these events to be compressed into a far shorter time frame. In fact, research evidence suggests that half-day sessions are becoming the norm, and in certain circumstances this is even reduced to a couple of hours. To paraphrase the comments of a facilitator explaining this trend: the average client has a horror of spending more than £5000 on me [as facilitator] and has a cynical view of the value of 16 people sitting round a table discussing the project (Ellis et al. 2005: 489).

The precise format and timing of workshops will vary from project to project. In theory, there is nothing stopping a value management review from taking place at any stage of the process. Indeed, the theoretical literature even recommends that a kind of post-occupancy review should be carried out to document the lessons learnt and provide feedback. In practice, however, it has been found that the process rarely proceeds beyond the early stages, as although the value of post-completion exercises is accepted, these are rarely carried out (Ellis et al. 2005: 486).

In Figure 6.2 we outline a typical project life cycle to emphasise that the earlier a value management study is undertaken the more opportunities arise to benefit from it. This is especially important as in the majority of cases there is only one value management intervention (Ellis et al. 2005: 486). So if value management is to be used to good effect, most participants agree that they need to become involved as early as possible in the project's cycle when clear choices have to be made. For example, holding a workshop to examine the outline business case allows time and discussion for all stakeholders to gain understanding of the project, improve its efficiency and overall value for money. Providing the value management event is held before the tender stage, there is still opportunity to discuss different options, seek better solutions, accommodate design changes and generally eliminate unnecessary costs. Ideally this would then be followed by a series of value management events as the project unfolds, and there are exceptional cases where several value management events are spread across a number of years.

Value management is clearly an organised effort to analyse functions, value, costs and sustainability. In short, it looks at construction projects as an integrated whole. It is not surprising therefore that the Government Procurement Service, which seeks to meet business requirements and deliver successful projects, is beginning to introduce value management reviews on large complex projects. Where the technique has been applied, there have been significant reductions in defects, maintenance charges and overall costs.

Figure 6.2 Project life cycle

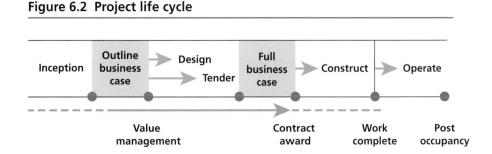

The United States Federal Highway Administration has been committed to value management since 1995 and has made it compulsory for all projects exceeding $25 million. As a consequence, its website lists some pretty impressive achievements. To take just 2010 as an example, the Federal Highway Administration's annual report refers to 402 value management studies that were carried out during the year. (In some instances these studies were undertaken at several stages of the same project, at various points from design through to completion. Figure 6.2 shows some of the key stages of a project.) These studies made a total of 3049 recommendations of which 1315 were taken on board. This generated a saving of around $2 billion on a spend of $34.5 billion. In some instances, the value management exercise also included an assessment of non-monetary benefits, such as travelling time and reducing the impact on nesting eagles, alongside the assessment of conventional monetary benefits such as construction and maintenance costs. At the current time there is clearly a far bigger commitment to the value management process in the United States than anywhere else in the world. But given the frequent recognition that 'value' is created on drawing boards not on building sites, and that designers and architects have found that conversations about how things are constructed can throw up many possibilities, value engineering could easily become commonplace on public sector projects in other countries during the next decade.

Finally, in the context of this chapter, it should not be forgotten that value management equally provides an excellent opportunity to build a competent team and achieve effective collaboration between clients and contractors.

Key Points 6.3

○ Compared to other industries, the processes associated with the traditional construction industry appear old fashioned, inefficient and unsustainable.

○ The Egan Report (1998) and the subsequent review by Wolstenholme (2009) emphasised that the industry needs to modernise by adopting efficient, integrated, innovative and sustainable construction practices.

○ Value management has the potential to address the relationships between cost, function and value. It is already used effectively in the United States to assess value for money on large-scale infrastructure projects.

○ The recognition that the value of a project commences at the design stage throws up many new possibilities over the next decade.

7 Costs of the Construction Firm

We now begin to examine more closely the supply side of an economy – that is, we attempt to explain how suppliers behave. To commence we draw on ideas traditionally associated with the theory of the firm, a theory popularised by Marshall (1920) to account for the production behaviour of firms involved in manufacturing. The theory proceeds from the relatively simple premise that providing a firm is aware of its production costs and revenue streams, it should be able to identify a specific profit maximising position. Clearly this requires a detailed analysis of the stream of costs and revenues that a firm faces as it reaches its optimum scale of production. In this chapter, we examine closely the nature of profits, productivity and costs relating to design, construction, maintenance, management, conservation and refurbishment, and in Chapter 8 we concentrate more on the revenue side of the picture. Before commencing this chapter, therefore, it may be useful to review Key Points 1.2, 2.4 and 5.1 to recap on basic ideas relating to models, markets, resource allocation, economic efficiency and the basic law of supply.

THE FIRM

We start by defining a business, or firm, in general terms.

> A firm is an organisation that brings together different factors of production, such as labour, land and capital, to produce a product or service which it is hoped can be sold for a profit.

The actual size of a firm will affect its precise structure. A common set-up for a larger firm involves entrepreneur, managers and workers. The entrepreneur is the person who takes the chances. Because of this, the entrepreneur is the one who will get any profits that are made. The entrepreneur also decides who runs the firm. Some economists maintain that the true quality of an entrepreneur is the ability to pick good managers. Managers, in turn, are the ones who decide who should be hired and fired and how the business should be generally organised. The workers are the people who ultimately use the machines to produce the products or services that are being sold by the firm. Workers and managers are paid contractual wages. They receive a specified amount for a given time-period. Entrepreneurs are not paid contractual wages. They receive no specified 'reward'. Rather, they receive what is left over, if anything, after all expenses have been paid. Profits are, therefore, the reward to the entrepreneur for taking risks. Note that roles can be combined, and in the many small firms that comprise a large part of the construction industry the entrepreneur is also the manager-proprietor.

According to EU data small and medium-sized enterprises – defined as firms employing less than 250 people – produce approximately 99 per cent of the construction output in Europe. In fact, more than 20 per cent of the European construction sector workforce is estimated to be self-employed; in the UK the comparable figure is closer to 40 per cent. Indeed, the construction industry may be characterised by risk-taking entrepreneurs.

Larger corporations, such as public limited companies, typically have many shareholders. They are the owners of the firm, which means that theoretically they are the entrepreneurs. In practice, they do not usually get involved in the day-to-day running of the firm. They do not, for example, determine the level of prices or output. These tasks are delegated to salaried managers, leaving shareholders with one sole interest – the level of profit that the firm manages to generate.

Profit

The costs of production must include an element of profit to provide a payment to the entrepreneur. If the level of profits falls in one area of activity, entrepreneurs may move their resources to an industry where the returns are higher. To illustrate this behaviour economists employ a concept of **normal profit**. Normal profit may be defined as:

> a minimum level of reward required to ensure that existing entrepreneurs are prepared to remain in their present area of production.

Normal profit is included in the cost of production, as it is an essential minimum reward necessary to attract the entrepreneur into economic activity. The concept of normal profit also highlights that all resources can be employed in several ways (that is, all resources have alternative uses). Note that what is meant by 'profit' by economists differs from its general meaning in everyday usage. To portray the general everyday meaning of profit, the following formula could be used:

> profits = total revenues – total costs

For economists, an alternative formula is required:

> economic profits = total revenues – total opportunity cost of all inputs used

The economic profits formula will become clearer by looking at two areas of resource allocation and the related cost accounting calculations. The first resource is capital, and the second is labour.

OPPORTUNITY COST OF CAPITAL

Firms enter or remain in an industry if they earn, at a minimum, a normal rate of return (NROR) – that is, normal profit. By this term, we mean that people will not invest their wealth in a business unless they obtain a positive competitive rate of return – in other words, unless their invested wealth pays off. Any business wishing to attract capital must expect to pay at least the same rate of return on the capital as all other businesses of similar risk are willing to pay. For example, if individuals can invest their wealth in almost any construction firm and get a return of 10 per cent per year, then each firm in the construction industry must expect to pay 10 per cent as the normal rate of return to present and future investors. This 10 per cent is a cost to the firm – formally referred to as the **opportunity cost of capital**. The opportunity cost of capital is the amount of income, or yield, forgone by giving up an investment in another firm. Capital will not stay in firms or industries if the rate of return falls below its opportunity cost. Clearly, the expected rate of return will differ from industry to industry according to the degree of risk and difficulty involved.

OPPORTUNITY COST OF LABOUR

The self-employed contractor or one-person business often grossly exaggerates profit, because the opportunity cost of the time that they personally spend in the business is often not properly measured. Take, as an example, the people who run small surveying offices. These surveyors, at the end of the year, will sit down and figure out their 'profits'. They will add up all their fees and subtract what they had to pay to staff, what they had to pay to their suppliers, what they had to pay in taxes, and so on. The end result, they will call 'profit'. However, they will not have figured into their costs the salary that they could have earned if they had worked for another company in a similar type of job. For a surveyor, that salary might be equal to, say, £20 an hour. If so, then £20 an hour is the opportunity cost of the surveyor's time. In many cases, people who run their own businesses lose money in an economic sense. That is, their profits, as they calculate them, may be less than the amount they could have earned had they spent the same time working for someone else. Take a numerical example. If an entrepreneur can earn £20 per hour, it follows that the opportunity cost of his or her time is £20 x 40 hours x 52 weeks, or £41,600 per year. If this entrepreneur is making less than £41,600 per year in accounting profits, he or she is actually losing money. This does not mean that such

Figure 7.1 Simplified view of economic and accounting profit

Here we see that on the left-hand side, total revenues are equal to accounting cost plus accounting profit – that is, accounting profit is the difference between total revenue and total accounting cost. On the other hand, we see in the right-hand column that economic profit is equal to total revenue minus economic cost. Economic costs equal explicit accounting costs plus a normal rate of return on invested capital (plus any other implicit costs).

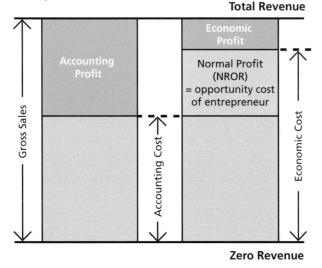

entrepreneurs are stupid: they may be willing to pay for the non-pecuniary benefits of being the boss.

We have only considered the opportunity cost of capital and the opportunity cost of labour, but the concept applies to all inputs. Whatever the input, its opportunity cost must be taken into account when figuring out true economic profits. Another way of looking at the opportunity cost of running a business is that opportunity cost consists of all explicit (direct) and implicit (indirect) costs. Accountants only take account of explicit costs. Therefore, accounting profit ends up being the residual after explicit costs are subtracted from total revenues.

The term profits in economics means the income that entrepreneurs earn over and above their own opportunity cost of time, plus the opportunity cost of the capital they have invested in their business. Profits can be regarded as total revenues minus total costs – which is how the accountants think of them – but economists include all costs. We indicate this relationship in Figure 7.1.

The Goal of the Firm

In developing a model of the firm, we will generally assume that the main business goal is maximisation of profits. In other words, a firm's goal is to make the positive difference between total revenues and total cost as large as it can. We use this **profit-maximising model** because it allows us to analyse a firm's behaviour with respect to the relationship between costs and units of output. Whenever this model produces poor predictions, we will examine our initial assumption about profit maximisation. We might have to conclude that the primary goal of some firms is not to maximise profits but rather to maximise sales, or the number of workers, or the prestige of the owners, and so on. However, we are primarily concerned with generalisations. Therefore, providing the assumption of profit maximisation is correct for most firms, the model serves as a good starting point.

Key Points 7.1

○ A firm is any organisation that brings together production inputs in order to produce a good or service that can be sold for a profit.

○ Accounting profits differ from economic profits.

○ Economic profits are defined as total revenues minus total costs, including the full opportunity cost of all the factors of production.

○ Small businesses and self-employed workers often fail to consider the opportunity cost of the labour services provided by the owner.

○ The full opportunity cost of capital invested in a business is generally not included as a cost when accounting profits are calculated. Thus, accounting profits overstate economic profits.

○ Profit maximisation is regarded as the main objective when considering a firm's behaviour.

PRODUCTION DECISIONS

When considering a firm's capacity for production, the time period is vitally important and throughout the rest of this chapter we will consider a 'short' time period as opposed to a 'long' time period. In other words, in the analysis that follows, we are looking at short-run production decisions.

Any definition of the short run is, necessarily, arbitrary. We cannot talk in terms of the short run being a specific period such as a month, or even a year. Rather, we must consider the short run in terms of a firm's ability to alter the quantity of inputs. For ease of understanding, we will simply define the **short run** as any time period when there is at least one factor of production that has a fixed cost. In the **long run**, therefore, all costs are variable; that is, all factors are variable.

How long is the long run? That depends on the individual industry. For retailing, the long run may be four or five months – because that is the kind of time period in which they can add new franchises. For manufacturing firms, the long run may be several years – because that is how long it takes to plan and build a new factory.

In most short-run analyses, the factor that has a fixed cost, or is fixed in quantity, is capital. We therefore state that in our short-run model, capital (as well as land) is fixed and invariable. This is not unreasonable – on a typical construction site the amount of land and the number of cranes in place will not change over several months. Likewise, the head office location and equipment can not be easily altered. The input that changes most is labour. The production relationship that economists use, therefore, holds capital and land constant, or given, and assumes labour is variable. This assumption is particularly pertinent in an industry such as construction in which labour accounts for a significant proportion of the costs.

The Production Function – a Numerical Example

The relationship between physical output and the quantity of capital and labour used in the production process is sometimes called a **production function**. The term production function in economics owes its origin to production engineers for it is used to describe the technological relationship between inputs and outputs. It depends therefore on the available technology.

Look at Table 7.1 (on page 102). Here we show a production function that relates total output in column 2 to the quantity of labour input in column 1. When there are no workers, there is no output. When there is the input of five workers (given the capital stock), there is a total output of 150 sq. m per week. (Ignore, for the moment, the rest of that Table 7.1.) In Figure 7.2a (on page 103), we show this hypothetical production function graphically. Remember that it relates to the short run and that it is for an individual firm.

Figure 7.2a shows a total physical product curve, or the amount of physical output that is possible when we add successive units of labour while holding all other inputs constant. Note that the graph of the production function in Figure 7.2a is not a straight line. In fact, it peaks at seven workers and starts to go down. Obviously the unique nature of each construction related project prevents us predicting an optimum number for each job, especially as the particular skills of the construction team also greatly influence the actual rate of output. In general terms, however, within any sector of the construction industry there is a recognisable trend: output increases as more units of labour are taken on but not at a constant rate. To understand why such a phenomenon occurs within any firm in the short run, we have to analyse in detail the law of diminishing (marginal) returns.

Table 7.1 Diminishing returns: a hypothetical case in construction

In the first column, we give the number of workers used per week on a project. In the second column, we give their total product; that is, the output that each specified number of workers can produce in terms of square metres. The third column gives the marginal product. The marginal product is the difference between the output possible with a given number of workers minus the output made possible with one less worker. For example, the marginal product of a fourth worker is 20 square metres, because with four workers 140 square metres are produced, but with three workers only 120 are produced; the difference is twenty.

Input of labour	Total product (output in sq. m per week)	Marginal physical product (in sq. m per week)
0	0	
1	20	20
2	60	40
3	120	60
4	140	20
5	150	10
6	160	10
7	165	5
8	163	−2

DIMINISHING RETURNS

The concept of diminishing marginal returns applies to many different situations. If you buckle a seat belt over your body in a car, a certain amount of additional safety is obtained. If you add another seat belt some more safety is obtained, but not as much as when the first belt was secured. When you add a third seat belt, again safety increases but the amount of additional safety obtained is even smaller. In a similar way, the **u-values** – a measure of heat loss – related to glazing do not decline steadily as you add more panes of glass within a window unit. The u-values typically associated with single, double and treble glazing are 5.7, 2.8 and 2.0 respectively. Therefore, assuming the wall construction and other factors remain constant, going from single to double glazing improves the u-value by 2.9, while adding a third pane of glass only improves the u-value by 0.8.

Figure 7.2a A production function

A production function relates outputs to inputs. We have merely taken the numbers from columns 1 and 2 of Table 7.1 and presented them as a graph.

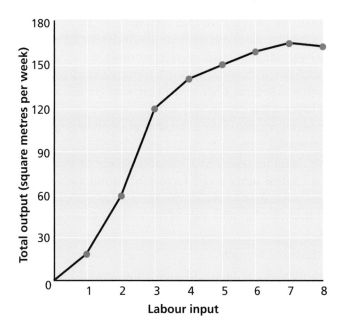

The same analysis holds for firms in their use of productive inputs. When the returns from hiring more workers are diminishing, it does not necessarily mean that more workers will not be hired. In fact, theoretically, workers should be hired until the returns, in terms of the value of the extra output produced, are equal to the additional wages that have to be paid for those workers to produce the extra output. Before we get into the decision-making process, let us demonstrate that diminishing returns can easily be represented graphically and subsequently used in our analysis of the firm.

Measuring Diminishing Returns

How do we measure diminishing returns? First, we will limit the analysis to only one variable factor of production (or input). Let us say that factor is labour. Every other factor of production, such as machinery, must be held constant. Only in this way can we calculate the marginal returns from using more workers and know when we reach the point of diminishing marginal returns.

Marginal returns for productive inputs are sometimes referred to as the **marginal physical product**. The marginal physical product of a worker, for example, is the change in total product that occurs when that worker joins an already existing team. It is also the change in total product that occurs when a worker resigns or is laid off from an already existing construction project. The marginal productivity of labour, therefore, refers to the change in output caused by a one-unit change in the labour input.

At the very beginning, the marginal productivity of labour may increase. Take a firm starting without any workers, only machines. The firm then hires one worker, who finds it difficult to do the work alone. When the firm hires more workers, however, each is able to specialise, and the marginal productivity of these additional workers may actually be greater than that achieved with the few workers. Therefore, at the outset, increasing marginal returns are likely to be experienced. Beyond a certain point, however, diminishing returns must set in; each worker has (on average) fewer machines with which to work (remember, all other inputs are fixed). Eventually, the site will become so crowded that workers will start running into one another and will become less productive.

Using these ideas, we can define the **law of diminishing (marginal) returns**. Consider this definition.

> As successive equal increases in a variable factor of production, such as labour, are added to other fixed factors of production, such as capital, there will be a point beyond which the extra or marginal product that can be attributed to each additional unit of the variable factor of production will decline.

We can express this more formally.

> As the proportion of one factor in a combination of factors is increased, after a point, the marginal product of that factor will diminish.

Put simply, diminishing returns means that output does not rise in exact proportions to increases in inputs, such as the number of workers employed.

AN EXAMPLE

The law of diminishing returns can be demonstrated in any sector of the economy. Take a building site as an example. There is a fixed amount of land (the building plot), and with the necessary supply of materials (bricks, timber, etc.) and certain tools, the addition of more workers eventually yields decreasing increases in output. A hypothetical set of numbers, based on discussions with project managers, that illustrates the law of diminishing returns has already been presented in Table 7.1 (and you may need to revisit the related caption). This is now represented graphically in Figure 7.2b.

Marginal productivity (the returns from adding more workers) at first increases, then decreases, and finally becomes negative. When one worker is hired, total output goes from zero to 20. Thus, the marginal physical product is equal to 20. When another worker is added, the total output, measured in square metres per week, increases to 60. Consequently the marginal physical product associated with the second unit of labour is 40. The third unit of labour adds a further 60 square metres to the total and, thereafter, the marginal product begins to decrease. In this example, therefore, diminishing marginal returns occurs after three workers are hired.

Diminishing Marginal Returns and the Theory of the Firm

If we now introduce business costs into the picture, we can begin to understand the central importance of the law of diminishing returns. For example, consider the relationship between marginal cost – that is, the cost of an extra unit of output – and the incidence of diminishing marginal physical returns as illustrated in Table 7.1. Let us assume that each unit of labour can be purchased at a constant price. Further assume that labour is the only variable input. We see that as more workers are hired, the marginal physical product first rises and then falls after the point at which diminishing returns are encountered. The marginal cost of each extra unit of output will first fall as long as the marginal physical product is rising, and then it will rise as long as the marginal physical product is falling. Consider again the data in Table 7.1. Assume that a worker is paid £500 a week. When we go from zero labour input to one unit, output increases by 20 square metres. Thus, the labour cost per square metre (the marginal cost) is £25. Now the second unit of labour is hired, and it, too, costs £500. Output increases by 40. Thus, the marginal cost is £12.50 (£500 divided by 40). We continue the experiment. The next unit of labour produces 60 additional square metres, so the marginal cost falls further to £8.33 per square metre. The next labourer yields only 20 additional square metres,

Figure 7.2b Diminishing marginal returns

Taking the data from Table 7.1, on the horizontal axis we plot the numbers of workers and on the vertical axis we plot the marginal physical product in square metres per week. When we go from zero workers to one worker, marginal product is 20. We show this at a point between 0 and 1 worker to indicate that the marginal product relates to a change in the total product as we add additional workers. When we go from one to two workers, the marginal product increases to 40. After three workers, marginal product declines. Therefore, after three workers, we are in the area of diminishing marginal physical returns. Total product, or output, reaches its peak at seven workers. In fact when we move from seven to eight workers, marginal product becomes negative.

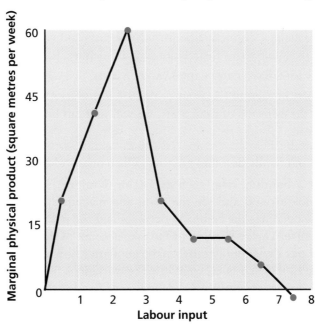

so the marginal costs starts to rise again back to £25. The following unit of labour increases marginal physical product by only 10, so that marginal cost becomes £50 per square metre (£500 divided by 10).

Marginal costs in turn affect the pattern of other costs, such as average variable costs and average total costs. Once these other costs have been considered, the importance of marginal cost analysis (and this whole section) will become clearer.

Key Points 7.2

○ In the short-run, all firms reach a point where diminishing marginal returns set in.

○ The law of diminishing returns states that if all but one of the factors of production are held constant, equal increments in that one variable factor will eventually yield decreasing increments in output.

○ A firm's short-run costs are a reflection of the law of diminishing marginal returns. Given any constant price of the variable input, marginal costs decline as long as the marginal product of the variable resource goes up. At the point of diminishing marginal returns, the reverse occurs. Marginal costs will rise as the marginal product of the variable input declines.

SHORT-RUN COSTS

In the short run, a firm's profit is largely determined by controlling various types of cost, which in general terms may be referred to as its total costs. Economists, however, like to make a distinction between fixed costs and variable costs, which we explain below. In simple terms the relationship, or identity, is:

total costs = total fixed costs + total variable costs

Examples of costs in each of these categories are introduced in Table 7.2. After we have looked at the elements of total costs, we will find out how to compute average and marginal costs.

Fixed Costs

Let us look at a building firm such as Taylor Wimpey. The managers of that business can look around and see the plant and machinery that Taylor Wimpey owns, the office buildings the company occupies and the permanent staff for whom they are responsible. Taylor Wimpey has to take account of the wear and tear of this equipment and pay the administrative staff, no matter how many houses it builds. In other words, all these costs are unaffected by variations in the amount of output. This leads us to a very straightforward definition of **fixed costs**.

All costs that do not vary – that is, costs that do not depend on the rate of production – are called fixed costs, or sunk costs.

According to Baumol's (1982) theory of contestable markets, when sunk costs are low, existing firms in a market have a continual concern that new entry is always possible. In these circumstances profit levels are constrained (see Chapter 8 for details). Contestable markets typify construction firms, as they usually have quite low fixed costs. Most contractors have no factory – as in effect each site represents the firm's new work location – and much of the necessary equipment, such as scaffolding, cranes, skips, toilets, office huts, security stores, floodlights and even water supplies, is usually hired as and when required.

Table 7.2 Typical construction costs

Type of Cost	Examples
Variable Costs	● **Labour used on site** ● **Materials used on site** ● **Equipment used on site** ● **Site management** ● **Tendering for future contracts**
Fixed Costs	● **Head office bills for energy, water and rates** ● **Wages for permanent head office staff** ● **Bank interest and leasing costs** ● **A sufficient (normal) level of return to keep the entrepreneur in the industry**

Variable Costs

The difference between total costs and total fixed costs is total variable costs: that is, total costs – total fixed costs = total variable costs. **Variable costs** are those costs whose magnitude varies with the rate of production. As a proportion of total costs, variable costs in construction tend to be much higher than in the manufacturing industry. One obvious variable cost is wages. The more a firm builds or makes, the more labour it has to hire and the more wages it has to pay. In fact a limiting factor to a construction firm's output is often its management: unlike machines, managers do not have an automatic safety cut-off point when they are operating at full capacity, and when things start to go wrong on site, costs can quickly spiral. As the size and number of their projects increases, construction firms need to employ good site and project managers, and these may well be in short supply. It is, therefore, difficult to determine when variable costs will rise. The same type of logic applies to the other main variable cost category – materials. As the demand for materials increases, they too may well become more expensive to acquire.

In construction, the distinction between fixed and variable costs can be difficult to make. For example, some firms may regard management salaries as fixed costs. It depends how the firm is organised. However, the way to avoid a high proportion of fixed costs is to meet any requirement to increase output by subcontracting and this is largely what happens in the construction industry across Europe.

Short-Run Average Cost Curves

In Figure 7.3a, we plot total costs, total variable costs and total fixed costs. You should note that the variable cost curve lies below the total cost curve by the vertical distance equivalent to total fixed costs. In manufacturing firms, the vertical distance representing fixed costs will be greater because its activities are based on a factory which probably houses expensive machinery and so it has relatively high fixed costs. Figure 7.3a is meant to represent the cost curves of a typical construction firm – the fixed costs are represented as proportionally low.

Next we want to look at the average cost. The average cost is simply the cost per unit of output. It is a matter of simple arithmetic to calculate the averages of these three cost concepts. We can define them simply as follows:

$$\text{average total costs} = \frac{\text{total costs}}{\text{output}}$$

$$\text{average variable costs} = \frac{\text{total variable costs}}{\text{output}}$$

$$\text{average fixed costs} = \frac{\text{total fixed costs}}{\text{output}}$$

Figure 7.3b shows these corresponding average costs. Let us see what we can observe about the three average cost curves in that graph.

AVERAGE FIXED COSTS (AFC)

As we can see from Figure 7.3b, **average fixed costs** fall throughout the output range. In fact, if we were to continue the diagram further to the right, we would find that average fixed costs would get closer and closer to the horizontal axis. This is because total fixed costs remain constant. As we divide this fixed amount by a larger number of units of output, AFC must become smaller and smaller.

AVERAGE VARIABLE COSTS (AVC)

We assume a particular form of the **average variable cost** curve. The form that it takes is a flattened U-shape: first it falls; then it starts to rise. The shape of the curve indicates that at first it costs less to build or make successive units, but as the law of diminishing returns sets in, it costs more and more to make successive units.

AVERAGE TOTAL COSTS (ATC)

This curve has a shape similar to the average variable cost curve. However, it falls even more dramatically at the beginning of the output range and rises more slowly after it has reached a minimum point. This is because **average total costs** is the summation of the average fixed cost curve and the average variable cost curve. So, when AFC plus AVC are both falling, ATC will fall too. At some point, however, AVC starts to rise while AFC continues to fall. Once the increase in the AVC curve outweighs the decrease in the AFC curve, the ATC curve will start to increase and it will develop its U-shape. An efficient firm will aim to achieve its output at the lowest point on the average cost curve – as this is where each unit of production is associated with its lowest possible cost, given the firm's existing level of capacity.

Figure 7.3a Total costs of production

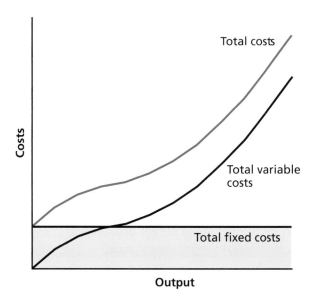

**Figure 7.3b Average fixed costs, average variable costs,
average total costs and the marginal costs of production**

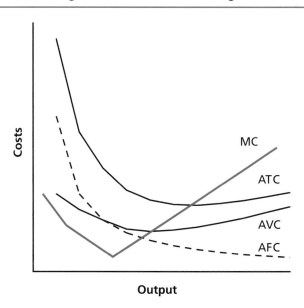

Marginal Cost

To highlight precisely when average costs reach their lowest point, economists are very interested in the principle of **marginal cost**. As we discussed earlier in this chapter in the section on diminishing returns, the term marginal means additional or incremental. In our previous example set out in Table 7.1, we considered the marginal physical product – that is, the additional output associated with taking on successive units of labour – and the marginal costs of that additional output. For convenience the data of Table 7.1 is reproduced in Table 7.3 together with columns setting out the marginal costs and average variable costs. These are calculated using the same assumptions – that each unit of labour costs £500 per week and that these are the only costs that alter – and the same method as on page 105. For example, when the second unit of labour is hired, costing £500, output increases by 40. Thus, the marginal cost is £12.50 (£500 ÷ 40) per square metre

Remember we find marginal cost by subtracting the total cost of producing all but the last unit(s) from the total cost of producing all units including the last one, and dividing the result by the additional output produced by the last unit. Marginal cost can be measured, therefore, by using the formula:

$$\text{marginal cost} = \frac{\text{change in total cost}}{\text{change in output}}$$

Table 7.3 Marginal and average costs

In the first column, we give the number of workers used per week on a project. In the second column, we give their total product; that is, the output that each specified number of workers can produce in terms of square metres. The third column gives the marginal product. The marginal product is the difference between the output possible with a given number of workers minus the output made possible with one less worker. In the fourth and fifth columns we calculate the marginal costs and average costs per square metre – assuming that each worker is paid £500 a week. For example, the marginal product of a fourth worker is 20 square metres, because with four workers, 140 square metres are produced, but with three workers only 120 are produced. The difference is 20. The marginal cost, therefore, is £25 (£500 ÷ 20), and the average variable cost at the same point is £14.29 (£2,000 ÷ 140).

Input of labour	Total product (output in sq. m per week)	Marginal physical product (in sq. m per week)	Marginal cost (£/sq. m)	Average cost (£/sq. m)
0	0			0
1	20	20	25.00	25.00
2	60	40	12.50	16.67
3	120	60	8.33	12.50
4	140	20	25.00	14.29
5	150	10	50.00	16.67
6	160	10	50.00	18.75
7	165	5	100.00	21.21
8	163	–2	–250.00	24.54

We show the marginal costs of production for our hypothetical example in column 4 of Table 7.3 and an equivalent marginal cost schedule is shown graphically in Figure 7.3b. By including marginal cost data on these graphs we can easily identify the minimum cost position.

Finding Minimum Costs

As Figure 7.3b demonstrates, the marginal cost curve first falls and then rises much like the average variable cost and average total cost curves. This should not be surprising: when marginal cost is below average total cost, average total cost falls; when marginal cost is above average total cost, average total cost rises. At the point at which average total costs are neither falling nor rising, marginal cost must then be equal to average total cost. When we represent this graphically, the marginal cost curve intersects the average total cost curve at its minimum.

The same analysis applies to the intersection of the marginal cost curve and the average variable cost curve.

THE CONTRACTOR'S COSTS

The theory of the firm suggests that a typical manufacturer of goods can determine an optimum selling price by analysing its costs. In the case of construction, however, few contractors know their real costs precisely, and most are completely unaware of their marginal cost positions. Furthermore, it is usually the client that initiates projects and the contractor is expected to determine the price before the work is complete. It is also often the case that there is only one party interested in buying the specific project – this type of market is known as a **monopsony** – which gives construction clients an unusually powerful bargaining position.

The Contractor's Bid

The usual bidding process involves the client inviting a selected number of contractors to compete for a project. Subsequently, the contractors that respond to this invitation will make various estimates before submitting a tender document. At the heart of the document is the price for the work and this is typically based on two components: a cost estimate, what it will cost the contractor to complete the project, and a mark-up to provide a profit. In many cases the key determinant to making a successful bid involves identifying the mark-up used by competitors. The implications of this will be fleshed out further in Chapter 8.

It is widely acknowledged that decisions about whether to bid or not to bid, and at what price, are incredibly complex. For example, Shash (1993) identified more than 50 factors that contractors take into consideration before submitting a bid. His top three factors – in rank order were as follows.

1 **Need for work** – The current state of the contractor's market, such as the firm's position on its cost curves, will be a significant factor in determining the nature of the bid. For example, if fixed costs are not spread over a sufficient volume of work, the firm will be willing to take on work at lower prices than when it is

working at capacity. In Shash's words (1993: 111): 'A contractor must secure a designated business volume...to cover operating costs and to realise a reasonable profit.' In other words, the more desperate a firm is for work, the lower its bid price and we will revisit the most desperate of bids – the so-called 'suicide bid' – in Chapter 8.

2   **Number of competitors** – Firms operating in a crowded, competitive situation have to accept a fair price dictated by others in the market. In contrast, if only a few firms dominate the market, a firm will be able to 'administer' its own price. This comparison between firms 'taking' a price or 'administering' a price will become clearer after the next chapter.

3   **Experience in similar projects** – The degree of complexity in the work required, compared to the firm's experience, is obviously a major determinant. (Shash's research related to the top 300 contractors, and had it broadened out to the lower end of the industry, location of the project would also have been a significant factor in deciding whether to bid.)

Key Points 7.3

○ The short run is that period of time during which a firm cannot alter its existing plant size.

○ Total costs equal total fixed costs plus total variable costs.

○ Fixed costs are those that do not vary with the rate of production; variable cost are those that do vary with the rate of production.

○ Average total costs equal total costs divided by output, that is $ATC = TC \div Q$.

○ Average variable costs equal total variable costs divided by output, that is $AVC = TVC \div Q$.

○ Average fixed costs equal total fixed costs divided by output, that is $AFC = TFC \div Q$.

○ Marginal cost equals the change in total cost divided by the change in output.

○ The marginal cost curve intersects the minimum point of the average total cost curve and the minimum point of the average variable cost curve.

○ A bid price has two components: an estimate of the costs, and a mark-up for profit.

○ Contractors consider more than 50 factors in deciding whether to bid for a project. Three important factors are: need for work, number of competitors and the contractor's experience in similar projects.

LONG-RUN COSTS

The long run is defined as the time during which *full* adjustment can be made to any change in the economic environment. In the long run, all factors of production are variable. The scope to vary all inputs allows a firm to produce at lower costs in the long run than in the short run (when some inputs are fixed). In other words, over time a firm has the opportunity to fine-tune the business. For example, in the long run a firm can alter plant size or increase the number of functions covered by its head office. There may be many short-run curves as a firm develops over the years, but only one long run. Long-run curves are sometimes called planning curves, and the long run may be regarded as the **planning horizon**.

We start our analysis of long-run cost curves by considering a single firm contemplating the construction of a single plant. The firm has, let us say, three alternative plant sizes from which to choose on the planning horizon. Each particular plant size generates its own short-run average total cost curve. Now that we are talking about the difference between long-run and short-run cost curves, we will label all short-run curves with an S; short-run average (total) cost curves will be labelled SAC, and all long-run average (total) cost curves will be labelled LAC.

Figure 7.4a Preferable plant size

If the anticipated permanent rate of output per unit time period is Q_1, the optimal plant to build would be the one corresponding to SAC_1 because average costs are lower. However, if the rate of output increases to Q_2, it will be more profitable to have a plant size corresponding to SAC_2, as unit costs can fall to C_3.

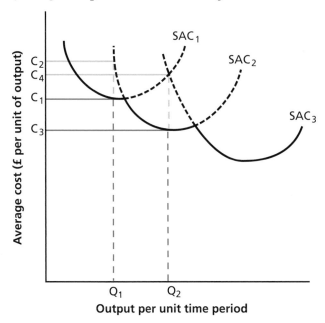

Look at Figure 7.4a Here we show three short-run average cost curves for the three (successively larger) plant sizes. Which is the optimal plant size to build? That depends on the anticipated rate of output per unit of time. Assume for a moment that the anticipated rate is Q_1. If plant size 1 is built, the average costs will be C_1. If plant size 2 is built, we see on SAC_2 that the average costs will be C_2, which is greater than C_1. So if the anticipated rate of output is Q_1, the appropriate plant size is the one from which SAC_1 is derived.

Note, however, what happens if the anticipated permanent rate of output per unit of time goes from Q_1 to Q_2. If plant size 1 has been decided upon, average costs will be C_4. However, if plant size 2 had been decided upon, average costs will be C_3, which is clearly less than C_4.

Long-run Average Cost Curve

If we make the further assumption that during the development of a firm the entrepreneur is faced with an infinite number of choices of plant size, then we can envisage an infinite number of SAC curves similar to the three in Figure 7.4a. We are not able to draw an infinite number, but we have drawn quite a few in Figure 7.4b.

Figure 7.4b Deriving the long-run average cost curve

If we draw all the possible short-run average curves that correspond to different plant sizes and then draw the envelope to these various curves, SAC_1 ... SAC_8, we obtain the long-run average cost curve.

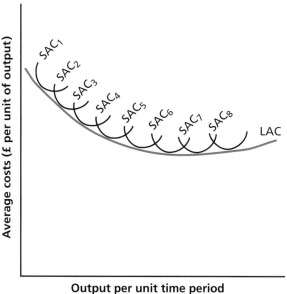

By drawing the envelope of these various SAC curves we find the **long-run average cost curve**. To be academically precise, the long-run average cost (LAC) curve should be wavy or scalloped, since it follows the path of the SAC curves enclosed. By tradition, however, it is portrayed as being tangent to the minimum point of the SAC curves from which it is derived. Either way, the long-run average cost curve represents the cheapest way to produce various levels of output – provided the entrepreneur is prepared to change the size and design of the firm's plant. Long-run average cost curves are sometimes referred to as **planning curves**.

Why the Long-run Average Cost Curve is U-shaped

Notice that the long-run average cost curve LAC in Figure 7.4b is U-shaped. It is similar to the U-shape of the short-run average cost curve developed previously in this chapter. However, the reason for the U-shape of the long-run average cost curve is not the same as that for the short-run U-shaped average cost curve. The short-run average cost curve is U-shaped because of the law of diminishing marginal returns. However, that law cannot apply to the long run – in the long run all factors of production are variable, so there is no point of diminishing marginal returns since there is no fixed factor of production. Why, then, does the long-run average cost curve have a U-shape? The reasoning has to do with changes in the scale of operations. When the long-run average cost curve slopes downwards, it means that average costs decrease as output increases. Whenever this happens, the firm is experiencing **economies of scale**. If, on the other hand, the long-run average cost curve is sloping upwards, the firm is incurring increases in average costs as output increases. The firm is said to be experiencing **diseconomies of scale**. There is a third possibility: if long-run average costs do not change with changes in output, the firm is experiencing **constant returns to scale**. In Figure 7.5 (on page 116) we show these three stages. In the first stage the firm is experiencing economies of scale; in the second stage, constant returns to scale; and in the third stage, diseconomies of scale.

Returns to Scale – In Three Stages

Savings (economies of scale) are possible as firms progress to larger production – that is, increases in output can result in a decrease in average cost. There are five types of scale economies.

- Technical economies: relating to the firm's ability to take full advantage of the capacity of its machinery.
- Managerial economies: as firms grow, they can afford to employ – and benefit from – specialised managers.
- Commercial economies: such as buying in bulk and advertising.
- Financial economies: larger firms have a greater variety of sources for funds and often at favourable rates.
- Risk bearing economies: larger firms may achieve distinct advantages by diversifying into several markets and researching new ones.

When economies of scale are exhausted, constant returns to scale begin. Some economists regard the commencement of this stage as the **minimum efficient scale** (MES) since it represents the lowest rate of output at which long-run average costs are minimised – and no further economies of scale can be achieved in the present time period. The MES is represented by point Q_1 in Figure 7.5.

Clearly, economies of scale are more easily associated with standardised manufactured products. Indeed, in many manufacturing industries a firm has to be big to survive. This is certainly not the case in construction. The unique nature of many construction projects, plus the relatively small size of many construction firms, prevents the industry from realising the full potential of economies of scale.

Figure 7.5 Economies and diseconomies of scale

Long-run average cost curves will fall when there are economies of scale, as shown in stage one up until Q_1. There will be constant returns to scale when the firm is experiencing output Q_1 to Q_2, as shown in stage two. And, finally, long-run average costs will rise when the firm is experiencing diseconomies of scale, beyond Q_2 in stage three.

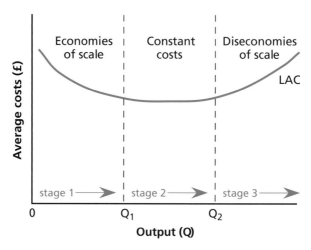

Cooke (1996: 134) shows very clearly the different capabilities of manufacturing and construction firms to benefit from economies of scale by using a diagram similar to Figure 7.6. In Cooke's analysis the long-run average cost curve LAC_1 represents the average costs of production of a typical manufacturing firm able to exploit economies of scale. In this case, the MES is at point Q_2. In contrast, LAC_2 – positioned to the left of and above LAC_1 – denotes the long-run average costs associated with projects with lower levels of standardisation, such as those experienced by the typical firm in construction. On LAC_2, a firm can only achieve economies of scale to output Q_1 – so clearly it would be beneficial to reorganise production to be on LAC_1. This is precisely what the contractor O'Rourke achieved by its acquisition of Laing Construction in 2001. Within a year O'Rourke was in a position to offer a range of standardised, prefabricated components that could be configured to create buildings to suit individual clients' requirements. This development was only possible due to O'Rourke's increased scale of operation. In effect it had moved from operating along LAC_2 to the more efficient LAC_1 schedule. Such outcomes are often the rationale that lies behind company mergers and acquisitions – but these achievements are still the exceptions rather than the rule in the construction industry.

Figure 7.6 Economies of scale in the construction sector

LAC$_1$ represents the average cost of production for a firm able to exploit economies of scale, such as a manufacturer. In comparison, LAC$_2$ is the average cost of production for a contractor unable to take full advantage of economies of scale, due to the unique nature of each unit of output.

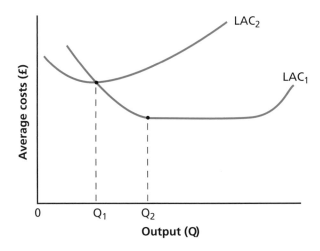

To sum up, the general rule seems to be that economies of scale, of any kind, have a limited role in construction. In fact, Patricia Hillebrandt consistently emphasised in each edition of her textbook that 'many construction firms are actually on a long-run *increasing* cost curve' (Hillebrandt, 1974: 121; 1985: 116 and 2000: 121). She attributes the problem to the localised nature of small traditional construction firms. For example, a firm needing to dig ditches might hire workers on a job basis. For jobs requiring up to ten workers, it simply hires the workers and gives them each a shovel to dig ditches. However, for jobs requiring ten or more workers, the firm may feel it is also necessary to hire an overseer to co-ordinate the ditch-digging effort. Thus, perhaps, constant returns to scale remain until ten workers are employed, then decreasing returns to scale set in. As the layers of supervision grow, the costs of communication grow more than proportionately; hence, the average cost per unit of output will start to increase. In the terms used in Figure 7.5, firms can run quite quickly into diseconomies of scale as they grow in size if they are not careful.

According to Hillebrandt, these diseconomies occur very soon in construction for two main reasons. First, most firms can only substantially increase their turnover by extending their catchment area, which in turn increases costs of transport and supervision. Second, there is the indivisible nature of entrepreneurial ability that causes the decision-making process to clog up as firms increase their scale of operations. This second argument is well rehearsed in the traditional business literature: many analysts have observed a crisis point in a firm's growth when it becomes too big for the directors of the firms to continue to exercise effective personal control yet too small to afford to recruit extra expertise. The precise point that this occurs depends on the management calibre of the existing directors.

A more current debate, captured in the research of Pan et al. (2008) and Chiang et al. (2008), focuses on the new opportunities offered by **off-site production (OSP)** where buildings, structures or parts are manufactured or pre-assembled prior to their installation on site. This offers contractors the opportunities to take advantage of a factory environment, with opportunities to improve on quality, time, cost, productivity and safety. Yet surveys of the industry in the UK and Hong Kong suggest that the difficulties in achieving economies of scale continue to be a significant constraint, restricting construction to traditional on-site approaches and cost schedules (Pan et al. 2008). The debate regarding whether traditional organisational problems or modern technological barriers limit a construction firm is examined in Reading 2 (see pages 137–9) where the option of off-site production to reduce average costs per unit is reviewed further.

The Minimum Efficient Scale and Levels of Competition

The manufacturing sector seems to be a constant reference point in these debates where, relatively speaking, state-of-the-art savings and economies of scale are more easily achieved. In fact, the conceptual idea of a minimum efficient scale (MES) first originated from a comprehensive study into the shape of long-run average cost curves in British manufacturing industry. This was defined as 'the minimum scale beyond which any possible subsequent doubling in scale would reduce total average unit cost by less than five per cent' (Pratten 1971: 278).

One of the uses of the minimum efficient scale (MES) is that it gives a rough measure of the degree of competition in an industry. If the level of output associated with the minimum efficient scale is small relative to the total industry demand, the degree of competition in that industry is likely to be high since there is room for many efficient firms. Conversely, when the level of output associated with the MES is large relative to industry demand, the degree of competition is likely to be small. For example, the markets in oil and gas extraction, air transport, automobile manufacturing and tobacco are dominated by a few very large businesses. These industries have a high **concentration ratio**, and many have the characteristics of oligopolies. In stark comparison, some industries are typified by many small businesses, some of which are run by the self-employed. These industries are said to have low concentration ratios. Construction is a prime example of this type of industry and these markets are characterised as contestable. (Types of market competition will be discussed further in Chapter 8.) Economists maintain that the number of firms comprising an industry is determined by the MES, as they claim that an industry is 'efficient' when the amount of resources used to produce the total output is at a minimum.

The importance of economies of scale to specific industries can be determined by the market share of an industry's largest firms, taking the gross value added (turnover), the percentage of total output, employment or profit as the benchmark. For example, imagine that an industry contains just ten firms that individually account for 25, 15, 12, 10, 10, 8, 7, 6, 4 and 3 per cent of turnover, respectively. The five-firm concentration ratio for this industry – the most widely used concentration measure – is 72 (calculated by 25 + 15 + 12 + 10 + 10). This means that the top five firms account for 72 per cent of this industry's sales. In the statistical material published by the Office for National Statistics (ONS) the variables used to calculate the five-firm concentration ratio are gross value added and output. In other words, the gross value added and the volume of output of the top five businesses are measured as a proportion of the total turnover and total output of the industry. The series is based on the *Annual Business Enquiry and Input-Output Analyses*. Note, however, before considering examples, that the precise definition of any specific industry is somewhat arbitrary and, consequently, as we narrow the definition of an industry the concentration ratio rises, and vice versa. Also conventional measures include only domestic production, and if foreign competition is a significant presence, the actual market concentration will be much lower than indicated by the ratios.

Table 7.4 Concentration ratios by industry (in 2004)

Industry	Five-firm concentration ratio	
	GVA (%)	Output (%)
Construction	5	5
Iron and steel	48	61
Cement, lime and plaster	74	71
Metal boiler and radiator	51	45

Source: Input-output analysis (ONS, 2006: Table 8.31)

Table 7.4 gives the five-firm concentration ratio for four industries. As noted, the more precise the definition of an industry the higher the ratio. Consequently, the broad industrial classification of construction produces a very low concentration ratio, as both total production and turnover (gross value added) are dominated by small businesses. In fact, as we pointed out in Chapter 6 there are more than 220,000 enterprises comprising the construction industry – responsible for all the gross value added and output of the UK construction firms. However, if we define house building as a separate industrial sector then it becomes a more concentrated market – the top five house builders account for more than 65 per cent of the gross value added by all the house builders across Great Britain. This increases the opportunities to benefit from economies of scale and changes the nature of competition. Indeed, house building at the regional and local level often allows businesses to act like firms operating in an oligopoly, as the number of competing firms at this level will be quite small. The bigger house building firms in an area should be able to control the price and quality expected in that marketplace.

Table 7.4 also shows that the five-firm concentration ratio for iron and steel output is 61 per cent, and such a high figure would be expected in an industry based around a specific production process that requires heavy plant and machinery. However, this concentration ratio has fallen in recent years due to structural changes experienced by the iron and steel sector, caused in part by an increase in foreign imports and the introduction of new technologies. Finally, it is interesting to observe that the ratio between the output and the gross value added by the top five firms in the iron and steel sector is significantly different to all the other industrial sectors shown in Table 7.4, suggesting that higher value, high-quality iron and steel is produced by smaller specialist firms in this industrial sector.

Key Points 7.4

○ The long-run average cost curve is derived by drawing a line tangent to a series of short-run average cost curves, each corresponding to a different plant size.

○ A firm can experience economies of scale, diseconomies of scale and constant returns to scale, depending on whether the long-run average cost curve slopes downwards, upwards or is horizontal (flat).

○ Economies of scale occur when all factors of production are increased and average costs fall. There are five reasons for economies of scale, relating to (a) managerial economies, (b) commercial economies, (c) financial economies, (d) technical economies and (e) risk-bearing economies.

○ The minimum efficient scale occurs at the lowest rate of output at which long-run average costs are minimised.

○ Construction firms may experience diseconomies of scale because of limits to the efficient functioning of management, the fragmented nature of the industry and the difficulties experienced in achieving economies of scale.

○ Concentration ratios estimate the importance of the largest firms in an industrial sector and give some indication of the level of competition and opportunities to benefit from economies of scale.

EXTERNAL ECONOMIES OF SCALE

The economies of scale we have discussed so far in this chapter are internal to the firm, and could be described as the direct result of individual company policy. In other words, economies of scale do not depend on what other firms are doing or what is happening in the economy. They are formally referred to as **internal economies** (or diseconomies) of scale. This is to distinguish them from **external economies of scale**, which benefit all firms in an industry – regardless of their individual size or policy. In the case of construction, external economies are more important than internal economies of scale. In most countries, the construction sector represents one of the largest parts of an economy but it usually comprises many small fragmented firms. External economies are, therefore, simply a by-product of being a firm involved in a large industrial sector.

When any industry expands, *all* firms in the industry benefit in ways that normally lead to savings for all the firms involved. Firms get the opportunity to buy in services more easily; firms can combine to fund research and/or training; firms often become more specialised; trade associations may form; professional bodies can emerge to represent their members; and specialised journals may be started to report on best practice. In construction, there are examples of all these developments.

As we mentioned in Chapter 1, there have been a succession of reports seeking to improve the efficiency of the industry as a whole. For example, the *Government Construction Strategy* published in May 2011 set a national agenda to improve the performance and culture of the construction industry. Targets relevant to the discussion in this chapter included a 20 per cent reduction in annual costs and greater integration across the industry. To achieve these targets, the report encouraged standardisation rather than bespoke designs, and it argued for more emphasis on whole-life costing in preference to the usual focus on up-front capital costs and an acceptance of new models of procurement, where all parties agree to a realistic cost benchmark before a project commences. To support these targets the government mandated the industry to adopt **building information modelling (BIM)** for all public contracts by 2016. In simple terms, BIM makes use of information technology to centralise all data relating to the design and cost of a project in such a way that it can be readily shared, interrogated and updated. This was formally justified on the basis that the integration of design, construction and operation stages of a building should lead to improved asset performance (Cabinet Office 2011: 14). We consider this modelling technique further in Chapter 15 as an opportunity that supports the sustainable construction agenda.

Another example of external economies of scale is the initiatives undertaken by the government on behalf of an industry. For example, Constructing Excellence is a government sponsored organisation that offers firms in the industry opportunities to learn how to improve practice through case studies and demonstration projects. Constructionline is a database that gathers information on contractors and consultants to provide public sector procurers (such as central government departments, local authorities, universities and colleges, and NHS trusts) with quick access to a list of fully accredited suppliers. The list covers the full spectrum of construction activities from architecture to demolition and includes businesses ranging in size from small specialists to the largest contractors. A final example of an external economy provided by the government is UK Trade and Investment, an organisation that is dedicated to helping UK companies grow their business internationally. This is done through a network of staff working in the UK and abroad to help any company wanting to export. Promotional exercises are undertaken in high growth markets and research is commissioned to identify opportunities for UK-based firms. For example, during 2011 construction sector briefings were published for Canada, Chile, China, Malaysia, Mexico, New Zealand, Russia and Singapore. Each of these reports highlighted the potential for contractors seeking to expand into these markets. To sum up, external economies, regardless of whether they are provided by trade associations, professional organisations or the government, help to improve overall industry performance.

All these developments represent external economies of scale, as in each case, industry-wide initiatives encourage the reduction of each participating firm's costs.

In terms of Figure 7.5, the industry as a whole should experience a downward shift in its long-run average cost curve reflecting a decrease in costs at every level of output.

Of course, external change will not always benefit the member firms of a particular industry. For example, as the construction industry grows larger, shortages of specific materials, land and/or skilled labour may occur. This would push up costs per unit of output across the whole industry – or, in the terms employed above, the industry would experience external diseconomies of scale.

A Final Note on Techniques

Before progressing further with the theory of the construction firm, it may be useful to emphasise three points which should help you avoid any confusion when studying other economic textbooks.

1   By now you should understand how economists use cost curve diagrams to provide a summary of their ideas. You should realise, therefore, that these diagrams are never intended to be more than a visual image – an aide-memoire. They never form a precise reflection of a particular firm.

2   Most microeconomic theory has been developed with manufacturing as the focus and it does not, therefore, always smoothly translate to the construction industry. As Ive and Gruneberg (2000: 150) emphasise, economic theory relates to the manufacturing of mass-produced products, whereas construction is oriented towards individual projects that tend to be 'one-off productions' (with the possible exception of housing). Indeed, in most mainstream textbooks on economic theory the construction industry is not even indexed.

3   Remember that we commence from the basic assumption that all firms in all sectors seek to maximise their profits. This is an important cornerstone of economic theory. Economists pursue this objective with an academic zeal that regards even the marginal detail as crucially significant. This will be developed further in the next chapter.

Key Points 7.5

○   External economies of scale relate to a whole industry; whereas internal economies of scale arise from the growth of one firm, regardless of what is happening to other firms.

○   Good examples of external economies of scale arise from the work of government departments that seek to deliver a marked improvement in the business performance of construction firms at home and abroad.

○   The cost-curve techniques associated with the theory of the firm have been developed for, and particularly apply to, the profit maximising manufacturer.

8 Types of Market Structure in the Construction Industry

In order to understand the precise relationship between output, revenue and price, a firm has to know the structure of the market or industry into which it is selling its product. There are various market structures. At one extreme, there is a **monopoly** where one producer dominates the market and controls the price and output decisions. At the other extreme, both buyers and sellers correctly assume that they cannot affect market price – this market structure is known as **perfect competition**. Most firms involved in construction are engaged in market structures between these two theoretical extremes. In the language of textbook economics, they are involved in **imperfectly competitive markets** – where the clients and contractors have to take into account how their individual actions will affect the market price. (Here, it might be useful to review Key Points 6.1.) We shall examine these real-life scenarios in more detail once we have set up a reference point for the discussion.

THE PURPOSE OF PERFECT COMPETITION

To begin, we consider in some detail the hypothetical scenario of a perfectly competitive market. Although no real industry actually operates in such a market, it provides an important reference point for economists. In the case of construction, there are also some interesting parallels in reality. The perfectly competitive market acts as a benchmark from which other market situations can be judged. As we will show, an optimum allocation of resources arises from perfect competition because every firm is producing at minimum unit cost. Consequently, we will be able to develop an understanding of what is meant by a 'fair' price, a 'normal profit' and an 'efficient' industry.

The Characteristics of Perfect Competition

The term 'perfect competition' relates to a specific model market structure with these defining characteristics.

- The product sold by firms in the industry is homogeneous. This means that the product sold by each firm in the industry is a perfect substitute for the product sold by every other firm. In other words, buyers are able to choose a product from a large number of sellers in the knowledge that it is essentially the same. The product is thus not in any sense differentiated regardless of the source of supply.
- Any firm can enter or exit the industry without serious impediments. Resources must also be able to move in and out of the industry unimpeded; without, for example, government legislation preventing any resource mobility.
- There must be a large number of buyers and sellers. When this is the case, no single buyer or seller has any significant influence on price. Large numbers of buyers and sellers also mean that they will be acting independently.

- There must be complete information available to both buyers and sellers about market prices, product quality and cost conditions.

Now that we have defined the characteristics of a perfectly competitive market structure, we consider the position of an individual firm. We define a **perfectly competitive firm** as:

> one that is such a small part of the total industry in which it operates that it cannot significantly affect the price of the product in question.

This means that each firm in the industry is a **price-taker** – it takes the price as something that is beyond its individual control.

How does a situation arise in which firms regard prices as set by forces outside their control? The answer is that even though every firm, by definition, sets its own prices, it must always consider the prices of its competitors. The firm in a perfectly competitive situation finds that it will eventually have no customers if it sets its price above the competitive price. Let us now see what the demand curve of an individual firm in a competitive industry looks like.

Single-Firm Demand Curve

We have already discussed the characteristics of demand schedules (for example, see Key Points 4.1). Figure 8.1 presents the hypothetical market demand schedule faced by any producer or contractor, who, we assume, controls only a very small part of the total market. This is how we characterise the demand schedule for a perfectly

Figure 8.1 The demand curve for an individual firm in a perfectly competitive market

We assume that the individual producer represents such a small part of the total market that it cannot influence the price. The firm accepts the price as given. At the going market price it faces a horizontal demand curve. If it raises its price, even by one penny, it will sell nothing. Conversely, the firm would be foolish to lower its price because it can sell all that it can produce at the market price. The firm is a price-taker and its demand curve is described as being perfectly elastic.

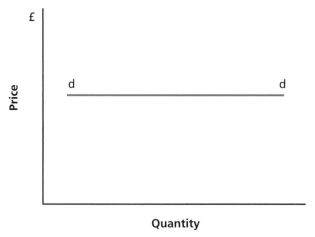

competitive firm – it is a horizontal line at the going market price. It is a completely elastic demand curve – by raising its price by one penny, the individual firm would lose all its business.

At the market price, demand is **perfectly elastic**; the firm can sell as much output as it wants, providing it does not alter the price. If the firm were to raise its price, consumers will buy the same skill, product or service from another producer.

Key Points 8.1

○ There are various market structures. Examples include a monopoly in which one producer alone controls price and output decisions, and perfect competition in which no single producer can control the market.

○ The perfectly competitive market is a hypothetical extreme that acts as a benchmark by which real (imperfect) markets can be judged.

○ The hypothetical model of perfect competition has four main characteristics: (a) homogeneous product, (b) freedom of entry and exit, (c) large number of buyers and sellers, and (d) full information.

○ A perfectly competitive firm is a price-taker. It takes price as given. It can sell all that it wants at the going market price. The demand curve facing a perfect competitor is a horizontal line at the market price.

HOW MUCH DOES THE PERFECT COMPETITOR PRODUCE?

We have established that a perfect competitor has to accept the given price of the product. If the firm raises its price, it sells nothing. If it lowers its price, it makes less money per unit sold than it otherwise could. The firm has only one decision variable left: how much should it produce? We will apply our model of the firm to answer this question. We shall use the **profit-maximisation model** and assume that firms, whether competitive or monopolistic, will attempt to maximise their total profits – that is, they will seek to maximise the positive difference between total revenue and total costs. (It may also help here to review the distinction between accounting and economic profits, see Key Points 7.1.)

Total Revenue

Every firm has to consider its total revenue. Total revenue is defined as the quantity sold multiplied by the price; or, expressed using the notation employed in some texts, TR = P x Q. This is also the same as total receipts from the sale of output.

In Figure 8.2 (see page 126) we assume that the firm is one of many comprising the total market, so it can sell all it produces at a given price. Thus, the total revenue curve is presented as a straight line. For every unit of sales, total revenue is increased by proportionally the same amount.

Total Costs

Revenue is only one side of the picture. **Total costs** must also be considered. Notice that when we plot total costs in Figure 8.2 the curve is not a straight line, but a wavy line, due to the existence of increasing and decreasing returns which we alluded to in Chapter 7. When the total cost curve is above the total revenue curve, the firm is experiencing losses. When it is below the total revenue curve, the firm is making profits. Where the two curves intersect represents breakeven points. Note that by reducing total costs, firms can make bigger profits.

Figure 8.2 Finding a profit-maximising position

The straight black line represents total revenue, as each unit is sold for the same price. Total costs, represented by the orange line, first exceed total revenues and a loss is made; then they become less than total revenue and a profit is made. We find maximum profits at the point where total revenues exceed total costs by the largest amount.

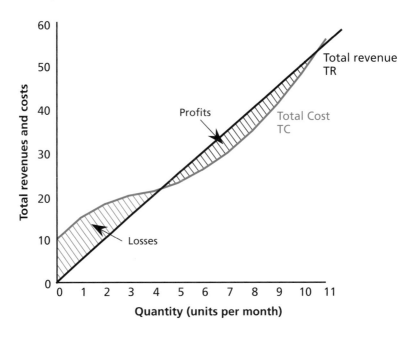

COMPARING TOTAL COSTS WITH TOTAL REVENUE

By comparing total costs with total revenue, it is possible to calculate the number of units that the individual competitive firm should aim to produce per month. Clearly, the firm will maximise profits at that place on the graph where the total revenue curve exceeds the total cost curve by the greatest amount. In Figure 8.2, that occurs at a rate of output and sales of either seven or eight units per month; this rate may be called the profit-maximising rate of production.

Marginal Analysis

Another way to find the profit-maximising rate of production for a firm is by marginal analysis. This method involves making a detailed study of marginal revenue and marginal costs. The concept of **marginal cost** has already been introduced in Chapter 7. It was defined as the change in total cost due to a one-unit change in production. The resulting schedule of costs was based on the law of diminishing returns: at first costs fall and then they begin to rise. Some example calculations for a marginal cost schedule were presented in column 4 of Table 7.3. This leaves **marginal revenue** to be clarified.

Marginal Revenue

Marginal revenue represents the increment in total revenue attributable to selling one additional unit of product. For example, if selling an extra unit of construction activity increases a contractor's total revenue from £1,800 to £2,100, the marginal revenue equals £300. Hence, marginal revenue may be calculated by using the formula:

$$\text{marginal revenue} = \frac{\text{change in total revenue}}{\text{change in output}}$$

In any market structure, therefore, marginal revenue is closely related to price. In fact in a perfectly competitive market, the marginal revenue curve is exactly equivalent to the price line or, in other words, to the individual firm's demand curve, since the firm can sell all of its output (including the last unit of output) at the market price.

COMPARING MARGINAL COST WITH MARGINAL REVENUE

Obviously, if the marginal revenue from a unit increase in output is greater than the marginal cost, it would seem rational for the profit-maximising firm to produce that unit of output. Conversely, if the marginal cost of an extra unit of output exceeds its marginal revenue, it would be produced at a loss and, therefore, it would be inappropriate for the profit-maximising producer to produce that unit of output. In fact, all firms have a clear incentive to produce and sell right up to the point at which the revenue received from selling one more unit of output equals the additional cost incurred in producing that unit. If the firm chooses to stop output before this point, it will not have maximised profits: it will not have squeezed the pips until they squeak. The profit maximiser should not be satisfied until the last penny of profit has been earned. This will only be achieved at the point where marginal costs equal marginal revenue. This decision rule is represented by point E in Figure 8.3 (see page 128).

Figure 8.3 Long-run perfectly competitive equilibrium

In the long run, perfectly competitive firms move towards a position at which marginal revenue equals marginal cost and average total costs. In short, 'where everything is equal' – represented by point E.

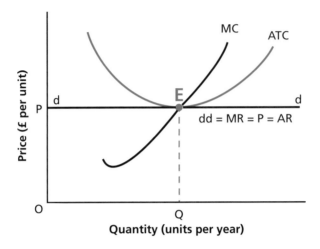

Quantity (units per year)

Key Points 8.2

○ Profit is maximised at the rate of output where the positive difference between total revenue and total costs is greatest.

○ Using marginal analysis, the profit-maximising firm will produce at a rate of output at which marginal revenue equals marginal cost.

○ Profit-maximising rules apply to all types of market structure.

TOWARDS THE NOTION OF AN EFFICIENT INDUSTRY

To consider an entire industry, the cost and revenue schedules of all its constituent firms need to be aggregated. For an industry with a perfectly competitive market structure, this is not a problem since the costs and revenue for each firm are identical. This theoretical extreme will prove relevant when we consider the notions of efficiency that exist in reality. It should also prove to be useful to those concerned with the performance of firms in the construction sector. Most firms in construction do seek to maximise their profits, they often have a large degree of freedom to enter and exit the various activities, and they rarely set their prices without regard to the terms expressed by their competitors.

Short-run Versus Long-run Profits

Economic theory suggests that in the long run all the firms in a competitive industry earn normal profits (as defined and discussed in Chapter 7). In the short run, however, which may represent a considerable length of time in construction markets, some firms will sell their product well above their minimum average costs and make **supernormal profits**. As a result, more firms will be enticed into the sector to get a slice of the action. In time, this increased competition will force the equilibrium price of the product down, until each firm is making only normal profit. This situation is shown in Figure 8.3.

In the long run, the perfectly competitive firm/industry finds itself producing at a rate Q. At that rate of output, the price is just equal to the minimum average total cost. Obviously, it is possible for supernormal profits to cause too many firms to enter the market, in which case the market price would fall below P and firms would make a loss – **subnormal profits**. Firms would then leave the industry and the decrease in supply would cause market prices to rise again to P. In this sense, perfect competition results in no 'waste' in the production system. Goods and services are produced using the least costly combination of resources. This is an important attribute of a perfectly competitive long-run equilibrium, particularly when we wish to compare the market structures that are less than perfectly competitive.

Key Points 8.3

- ○ In the short run, the perfectly competitive firm can make supernormal profits.

- ○ In a perfectly competitive market, new firms entering the industry will absorb supernormal profits.

- ○ In the long run, a perfectly competitive firm (industry) produces at the point where P = MR = MC = ATC.

MARKET STRUCTURES THAT TYPIFY THE CONSTRUCTION INDUSTRY

So far we have discussed in some detail the hypothetical market structure of perfect competition, in which there are numerous firms that produce the same product and have no influence over price: they are price-takers. In this section, we analyse the actual markets in which construction firms are engaged. These are also generally competitive, in the sense that there are usually many firms producing an insignificant part of total construction output, and there are certainly no formal restrictions preventing firms from participating. Most firms comprising the construction industry do, however, have some control over price. In fact, we shall conclude that, in general terms, markets representing construction are often representative of either monopolistic competition or oligopoly and, in practice therefore, there is an element of control over price-making.

In making this comparison between the world of perfect theory and actual practice, it is important to remember that the construction industry, as a whole, consists of many different markets – some are defined by a specific service or product; others by the size and complexity of contracts awarded in the market, or by the geographical location of the market. Consequently, we should not aim to pigeonhole all firms into one model of market behaviour. However, the following analysis represents a quick tour through some generalised principles that apply to significant parts of the construction industry. It may help your comprehension if you have some specific construction firms in mind as you proceed.

Monopolistic Competition

In reality, most markets are far from perfect. For example, in any construction market contractors, subcontractors and material producers will try to obtain some monopoly advantages by distinguishing their firm's product from that of their competitors. They may do this by somehow implying – or, indeed, achieving – better quality and/or reliability. In these types of market, firms can earn above normal profits – but only for a short while, because other firms in the market will respond by producing similar products. This keeps the market very competitive, and constrains long-term profits.

This model of behaviour is known as **monopolistic competition** as each firm can easily achieve a degree of local monopoly but is ultimately restricted by the presence of many competing firms. As we have suggested, this type of competition is well exemplified by firms in the construction industry: firms tend to be highly fragmented across the country but are somehow constrained by the potential competition of similar firms in neighbouring towns.

Oligopoly

In the strictest sense of the word, **oligopoly** is where a few sellers compete for the entire market. Blue Circle, for example, accounts for approximately half of the supply of cement to building firms in the UK and two other firms, Rugby and Castle, make most of the remaining cement. In this kind of market each firm has enough power to avoid being a price-taker – but they are still subject to a sufficient amount of competition to know the market is not entirely under their control. In other words, firms in oligopolistic markets will price their produce or service according to how they think competitors will react. This leaves them faced with the dilemma of not knowing whether to compete or co-operate. In short, the world of oligopoly is one of uncertainty.

To paraphrase Lipsey and Crystal (1995: 264), firms in an oligopolistic industry will make more profits if they agree to co-operate as a group; however, if one firm deviates from the agreement and/or becomes aggressively competitive, it stands to make more profit for itself.

When firms agree to co-operate to raise profits it is called **collusion**. Anecdotal evidence from managers and postgraduate students working in the industry suggests that collusion is common practice across the whole breadth of construction wherever contractual agreements are formed. Hillebrandt's theoretical analysis also suggested that forms of oligopolistic behaviour would be pervasive throughout the industry (1974: 155–7; 2000: 153–5). More recently these assumptions were confirmed by an extensive investigation carried out by the Office of Fair Trading (OFT). In 2008, the OFT found that more than 100 British construction companies were participating in the rigging of bids. This inquiry is described further below (see page 132), but first, as collusive agreements appear to typify so many transactions in the industry we need to examine them in more detail.

THE HYPOTHESIS OF QUALIFIED JOINT PROFIT MAXIMISATION

As long ago as 1776, Adam Smith, the founder of modern economics, alluded to collusion. In commenting upon contracts between rival firms, he showed deep suspicion. He observed that 'people of the same trade seldom meet together for fun and merriment, but the conversation ends in a conspiracy against the public, or in some contrivance to raise prices'.

It was not until 170 years later that this statement was elaborated into a hypothesis by Professor William Fellner. He explained how some kind of agreement between competing firms of the same trade was inevitable. The agreements, however, were not necessarily formal. In fact he distinguished between two types of agreement: 'explicit agreements' and 'quasi-agreements'. Explicit agreements would today be regarded as 'overt' or 'covert' collusion, depending on whether the agreement is open or secret. In contrast, a 'quasi-agreement' is far less formal – it is what Fellner describes as somehow representing a 'spontaneous co-ordination'. An example of a quasi-agreement would be the kind of unwritten rule that says all firms should take their price from the most dominant company in the market to avoid rounds of price-cutting. This type of arrangement, where no formal agreement actually occurs, is commonly referred to as tacit collusion.

Although collusive behaviour of any form is usually frowned upon, it is entirely understandable that firms may try to act in a common manner to protect or promote common interests. The importance of Fellner's work is that it shows that agreements to protect common interests do not require an explicit arrangement. As Fellner (1949: 16) noted: 'The difference between "true" agreement and quasi-agreement is that the former requires direct contact while the latter does not.' The basic aim of his hypothesis was to explore how rival firms in an oligopolistic type market resolved the conflict of knowing whether to:

- compete with rivals to gain as large a share of the potential profits as possible
- co-operate with rivals to maximise their joint profits.

The relative strength of these two forces – which operate concurrently – varies from industry to industry as the following five market characteristics suggest. As you read through them try to think about how many may apply to a specific market within the construction industry. In formal terms, that is, how many of these five market characteristics could lead to **joint profit-maximisation** in a specific construction market.

1 There are very few firms. They know each other well enough to understand that one of them cannot gain sales without inducing retaliation. So some agreement to co-ordinate their policies may be reached.
2 The firms produce similar products. As a result, it is difficult to gain a specific advantage in the market. In such a situation, firms may prefer some form of joint effort in preference to the cut-throat behaviour necessary to take customers away from each other.
3 There is a dominant firm. Other firms may look to the dominant one for its judgement about market conditions and take its lead on prices. In short, the dominant firm becomes a reference point and the focus for tacit agreement.
4 The firms have very similar average costs. In this case it is unlikely that firms will enter into price competition. Rivalry could break out in other forms, unless some joint agreement is reached to maximise profits.
5 New entrants face significant barriers to entry. The theory of perfect competition suggests that high profits in an existing market will attract new entrants and, as a result, prices and profits reduce. This profit-damaging activity is less likely to occur if some agreement between the existing firms has been made to prevent other firms breaking into the market.

COLLUSION – RIGGED PRICES

It is difficult to gauge the extent of qualified joint profit-maximising agreements, as in most countries they are illegal. There are, however, countless opportunities that one can envisage and the examples that are talked about are no doubt only a tip of the iceberg. Certainly there is widespread evidence of collusion across the construction sector in Australia, Canada, Germany, Netherlands, South Africa, South Korea, Philippines, USA and the United Kingdom (Brockmann, 2011: 31–2). For example, in the UK, the investigation carried out by the OFT during 2008 into anti-competitive agreements resulted in 112 construction companies being fined and prosecuted for a total of 240 incidents of **cover pricing**.

Cover pricing describes an age-old practice where one contractor asks another contractor on the same tender list to quote a price which will be above that quoted by the first contractor. In other words, it involves one or more bidders colluding with a competitor during the tender process to put in prices which are intended to be too high to win the contract. As a result, the client is left with a false impression of the level of competition and this could easily result in the client paying inflated prices for the work. In short, those involved in a bid rigging cartel work together to decide who will win a contract and at what price. In extreme cases firms may even 'pay off' those who agree not to tender a bid, although the more common arrangement seems to be where the firm providing the 'cover price' does so in exchange for a similar favour when future contracts arrive in the market. In these cases, the group of contractors are virtually, by concerted action, forming themselves into the structure of an oligopoly so that they can raise the level of profits and share them around on a rota basis. As an example of this process consider the data shown in Table 8.1 which represents four bids submitted to Newcastle City Council (the client) for the complete refurbishment of 194 council houses. As you can see each of the rival bids were at least 5 per cent higher than the winning submission, but this is not surprising as the OFT's evidence suggests that the winning contractor was 'assisted' (covered) by three other organisations.

Table 8.1 An example of bid rigging

Company	Date & time tender received	Amount of tender	Awarded contract
Cityworks	11 October 2002 before deadline	£2,518,150	
Connaught	11 October 2002 before deadline	£2,175,484	
Frank Haslam Milan	11 October 2002 before deadline	£2,065,788	Yes
Mansell	11 October 2002 before deadline	£2,189,244	

Source: Adapted from OFT (2009: 1011)

Intriguingly, the 112 firms accused by the OFT defended themselves on the grounds that they often submit bids to provide identical services and naturally respond by seeking covert and informal agreements to manage their workloads efficiently. This is particularly necessary when construction firms are operating near full capacity but want to remain on a list of recognised contractors for future projects. The defending firms were also keen to highlight that full tendering is a rigorous and costly exercise achieved on relatively tight profit margins. It is also worth noting that although in statistical terms it might appear that the construction industry comprises many thousands of competitive firms, it is in fact fragmented by trades and regions, and in many of these market segments there are examples of a local monopoly or a local oligopoly with a price leader.

THE THEORY OF GAMES

In recent years economists have begun to use a branch of mathematics called **game theory** to study collusion. In this work, the competition between firms is analysed as a game. Each firm decides its own game plan in terms of price and output, but realises that the result or success of its strategy depends upon the action of its opponents (the competing firms).

In 1994 three game theorists shared the Nobel Prize for economics and since then examples of the application of game theory have proliferated in the mainstream texts. Two common examples are the prisoners' dilemma and the zero-sum game. The prisoners' dilemma is used to demonstrate that competing firms have conflicts of interest (like the conspirators in a crime who are interviewed separately) that may or may not be resolved by some kind of tacit agreement. The zero-sum game is used to describe a situation in which the total winnings are fixed – some must lose, if others win. This is similar to competing for work through the tendering process. (Unless, of course, there is some agreement to share the winnings on a rota basis!)

Key Points 8.4

○ The construction industry comprises such a wide diversity of firms that it is impossible to categorise them all within one type of market structure.

○ Monopolistic competition is a market structure that lies between pure monopoly and perfect competition.

○ Oligopoly refers to a market structure in which there are just a few firms that are highly interdependent. In very generalised terms, construction firms can be seen to follow this type of market behaviour.

○ Because the world of oligopoly is one of uncertainty, there are incentives to try to collude, and there are several types of agreements that can exist between firms, including tacit, covert or overt arrangements.

○ The actions of rival firms may affect the size of the profits of all firms in a market. This is highlighted by the hypothesis of qualified joint profit maximisation.

○ It is difficult to gauge the actual extent, and nature, of collusive agreements as they are illegal, but there is plenty of evidence that alludes to an international practice of cover pricing.

CONTESTABLE MARKETS

Most of the market behaviour we have described in this chapter has been concerned with the actions of firms *inside* the market. The theory of **contestable markets** focuses on the possibility of firms entering the market from the *outside* and the effect that this potential competition has on the behaviour of the firms already inside the market. The theory highlights that markets are particularly contestable when there are no 'sunk costs' (that is, costs of entry or exit), as this allows firms to freely enter or leave a market sector without incurring too much financial risk.

The relevance of this theory is that markets do not have to contain many firms for profits to be held near the competitive level – the threat of a potential new entrant is sufficient to constrain prices. Research based on a study of the industry in 1990–1994 concluded that construction markets were contestable and that periods where firms could exploit market dominance to make high profits were unlikely to persist (Ball et al. 2000). During the subsequent boom (1994–2007), however, the contestability of construction markets became questionable. At the bottom end of the market, registration schemes made it more difficult for firms to enter the industry; at the top end, the increasing trend for partnering reduced the number of firms able to compete. In short, as **barriers to entry** emerge, the level of competition in an industry reduces.

The financial crisis of the late 2000s halted the construction boom and radically changed the competitive dynamics in the industry. The amount of spare capacity has significantly increased and the opportunity, and inclination, for collusive behaviour has declined. In effect each firm has had to become more competitive, and the construction marketplace has the character of a dog-eat-dog type of existence.

As explained in Chapter 7, a bid for construction work is typically based on two components: a cost estimate to complete the project and a mark-up to provide a profit. The firm that offers the lowest price typically wins the contract. According to the theory presented in this chapter, the greater the number of bidders the smaller the margins. So the more contestable a market becomes, the more problematic it is to secure any level of profit. This trend for reducing profit levels is what Harris and McCaffer (1989) refer to as 'the margin lost in competition'. Taken to its extreme, a bid could even be set at a price that would make the firm a loss. This is known as a **suicide bid**, but as firms experience difficult economic times some may tender for contracts simply to secure work in preference to making profits. In short, firms can find themselves forced into situations where they actually 'buy' work. Although this makes little sense in theory as it is not economically viable, the practice has occurred many times in the past. Latham (1994) warned that 'many clients do not understand that fiercely competitive tenders and accepting the lowest bid do not provide value for money in construction. Lowest priced tenders may well contain no margin of profit for the contractor, whose commercial response is then to try to claw back the margin through variations, claims and Dutch auctioning of subcontractors and suppliers.' Yet nearly 20 years later a survey of 390 RICS members suggested that more than half of them had recently seen tenders that were not priced at a sufficient level to cover the cost of the work (Gardiner 2011).

RESOURCE ALLOCATION AND SUSTAINABILITY

This chapter has demonstrated the importance of free markets. It has highlighted how resources should, in theory, be used more efficiently if there is freedom of information, and if firms know about each other's costs and procedures. The analysis suggests that as markets become more competitive there is a greater likelihood that society's welfare will be maximised. To paraphrase Adam Smith's writing of two centuries ago, as people follow the signals of the invisible hand of the competitive market in pursuit of their own interest they unintentionally also promote public interest. In the past 30 years many governments have drawn on this classic wisdom to reduce public sector provision and government regulation in favour of a greater reliance on decentralisation for the greater good. Arguably, however, this has led to the current economic crisis and the case is being made for a revision back towards a more mixed economy; one that combines the incentives of the market within a regulatory government framework. In short, the current crisis could revive a more 'centrist' approach to the question of allocating resources efficiently.

Indeed the Government Procurement Service (GPS), the National Audit Office and the Office of Fair Trading each has a responsibility to assure that taxpayers get value for money. The public sector is certainly a major player in the construction market, and local authorities and government departments are therefore encouraged to be aware of the benefits of competitive tendering. Competition in the supply chain is regarded as healthy, whereas arrangements within the supply chain to restrict competition and control prices are unhealthy. Consequently, wherever possible, public (and private) sector clients are encouraged to vet tenders and bids for evidence of collusive agreements.

It is understandable that competition needs to be carefully monitored by government and there is a substantial history of **competition policy** that applies right across the economy. For example, in the last ten years there have been investigations into allegations of anti-competitive activity covering industries as diverse as toy retailing, horse racing, construction, newspaper distribution, insurance, crematoria, private schools, bus transportation, groceries, healthcare, mobile phones and airports. Competition policy attempts to restrict unethical business behaviour that acts against the public interest. It is represented by statutory measures operating at the national and European level. In the UK, competition policy is managed by two agencies: the Office of Fair Trading (OFT) and the Competition Commission. Their respective powers as independent enforcers of consumer legislation and competition law are enshrined in the Enterprise Act 2002. For the most recent information on the promotion of competition, readers should visit the respective websites of these organisations which were reviewed on page 30.

As implied above, and explicitly stated in Chapter 2, competitive markets are a means of securing efficient production that does not waste inputs and promotes economic growth. Consequently, an effective competition policy that prevents the development of unethical behaviour or unfair market transactions is important. The closing thought here, however, should remind us of a suggestion made towards the start of this text – namely that, in the final analysis, questions relating to sustainability can be reduced to effective resource allocation – but the real question facing policy-makers today is should this be pursued through markets that are allowed to operate freely or should there be a greater emphasis on regulation and levers imposed and controlled by governments.

Key Points 8.5

○ A market is perfectly contestable if there are no barriers to entry or exit. The absence of barriers means that any new entrant can compete with existing firms in the market.

○ Increased contestability makes it more difficult for firms to achieve healthy profit margins.

○ Theoretically the efficient allocation of resources is associated with free competitive markets.

○ The current economic crisis raises questions about the level of freedom that markets should be allowed. A case is developing for making stronger policy interventions to monitor and ensure free competition.

Reading 2

Two important aspects of Part A have been to explain the importance of resource efficiency and to overview the characteristics of various market structures. We suggested that, in most cases, the market structures in the construction sector are different from those found in the manufacturing sector. Manufacturing is mostly dominated by a concentration of very large companies that are able to utilise capital-intensive modes of mass production and benefit from economies of scale. Whereas, in direct contrast, the construction industry is traditionally characterised by a large number of small firms, with very few barriers to entry, a dispersed market structure and a relatively low level of fixed costs. Which sector achieves greatest resource efficiency (value for money) is left open to debate.

In the first of two extracts relating to Part A, Blismas, Pasquire and Gibb argue that cost benefit evaluations of off-site production (OSP) are currently too focused on costs, and that they need to become far more holistic to take full advantage of the resource efficient benefits offered by OSP. While studying the extract, therefore, a useful exercise would be to try to identify some of the cost-based evaluations to account for the slow uptake of OSP and some of the broader resource efficient advantages that would make it more economically viable.

Nick Blismas, Christine Pasquire and Alistair Gibb (2006) 'Benefit evaluation for off-site production in construction', *Construction Management and Economics* 24: 121–30

Introduction

Recent UK government reports, including the Egan Report Rethinking Construction (1998), produced by the Construction Task Force, discussed the need for performance improvements in the UK construction industry. Egan (1998) identified supply chain partnerships, standardisation and off-site production (OSP) as having roles in improving construction processes. The Australian construction industry has also recently identified OSP as a key vision for improving the industry over the next decade (Hampson & Brandon, 2004).

The uptake of OSP in construction is limited however, despite the well documented benefits that can be derived from such approaches (Neale et al., 1993; Bottom et al., 1994; BSRIA, 1999; CIRIA, 1999, 2000; Housing Forum, 2002; Gibb & Isack, 2003). A major reason posited for the reluctance among clients and contractors to adopt OSP is that they have difficulty ascertaining the benefits that such an approach would add to a project (Pasquire & Gibb, 2002). The use of OSP, by many of those involved in the construction process, is poorly understood (CIRIA, 2000). Some view the approach as too expensive to justify its use, whilst others view OSP as the panacea to the ills of the construction industry's manifold problems (Groak, 1992; Gibb, 2001). Neither of these views is necessarily correct.

The benefits of OSP are largely dependent on project-specific conditions, and the combination of building methods being used on a project. Decisions, regarding the use of OSP, are consequently unclear and complex. Direct comparison of components is not usually possible due to interdependencies between elements, trades and resources. These complexities make the derivation and use of holistic and inclusive evaluation methods difficult. The unlimited combinations of components, site conditions and degrees of OSP do not permit the derivation of a comprehensive evaluation system; however sufficient common factors exist for a degree of valid comparative analysis.

A pilot study by Pasquire and Gibb (2002) demonstrated that decisions to use OSP are still largely based on anecdotal evidence rather than rigorous data, as no formal measurement procedures or strategies are available.

In other words, current evaluation methods used to compare traditional and off-site produced building solutions do not adequately account for 'value' and therefore cannot 'record' the benefits that OSP can promote.

This paper investigates the proposition that current evaluation methods for OSP are cost – and not value – based, and therefore cannot account for the recognised benefits of OSP. The consequence of this is that OSP invariably appears as an expensive alternative to traditional on-site options. The next section identifies the main benefits of OSP from previous research...

Benefits of site production

The benefits attributed to OSP are numerous and well documented. Gibb and Isack (2003) conducted a large interview-based survey in which they determined construction clients' views on the benefits of OSP. Their findings showed that clients' perceived the benefits of OSP as being mainly time – and – quality based. Table 1 (in the next column) summarises their findings in descending order of benefit.

Further, interviewees of Gibb and Isack (2003) were asked to rank a list of key benefits from the initial interviews and literature, noting both the importance of the benefit and the likelihood of realising the benefit (see Table 2).

The interviewees rated benefits in non-direct cost terms, such as minimisation of on-site operations; reduction of site congestion; reduction of on-site duration; improved health and safety, etc. Direct cost benefits did not feature in these ranked lists (Table 2), although identified within Table 1.

These findings clearly demonstrate that, although OSP can offer direct monetary benefit (in terms of costs), the main benefits are from indirect savings (so-called non-cost items).

Table 1 Clients' perceptions of the benefits of OSP (from Gibb & Isack, 2003)

Benefit	Description
Time	Less time on-site—speed of construction*
	Speed of delivery of product
	Less time spent on commissioning
	Guaranteed delivery, more certainty over the programme, reduced management time
Quality	Higher quality—on-site and from factory*
	Product tried and tested in factory
	Greater consistency—more reproducible
	More control of quality, consistent standards
Cost	Lower cost*
	Lower preliminary costs
	Increased certainty, less risk
	Increases added value
	Lower overheads, less on-site damage, less wastage
Productivity	Includes less snagging
	More success at interfaces
	Less site disruption
	Reducing the use of wet trades
	Removing difficult operations off-site
	Products work first time
	Work continues on-site independent of off-site production
People	Fewer people on-site
	People know how to use products
	Lack of skilled labour
	Production off-site is independent of local labour issues
Process	Programme driven centrally
	Simplifies construction process
	Allows systems to be measured

* indicates high incidence

Table 2 Rating of benefits from highest to lowest according to importance and likelihood (from Gibb & Isack, 2003)

Benefit	(from highest rated to lowest rating)	Cost-related
1.	Minimises on-site operations	
2.	Reduces congested work areas and multi-trade interfaces	
3.	Minimises on-site duration	b
4.	Improved health & safety by reduction and better control of site activities	
5.	Produces high quality or very predictable quality finishes	
6.	Minimises number of site personnel	a
7.	Benefits when only limited, or very expensive on-site labour	
8.	Enables existing business continuity	
9.	Can cope with restricted site storage area	
10.	Enables inspection and control off-site works	a
11.	Provides certainty of project cost outcomes	
12.	Provides certainty of project completion date	
13.	Less environmental impact by reduction and better control of site activities	

a = cost-related; b = impact on cost.

Based on these findings, pure direct cost comparisons will favour traditional on-site operations that are costed on a rate-based system, with overheads, access, cranage, repairs and reworks hidden within preliminary costs. OSP costs are usually presented as all-inclusive amounts with a premium for off-site capital costs. Having established that the benefits of OSP are largely identified as non-cost items, the paper continues to analyse several cases to ascertain the emphasis of current OSP evaluation methods....

References

Bottom, D., Gann, D., Groak, S. and Meikle, J. (1994) *Innovation in Japanese Prefabricated House-Building Industries*, Construction Industry Research and Information Association, London.

BSRIA, compiled by Wilson, D.G., Smith, M.H. and Deal, J. (1999) *Prefabrication & Pre-assembly—Applying the Techniques to Building Engineering Services*, Briefing Note ACT 2/99, Building Services Research and Information Association, Bracknell.

CIRIA, compiled by Gibb, A.G.F., Groak, S., Neale, R.H. and Sparksman, W.G. (1999) *Adding Value to Construction Projects through Standardisation & Pre-assembly in Construction*, Report R176, Construction Industry Research and Information Association, London.

CIRIA, and principal author Gibb, A.G.F. (2000) *Client's Guide and Toolkit for Optimising Standardisation and Pre-assembly in Construction*, Report CP/75, Construction Industry Research and Information Association, London.

Egan, J. (1998) *Rethinking Construction*, The Egan Report, Department of the Environment, Transport and the Regions, London.

Gibb, A.G.F. (2001) Standardisation and pre-assembly—distinguishing myth from reality using case study research. *Construction Management & Economics*, 19, 307–15.

Gibb, A.G.F. and Isack, F. (2003) Re-engineering through pre-assembly: client expectations and drivers. *Building Research and Information*, 31(2), 146–60.

Groak, S. (1992) *The Idea of Building*, E and FN Spon, Routledge, London.

Hampson, K. and Brandon, P. (2004) *Construction 2020: A Vision for Australia's Property and Construction Industry*, Cooperative Research Centre for Construction Innovation for Icon.Net Pty, Brisbane, Australia.

Housing Forum (2002) *Homing in on Excellence—A Commentary on the Use of Off-site Fabrication Methods for the UK House Building Industry*, Housing Forum, London.

Neale, R.H., Price, A.D.F. and Sher, W.D. (1993) *Prefabricated Modules in Construction: A Study of Current Practice in the United Kingdom*, Chartered Institute of Building, Ascot.

Pasquire, C.L. and Gibb, A.G.F. (2002) Considerations for assessing the benefits of standardisation and pre-assembly in construction. *Journal of Financial Management of Property & Construction*, 7(3), 151–61.

....

Extract information: Edited and adapted from pages 121–3 of original plus relevant references from pages 129–30.

Reading 3

This reading reviews the process of winning and pricing work in the construction sector. It is based on Bee-Lan Oo's doctoral thesis that researched contractors' bidding decisions in different competitive market environments. As such, it draws on the classic literature on bidding for construction contracts. It has been selected as it investigates the variables that inform the decision to bid for work. The basic variables have been discussed in Part A and clearly the ideas relating to partnering, market structures, economic conditions and the possibilities to collude all appear relevant. As indicated in the *Government Construction Strategy* (2011) and earlier reports by Egan (1998) the UK construction industry tends to compare badly with other industrial sectors in terms of profitability, product quality, client satisfaction and investment in research and development.

The general understanding is that contractors are usually selected through competitive tender on the basis of lowest price. As a consequence, during a recession price competition tends to increase and profit margins reduce as contractors become desperate to win projects.

The extract offers the opportunity to extend these ideas and think them through. Clearly winning the 'right' project at the 'right' price is a crucial aspect of survival in the construction industry, and it does not really take a course in economics to understand this. However, it might be interesting to briefly reflect on a question that has already recurred several times in the text – namely, does economic theory satisfactorily explain the nature of competition in the construction industry?

Bee-Lan Oo, Derek S. Drew and Goran Runeson (2010) 'Competitor analysis in construction bidding', *Construction Management and Economics* 28 (12): 1321–29

Introduction

In the construction industry, competitive bidding is used for a variety of procurement routes available for satisfying clients' construction needs. These include both the traditional procurement via design-bid-construct, and the non-traditional ones such as the design-and-build, management contract, and build-own-operate-transfer. While clients will naturally aim to strike the best bargain by maximizing competitive intensity, contractors would ideally submit a bid offer that is likely to provide the best pay-off, allowing for the cost and potential risks of undertaking a particular project. It should be noted, though, that contractors do not always bid for every job that comes along but select from a continually changing array of potential projects (Odusote and Fellows, 1992). Competitive bidding in construction is therefore concerned with contractors making strategic decisions in respect of: (i) project selection whether or not to bid for a job; and (ii) determination of bid price if contractors opt to bid (Skitmore, 1989).

To meet specific firm objectives, bidding strategies vary from contractor to contractor, and each will have different degrees of preference or sensitivity towards the factors affecting their bidding decisions. It has been found in many studies that there are differences in ranking of factors which contractors consider when making bid/ no-bid and mark-up decisions; see for example, Ahmad and Minkarah (1988),

Odusote and Fellows (1992), Shash (1993) and Fayek et al. (1999).

This suggests that contractors' bidding decisions are dependent on many individual firm-specific characteristics, including some that are unobserved, i.e. the notion of heterogeneity across contractors. Gonzalez-Diaz et al. (2000) suggest that one may think of the unobserved heterogeneity as the management style of a construction firm, which may include the capability of its manager, the quality of its output and its competitive strategy. By adapting the definition of heterogeneity in Jain et al.'s (1994) economic behaviour study to the context of construction bidding, it could be expected that individual contractors, when confronted with a given set of bidding variables (e.g. market conditions and number of bidders) exhibit different bidding behaviour due to (i) differences in overall bidding preferences—preference heterogeneity; and (ii) variations in their responses to the given set of bidding variables—response heterogeneity (Oo, 2007).

Empirical studies have been conducted to analyse the bidding behaviour of competing contractors according to various bidding variables such as type and size of construction work (Drew and Skitmore, 1997), market conditions (de Neufville et al., 1977; Runeson, 1988; Chan et al., 1996) and number of bidders (Carr and Sandahl, 1978; Wilson et al., 1987). These models were, however, being built on the assumption that individual contractors can be treated as behaving collectively in an identical (statistical) manner—the bidder homogeneity assumption. It is likely that models at the level of individual contractors, instead of collective models, will be needed if there is heterogeneity across contractors. Recognizing the need to consider this, there are only a few studies aimed at establishing the extent to which heterogeneity across contractors exists in practice. Skitmore (1991) has detected the existence of heterogeneity across bidders in his attempt to derive a probability distribution of bids to represent bidding behaviour of all bidders in three datasets. At the level of the effects of bidding variables on contractors' bidding

strategies, it was found that there is significant heterogeneity across individual Hong Kong and Singapore contractors in their bid/no-bid (Oo et al., 2007, 2008) and mark-up decisions (Oo et al., 2009) in response to a given set of four bidding variables. The significant implication of these empirical studies is that future bidding modelling attempts should take into account the possible heterogeneity that exists across contractors. As Hsiao (2003) points out, ignoring such heterogeneity or individual effects could lead to (i) parameter homogeneity in the model specification; and (ii) inconsistent or meaningless estimates of interesting parameters.

The approach taken here was to apply a heterogeneous approach to modelling individual competitors' bidding behaviour. The competitor analysis focuses on individualized models that consider bidding competitiveness of a large Hong Kong contractor relative to a group of its key competitors according to four bidding variables, namely: (i) project size; (ii) work sector; (iii) work nature; and (iv) number of bidders. It offers a more informed approach in identifying key competitors, and shows that the identified competitors' bidding behaviour provides an aid to greater understanding and opportunities for possible future exploitation by a contractor concerned, particularly for the formulation of bidding strategies targeting key competitors.

Competitor analysis

Competitor analysis in construction bidding is essentially about comparing competing contractors on the basis of bid prices. For most practical purposes, it is sufficient to consider bids in relation to a baseline in considering competitiveness between bids (Drew and Skitmore, 1993). In this paper, the lowest bid was used as a baseline that has the advantage of representing maximum level of competitiveness at the time of bidding. It is the lowest bid that determines not only the identity of the winning contractor, but also the legally binding contract value of a particular project in the vast majority of cases (Merna and Smith, 1990)....

References

Ahmad, I. and Minkarah, I. (1988) Questionnaire survey on bidding in construction. *Journal of Management in Engineering*, 4(3), 229–43.

Carr, R.I. and Sandahl, J.W. (1978) Bidding strategy using multiple regression. *Journal of Construction Division*, 104, 15–26.

Chan, S.M., Runeson, G. and Skitmore, M. (1996) Changes in profit as market conditions change: an historical study of a building firm. *Construction Management and Economics*, 14(3), 253–64.

De Neufville, R., Hani, E.N. and Lesage, Y. (1977) Bidding models: effects of bidders' risk aversion. *Journal of Construction Division*, 103(1), 57–70.

Drew, D.S. and Skitmore, M. (1993) Prequalification and c-competitiveness. *Omega*, 21(3), 363–75.

Drew, D.S. and Skitmore, M. (1997) The effect of contract type and size on competitiveness in bidding. *Construction Management and Economics*, 15(5), 469–89.

Fayek, A., Ghoshal, I. and AbouRizk, S. (1999) A survey of the bidding practices of Canadian civil engineering construction contractors. *Canada Journal of Civil Engineering*, 26, 13–25.

Gonzalez-Diaz, M., Arrunada, B. and Fernandez, A. (2000) Causes of subcontracting: evidence from panel data on construction firms. *Journal of Economic Behavior & Organization*, 42(2), 167–87.

Hsiao, C. (2003) *Analysis of Panel Data*, 2nd edn, Cambridge University Press, Cambridge.

Jain, D.C., Vilcassim, N.J. and Chintagunta, P.K. (1994) A random-coefficients logit brand-choice model applied to panel data. *Journal of Business and Economic Statistics*, 12(3), 317–28.

Merna, A. and Smith, N.J. (1990) Bid evaluation for UK public sector construction contracts. *Proceedings of the Institute of Civil Engineers: Part 1*, 88, 91–105.

Odusote, O.O. and Fellows, R.F. (1992) An examination of the importance of resource considerations when contractors make project selection decisions. *Construction Management and Economics*, 10(2), 137–51.

Oo, B.L. (2007) Modelling individual contractors' bidding decisions in different competitive environments, unpublished PhD thesis, Hong Kong Polytechnic University.

Oo, B.L., Drew, D.S. and Lo, H.P. (2007) Applying a random coefficients logistic model to contractors' decision to bid. *Construction Management and Economics*, 25(4), 387–98.

Oo, B.L., Drew, D.S. and Lo, H.P. (2008) A heterogeneous approach to modelling the contractors' decision to bid. ASCE *Journal of Construction, Engineering and Management*, 134(10), 766–75.

Oo, B.L., Drew, D.S. and Lo, H.P. (2009) Modeling the heterogeneity in contractors' mark-up behaviour. *ASCE Journal of Construction, Engineering and Management*.

Runeson, G. (1988) An analysis of the accuracy of estimating and the distribution of tenders. *Construction Management and Economics*, 6(4), 357–70.

Shash, A.A. (1993) Factors considered in tendering decisions by top UK contractors. *Construction Management and Economics*, 11(2), 111–8.

Skitmore, M. (1989) *Contract Bidding in Construction*, Longman, Harlow.

Skitmore, M. (1991) The construction contract bidder homogeneity assumption: an empirical test. *Construction Management and Economics*, 9(5), 403–29.

Wilson, O.D., Atkin, A.S., Sharpe, K. and Kenley, R. (1987) Competitive tendering: the ideal number of tenders, in Lansley, P.R. and Marlow, P.A. (eds) *Managing Construction Worldwide*, E & FN Spon, London, pp. 175–86.

Extract information: Edited and adapted from pages 1321–2 of original plus relevant references from pages 1328–9.

Part B

Protection and Enhancement of the Environment

WEB REVIEWS: Protection and Enhancement of the Environment

On working through Part B, the following websites should prove useful.

www.bre.co.uk

The Building Research Establishment is a research-based consultancy with offices in England and Scotland. It has a particular expertise in the area of sustainable construction and its website advertises its latest products. Up-to-date information on the BREEAM schemes and whole life costing, introduced in Chapter 9, is available at this site.

www.ends.co.uk

ENDS is an environmental data service, providing a daily news service on European environmental affairs. The homepage provides the opportunity to sample the organisation's authoritative monthly report and has links to other environmental resources on the web. The information can be used in conjunction with all the chapter themes in this section – visit and see why it claims to be the best environmental website.

www.worldgbc.org

For a different perspective, it is often interesting to look at information presented by non-government organisations. The World Green Building Council site has been working since 2002 to assist the property industry towards sustainability. It currently has a presence in more than 80 countries and you might find it interesting to search for a Green Building Council in a specific country, or download the directory to check coverage.

www.buildoffsite.org

As the name implies, Buildoffsite is an organisation formed to promote modern methods of construction. It was established with government backing in 2005 to create a step change in the application of off-site techniques within the construction sector. It was initially set a target of raising expenditure on these techniques tenfold to £20 billion by 2020. Its members are drawn from all sectors of the UK industry, including developers, designers, contractors, manufacturers, clients and government; the current membership list exceeds 60 organisations. The website contains links to events, publications, case studies and a quarterly newsletter. As you will sense from Chapters 7 and 9, this site contributes significantly to the promotion of sustainable construction.

www.carbonbuzz.org

CarbonBuzz was set up in November 2008 by CIBSE and RIBA (the respective professional organisations of UK building services engineers and architects) to raise awareness of the fact that most buildings generate too much carbon. The organisation's website collects building energy consumption data to highlight the 'performance gap' between design figures and actual readings. To date, it has collected data on more than 300 buildings. Although it has only published detailed data on 20 projects, CarbonBuzz has contributed to a range of post-occupancy reviews, in particular the government-funded *Building Performance Evaluation* that was launched in 2010. It is a site that will help students studying Section B and that will continue to inform the low carbon agenda.

9 Markets for Green Buildings and Infrastructure

An important consideration for any firm seeking to control the market and stand out from its competition is to satisfy, or create, a niche market – to produce a service or product that is in some way different from its rivals. In economic terms this is referred to as **product differentiation**. We have already discussed how in the extreme case of perfect competition we assume that the market consists of homogeneous products, in which each individual firm in the market produces an identical product (or service) and has a horizontal demand curve. To express it another way, in a perfectly competitive market there is only one specific 'undifferentiated' product (see Key Points 8.1).

Providing a firm can manage to differentiate its product or service from other similar products – even if only slightly – it can gain some control over the price it charges. Firms producing a differentiated product are able to achieve some independence from their competitors in the industry. They should be able to raise their prices, and thereby increase profits, without losing all their customers. Unlike firms operating at the perfectly competitive extreme, they face a slightly downward sloping demand curve. In fact, the greater a firm's success at product differentiation, the greater the firm's pricing options – and the steeper the demand curve.

OPPORTUNITIES TO DIFFERENTIATE CONSTRUCTION PRODUCTS

Economics textbooks usually emphasise that the opportunities to differentiate a product or service in the construction industry are limited. Firms may be able to market themselves as somehow superior to their competitors in terms of quality or reliability, but they are always constrained by the large number of firms that compete and produce close substitutes. Consequently, the ability of one firm to significantly raise its prices above that of its competitors is restricted. Gruneberg and Ive (2000: 92) extend this hypothesis. They argue that the tendering process creates a further complication, as it is usually assumed that all those selected to submit tenders are undifferentiated – equal, in terms of the service they are offering.

An important aim of this chapter, however, is to identify the economic arguments that may encourage construction firms to take up the green challenge. This depends upon firms in the industry taking the opportunity to differentiate their product by moving away from traditional techniques to those that demonstrate environmental awareness. It also involves paying attention to global, local and user concerns if firms are to develop sustainable buildings and infrastructure.

At the time of writing, a construction firm producing environmentally sensitive products would be able to distinguish itself so effectively from the majority that it could secure short-term monopoly profits – that is, until the time when competitors recognise the benefits of following the same mould, bringing the market back to something nearer to perfect competition and, in this case, bringing the market closer to the idea of sustainability. A trend for sustainable construction is slowly emerging and being taken up by some contractors and clients, and traditional specifications are being challenged in favour of those that demonstrate environmental benefits. The common characteristics of environmentally sensitive specifications are discussed in the next sections.

Emerging Green Markets

Even in manufacturing – with supply based on factory techniques and where products are demanded and used by a single customer – it is difficult to develop a market for environmentally superior products. In construction the challenge is even more complex, as there are fewer standard prototypes and often the 'users' of construction products are not the owners. As we have suggested in preceding chapters, each construction product can be regarded as unique. Products are assembled on site by a team of subcontractors. The large labour force is often one stage removed from the agreement made between the client and the contractor. And, as a final twist, the interests of the users are often different from those of the investors that produce the original specification. This makes it difficult for those supplying the products to final users to communicate effectively through market signals. Yet it is in the marketplace where people display their green credentials.

It is therefore not surprising that green development in the construction industry has been relatively slower than in manufacturing – but it is emerging. The most activity has been seen in the commercial sector, with owner-occupiers beginning to specify bespoke headquarters that reflect their corporate ethos. The level of green activity within the residential sector, however, is not so evident, as the volume builders engaged in speculative residential developments have been slow to see the market potential of adopting an environmentally aware corporate image. There are some innovative exceptions, particularly within the government-funded social housing sector, and some examples of architect-designed homes – eco-homes – for environmentally conscious clients. Finally, awareness is emerging in the sector specialising in infrastructure, which could make an important contribution once it takes off. We now look at each of these sectors in turn.

THE COMMERCIAL SECTOR

Each year, the largest amount of new building work is in the commercial sector (broadly defined as offices, factories, warehouses and shops – see Table 5.2, page 75). Most activity in this sector continues to produce a standard undifferentiated product that tends to be over specified, mechanically ventilated and energy guzzling. However, an increasingly significant proportion – say 25 per cent – of the new additions are able to boast environmentally friendly features. Sensitivity to the environment is becoming an ever more important issue, and commercial organisations want to reflect their environmental credentials in the types of building that they rent and own. There appear to be good corporate arguments in favour of situating commercial outlets in buildings that minimise global and local impacts, reduce energy bills and facilitate greater worker productivity.

According to the Building Research Establishment Environmental Assessment Method (BREEAM), and its US equivalent Leadership in Energy and Environmental Design (LEED) developed by the United States Green Building Council (USGBC), it is possible to audit and assess a broad range of issues within the design, procurement and management of an office building. For example, a detailed evaluation can be made of the materials selected and the energy systems employed to light, heat and cool the building. Interestingly, both of these assessment methods identified the new commercial office market as having most potential and this sector became the testing ground for various BREEAM and LEED schemes. The BREEAM scheme for new office designs was launched in 1990, and the LEED equivalent followed eight years later in 1998. Subsequently, schemes to evaluate existing commercial buildings, homes (both new and old) and various other outlets such as shops, schools, health centres and industrial units followed.

BREEAM and LEED have the advantage of sharing many years of experience and their websites currently boast that more than 250,000 buildings have been certified across 121 different countries. A significant majority of these, however, are still in the UK and the United States. Furthermore, these figures simply represent the number of environmental assessments that have been carried out by the Building Research Establishment, the United States Green Building Council or their authorised assessors. It would be more interesting and informative to know how many other green buildings exist that have not been put through a formal environmental assessment scheme. Either way, the number of green buildings is certainly on the increase.

Construction firms seeking to differentiate their products on the basis of their environmental performance need to deploy their assets in a distinctive way. There is a new breed of commercial client emerging that needs to know that their requirements can be competently fulfilled by the contractor. There are a range of features that typify state-of the-art green developments, and the common ones are listed in Table 9.1.

Table 9.1 The characteristics of green commercial buildings

✓	Makes maximum use of natural daylight
✓	Minimises consumption of fossil fuels, by techniques such as natural ventilation, combined heat and power, and orientation of site to benefit from passive solar energy
✓	Reduces the use of fresh water by using grey water recycling for landscape irrigation, flushing toilets, etc.
✓	Minimises site impact by careful landscaping and the preservation of local ecosystems
✓	Reduces the quantity of 'virgin' materials used and selects those that have the least negative environmental impact
✓	Reuses and recycles existing buildings and sites
✓	Minimises material waste during construction and demolition

Source: Adapted from Shiers (2000: 354)

Architecture that is based on (some of) the features outlined in Table 9.1 is slowly emerging. Some of these examples of green buildings are listed in Table 9.2. The ones selected in the table are on, or near, a university campus – so you might have the opportunity to take a closer look.

Table 9.2 Examples of green buildings in the UK

These buildings have been developed since the early 1990s. They are listed in chronological order, with the most recently opened building at the foot of the list.

Queens Building (School of Engineering), De Montfort University, Leicester

The Inland Revenue Building, Nottingham

Elizabeth Fry Building, University of East Anglia, Norwich

Learning Resource Centre, Anglia Polytechnic University, Chelmsford

Wessex Water Headquarters, Bath

Architectural and Planning Studios, University of the West of England, Bristol

The Gherkin, 30 St Mary Axe, London

The National Assembly for Wales, Cardiff

PricewaterhouseCoopers Headquarters, London

THE RESIDENTIAL SECTOR

Everybody needs some kind of shelter to live in, and there is a large and somewhat varied stock of dwellings across the globe. Due to the limitations of national data it would be difficult to estimate the number of houses that have been built to a green or energy efficient standard. Some of the buildings built to the UK's Code for Sustainable Homes levels 5 and 6 and the European *Passivhaus* standards represent good examples of what can be achieved in today's markets, but these exceptions represent only a tiny proportion of the global housing stock.

The residential sector as a whole is responsible for about 25 per cent of global energy usage. In the United Kingdom alone, the existing stock of houses exceeds 27 million units, and these buildings are responsible for about 30 per cent of the national energy used each year. Most of these buildings were designed in the days when energy was relatively cheap and problem free, so over time they have become problematic in terms of sustainability.

To some extent, it is inevitable that houses built in the past do not meet today's exacting requirements, but rather worryingly the majority of new additions to the housing stock are also relatively inefficient in terms of energy usage. Furthermore, in the UK, at least, much of the new housing stock comprises low density developments on greenfield sites that are car dependent. This contrasts strongly with government policy. Governments have sought to promote development on brownfield sites, designed around good public transport and utilising high-density designs that exceed the minimum expectations for energy efficiency. Governments would also prefer to see developments that include provision of social housing or housing that can be afforded by people on relatively low incomes. This contrast between public policy and actual building practice highlights the challenge that governments face in supporting sustainable construction.

To compound the government's frustration, resource efficient, environmentally friendly housing is by no means 'rocket science' – indeed, technically it can be achieved by most contractors. Take energy efficiency as an example: all that is needed is greater levels of insulation, the careful sealing of all joints, the positioning of windows to make the most of sunlight, and use of a heat exchange system that permits the air going out to preheat the air coming in. Most volume developers, however, have been reluctant to adopt such energy efficient measures because of the extra cost (and care) involved. They claim that energy efficiency comes at a cost that cannot be passed onto the consumer. Of course, house builders are private companies whose purpose is to make money for their shareholders, they are not green charities with a mission to promote sustainable construction. Few economists were surprised to note that significantly fewer houses, of any description, were built in the UK in 2010 and 2011 but the profits of the big ten volume builders greatly increased (see Gardiner 2012).

Zero-energy or carbon neutral buildings, where energy is provided from renewable sources (such as solar, wind, geothermal, biomass, or a combination of these sources) supplied into highly insulated and tightly sealed units, are commercially achievable. In fact, several low energy and zero carbon projects have been built in the last few years or are currently in development; 15 prime examples from 15 different countries across the developed world were reviewed in 2010 by the National House Building Council. This review not only showcased the current state of the art, but it also highlighted in global terms how slow these exceptional developments have been to evolve. In most cases the developments are small, typically around 40 units, and some are no more than one-off demonstration projects (NHBC 2010). By the end of 2011, there had only been 128 units completed to precise zero carbon standards in the United Kingdom, and a further 209 units at code level 5 – a slightly less demanding standard (DCLG 2012). This poor record is despite the fact that the UK government has been working since 2006 towards a target for all new homes to be carbon neutral by 2016. As this date comes closer on the horizon, the government has begun to back pedal and redefine its interpretation of zero carbon.

The approach is only slightly more advanced in the rest of Europe. The *Passivhaus* (passive house) movement has gained some momentum, but it is still an exception rather than the rule. A preference is slowly emerging for high-quality, well-insulated buildings that require relatively small amounts of energy for space heating, but they do not yet dominate the output. Although there are currently thousands of passive houses in Europe, mainly in Germany, Austria, Denmark and Switzerland – it is estimated that there are in the region of 25,000 to 30,000 homes built to this standard – this only represents something like 0.04 per cent of the related stock (NHBC 2010).

The five basic principles of sustainable housing are shown in Table 9.3. If developments of this type became commonplace in the property market, we

Table 9.3 Five principles of sustainable housing

1	Improve thermal efficiency to a point where homes can achieve zero carbon energy usage
2	Reduce mains water consumption by collecting rainwater and recycling grey water
3	Maximise the use of local, reclaimed and recycled materials
4	Promote public transport and car pools to create a lifestyle that is less car dependent
5	Design into the estate services to enable on site composting, home delivery of grocery and recycling

Source: Adapted from Desai and Riddlestone (2002: 20)

could become less dependent on fossil fuels and reduce carbon dioxide emissions. An authoritative survey of the literature (combining more than 80 national and regional studies) indicated that there is a potential to reduce global carbon dioxide emissions by approximately 29 per cent by 2020 in the residential and commercial sectors (Urge-Vorsatz et al. 2007: 388). These calculations were drawn upon by the Intergovernmental Panel on Climate Change (IPCC) in its fourth assessment report, where it concluded that the biggest potential saving in any sector (including transport) related to energy generated for use in buildings.

This discussion suggests that firms specialising in house building (or other structures) could benefit by differentiating their product in several ways and by demonstrating a greater awareness of the techniques and specifications that support sustainable construction. In this way, they could win business in the marketplace by beating their rivals at a new game. As the sources behind the IPCC report made clear, achieving a low carbon future is dependent on new programmes and policies for energy efficiency in buildings that go well beyond what is happening today (Urge-Vorsatz et al. 2007: 395).

INFRASTRUCTURE

Some economists claim that current global investment in infrastructure projects represents a 'drop in the ocean' compared to the massive worldwide need particularly in the developing world, which is estimated in trillions of dollars (Miller 2011: 72). In the UK the infrastructure sector represents, in value terms, approximately 25 per cent of construction output each year. It encompasses the construction and maintenance of roads, railways, airports, tunnels, bridges, telecommunications networks, power stations, coast and river works, and water supply and wastewater treatment facilities.

In a developing country investment in infrastructure is essential as it is positively and significantly correlated with economic growth. In crude terms infrastructure provides the 'wheels' of economic activity as telecommunications, power, water, and transport are key parts of the production process in nearly every economic sector. It is not only the quantity of infrastructure that is important, but also the quality of the service it provides. So the maintenance of infrastructure is equally vital, as low operating efficiency and lack of attention to the needs of users can significantly reduce its impact. In short, adequate infrastructure helps to determine one country's success and another's failure in economic, social and environmental terms.

Infrastructure projects can be specified with sustainability in mind. Indeed, the Institution of Civil Engineers presented its first awards to recognise environmental excellence in the summer of 2003. CEEQUAL (Civil Engineering Environmental Quality Assessment and Award Scheme) is an audit-based assessment similar to the Building Research Establishment Environmental Assessment Method (BREEAM) but appropriate for non-building projects. It shows how infrastructure may be constructed in an environmentally friendly manner. The characteristics that identify green infrastructure are in many ways similar to those listed in Tables 9.1 and 9.3 – the minimisation of waste, use of recycled aggregates, protection of landscape, ecology and archaeology, management of noise, and efficient use of water and energy. By 2012, more than 130 projects have been assessed under the scheme, with a further 240 projects in the pipeline. The cumulative value of all projects assessed to date is reported to be approaching close to £20 billion (CEEQUAL 2012). In the UK and Ireland, where most of these assessed projects are located, this represents a significant percentage of infrastructure spend, and it could be argued that 'green' construction has made more impact in infrastructure than in other sectors of the industry.

EXISTING BUILDINGS

By far the biggest challenge of sustainability in the built environment relates to existing homes, offices and infrastructure. In fact, most estimates based on UK data suggest that on average only 1 per cent of buildings are replaced each year, and similar replacement levels are recorded in most developed nations. It would therefore take more than 100 years to replace the existing stock with efficient buildings that have been constructed to today's environmental standards. However, although the technology exists to design the new stock in ways that could improve sustainability by 50 per cent or more, most new builds (about 0.9 per cent) achieve little more than the minimum standards laid down in the building regulations. So, without interventions from governments, it could take 1,000 years to replace the stock with the best energy performance currently achievable.

Another way of expressing the challenge is to recognise that around 60 per cent of the building stock that will exist in 2050 has already been built. In effect, the largest potential market for green building sits in the refurbishment and retrofit sector, and this will be important if carbon emissions reductions are to hit the ambitious targets being set by governments. Several countries have introduced grant schemes and tax incentives to encourage property owners to make their buildings more energy efficient. For example, schemes to encourage or subsidise improvements through better insulation and the greater use of renewable energy technology have been supported by governments in the United States, Australia, Belgium, Canada, Denmark, France, Germany, Japan and the United Kingdom. The German scheme is particularly interesting as this has already managed to improve the energy performance of more than 3000 existing buildings to the extent that, in energy terms, these now significantly outperform new buildings built to the highest energy conservation standards. On the basis of this evidence, the Federal Housing, Urban and Transport Ministry has announced an ambitious energy reduction programme to upgrade all pre-1984 properties in Germany by 2020. Using a system of loans, grants and tax incentives, the scheme will cover an estimated 30 million units of all kinds (including schools, public offices and residential buildings). The retrofit building upgrade programme will make a major contribution to Germany achieving its ambition to reduce overall carbon dioxide emissions by 40 per cent by 2020 (Powers 2008).

In the UK, through the Energy Act 2011, the government set up a similar, but less ambitious, scheme to improve the energy efficiency of British properties. The so-called 'Green Deal' is forecast to bring about a cut of over 2 million tonnes in carbon dioxide emissions a year. The underlying principle is simple enough – energy efficiency measures, such as loft insulation and heat pumps, will be undertaken by a private firm at no upfront cost to the consumer, and the payment for the work will be recouped over time through charges added to that consumer's energy bills. The policy's 'golden rule' is that energy efficiency measures can only be provided if the expected savings on the fuel bills are greater than the charges to pay back the cost of the work within a maximum 25 year time frame.

So at the heart of the government's Green Deal is an innovative financing mechanism that allows consumers to pay back the cost of energy efficiency improvements through their future energy bills. If the resident moves out and ceases to be the energy bill payer at that property, the financial obligation transfers to the next bill payer at the property. In this way, the Green Deal differs from the conventional top-down government programmes based on tax incentives and grants as it is in effect a loan funded by private capital.

Alongside the Green Deal sits another legislative proposal that, from April 2018, private rented properties must be brought up to a minimum energy efficiency rating of E. It will be unlawful to rent out residential or business premises that do not reach this minimum standard. In short, owners of energy inefficient buildings (those that are F or G-rated) will need to undertake measures to improve the performance of their properties. The Green Deal, together with the E rating for rented property, should impact on the nature and scale of building refurbishment projects as the measures incentivise residents and landlords to improve the energy efficiency of existing buildings.

This increasing focus on the energy performance of buildings reflects the widely accepted view that the property sector can play a key role in reducing carbon emissions. The International Panel for Climate Change (IPCC 2007: Chapter 6) noted that in comparison to other economic sectors buildings offer significant opportunities for cost-effective improvements to energy efficiency, thereby reducing demand for heat and power in line with international energy strategies. A focus on buildings would allow governments to exploit what the panel referred to as the 'low hanging fruit'. Extending this analogy further, some reviewers have noted that: 'Once the Green Deal "carrot" is on the table, landlords can face the threat of being unable to let their properties if they refuse to eat it' (Mactavish et al. 2012: 50).

In summary, this approach seems fairly straightforward providing, of course, that old buildings are refurbished accordingly. But the practicalities of implementing this strategy (regardless of who carries the initial burden of the cost of the work – public or private sector, landlord or tenant) could prove to be complex and challenging. Government grants, tax credits and low interest loans for retrofitting may well be on the table, but the work still needs to be commissioned, approved, funded and carried out to a high standard before the policy goals are secured. Consumers need to be convinced that the hassle of organising or applying for the work to be done is worth the economic and environmental benefit. In Chapter 10, we explain how governments do not always achieve their goals as numerous barriers, such as market failures and behavioural issues, can get in the way. These barriers can limit the level of take-up, and this raises questions about the potential effectiveness of schemes such as the Green Deal.

Key Points 9.1

○ The development of green buildings is important to sustainable construction.

○ Product differentiation can lead to short-term monopoly profits.

○ A construction firm may differentiate its product by introducing environmental specifications, and opportunities to achieve this are emerging in the commercial, residential and infrastructure sectors.

○ In relative terms, only a small number of green buildings exist (some examples are listed in Table 9.2). The characteristics they display are shown in Tables 9.1 and 9.3.

○ Policies designed to promote environmental design and energy efficient refurbishment are difficult to implement. This issue is discussed further in Chapter 10.

RESOURCE EFFICIENCY

Implicit in the characteristics of green buildings and infrastructure is a better use of resources. This is particularly well illustrated by low carbon projects and by the passive house developments in Europe, which can achieve up to 90 per cent reduction in energy consumption. Similar levels of resource gains are evident when construction firms reuse and/or recycle materials, develop brownfield sites, minimise waste, promote public transport and employ local labour. Indeed, achieving greater levels of output with fewer resources lies at the very heart of achieving sustainable construction.

Some analysts argue that much greater resource efficiency is achievable. In the 1990s, an important optimistic report – *Factor Four: Doubling Wealth, Halving Resource Use* (Weizsäcker et al. 1998) – claimed that resource productivity could be increased by a factor of four. Obviously such an increase in efficiency would reduce the demands placed on the natural environment. To demonstrate that a quadrupling of resource productivity was technically possible the report included fifty examples. Twenty were related to energy productivity in various contexts, from refrigerators to hypercars; a further twenty were concerned with material productivity, ranging from residential water efficiency to timber-framed building. A decade later an even more ambitious target was set in the sequel *Factor Five: Transforming the Global Economy* (Weizsäcker et al. 2009). This work focused on four sectors, namely buildings, agriculture, transport and industry (steel and cement), and presented several best practice case studies. These included a whole systems approach to commercial buildings and a detailed analysis of the passive house movement. Encouragingly, in the context of construction economics, many of the examples were relevant to the markets for green buildings and infrastructure, and some of these are listed in Table 9.4.

Table 9.4 Examples of resource productivity (Factors four and five)

✓	Steel or timber frame versus concrete
✓	Compact fluorescent and LED lighting
✓	Air conditioning versus passive cooling
✓	Liquid crystal display (LCD) TVs and computer screens
✓	Renovating old terraced derelict slums
✓	Superwindows and large office retrofits
✓	Solar energy systems
✓	Conservation versus demolition

Source: Adapted from Weizsäcker *et al.* (1998) and (2009)

In discussing the various examples, Weizsäcker et al. (1998 and 2009) highlight the competitive advantages that could be achieved by exploiting resource efficiency. The possibilities and opportunities they describe are achievable by most firms in any part of the world seeking to differentiate their products. In most industries, if producers are offered the opportunity to adapt production to make it significantly quicker, of consistently higher quality and with a potential of saving up to 80 per cent of resources, they would give it a try. Construction, however, is notoriously slow to take advantage of any new opportunities that present themselves. The debate concerning the slow uptake of off-site production was introduced in Chapter 7 and Reading 2. This debate has been ongoing for more than twenty years and forms an important part of the sustainable construction agenda. It has been informed and given impetus by several government reports. For example, the Innovation and Growth Team report published in 2010 encouraged the industry once again to realise the benefits that the controlled environment of a manufacturing plant could offer to achieve reductions in construction costs, delivery times and in building defects. Off-site manufacture, the report argued, would enable project teams to deliver higher quality products at significantly lower cost (IGT 2010: 49). Similarly the National Audit Office (2005) highlighted that modern methods of construction made it possible to build up to four times as many homes with the same amount of on-site labour while reducing on-site construction time by up to half. The next section explores some of the arguments for and against the uptake of resource efficient methods in the construction sector.

OFF-SITE CONSTRUCTION METHODS

Modern methods of construction utilise a number of innovations that transfer work from the construction site to the factory. They embrace a variety of approaches referred to by several different terms, such as off-site manufacturing (OSM), off-site production (OSP), prefabrication, standardisation, lean construction and modular build. Common examples of building elements that are produced using off-site production techniques are bathroom and toilet pods, and timber and steel frame structures and roofs. The wide range of benefits that follow as result of adopting these methods are summarised in Table 9.5.

Table 9.5 The benefits of modern methods of construction

✓	Off-site working leads to improved safety
✓	Controlled factory environment leads to increased productivity
✓	Running costs are reduced as air tightness and energy efficiency are improved
✓	Reduced on-site construction time
✓	Significant reduction in costs
✓	Less waste from surplus and damaged materials
✓	Fewer defects and fewer environmental impacts

Clearly these benefits would assist a firm to achieve resource efficiency, improve product quality and secure a greater level of profit. Indeed, there are some exceptional examples in China and Japan and those that are interested are encouraged to watch the remarkable time-lapse video of a 30 storey (16,700 square metres) hotel built almost entirely of prefabricated components in China (by the Broad Group) in 360 hours (just 15 days). Yet, despite the obvious advantages of these modern off-site technologies, several barriers are reported by construction firms in the United States and Europe, such as higher capital costs, difficulties of achieving significant economies of scale, concerns relating to manufacturing capacity, the fragmented nature of the industry's structure, skills shortages and a risk-averse culture (Pan et al. 2008: 61).

As Weizsäcker et al. (2009) pointed out, the constraints to achieving gains in resource productivity are not technological but institutional. This finding was reinforced by the results of a survey of 100 house builders in the UK. Several firms responded that a lack of previous experience prevented them from a wider take up of off-site construction methods (Pan et al. 2008: 62). In short, inertia and cultural issues are underlying barriers to change. As the chairman of Buildoffsite (a UK organisation that campaigns for prefabrication – see Web Review on page 144 for details) sarcastically remarked: 'This is the twenty-first century; we've flown to the moon. Surely people can get their heads around a house that was built in Poland and shipped to the UK' (cited in Wright 2010). This line of argument suggests that reforming the processes of construction, or economic development generally, and introducing a more sustainable approach is as much a challenge to our personal values as to our political and economic systems.

Capital Costs Versus Running Costs

A significant example of inertia is the way markets tend to favour the short term in preference to the long term. To some extent this is exemplified by the reluctance of volume builders to construct energy efficient homes for fear that it might reduce their profits (see earlier discussion on page 149). This type of short-termism is particularly common whenever one person pays for the efficiency gains and another party reaps the benefits. This is easy to see in the commercial sector, in which the priorities of landlords and tenants are frequently regarded as distinct. An often-quoted general rule for traditional commercial buildings is that running costs outstrip capital costs by a ratio of 10:1 over a 25 year period. More specifically, a study carried out on behalf of the Royal Academy of Engineering (Evans et al. 1998) estimated that the costs for a typical commercial building over a 20 year period are in the ratio of 1 (for construction costs): 5 (for maintenance costs): 200 (for staff costs). Yet the present culture in the construction industry still tends to place far greater emphasis on the initial capital cost, while demonstrating little regard for the costs incurred by end users. In terms of efficiency, this attitude creates major resource cost implications – indeed the figures suggest that we may be more than ten times better at wasting resources than using them.

This line of analysis creates another opportunity for construction firms to differentiate their products and service. An integrated approach – which fully takes into account the end user – makes it far easier to suggest that the construction firm is adding value to a client's future business. Yet a client seeking to place an order for a business headquarters that makes use of natural materials, sunlight, energy efficiency, low noise, green plants and a genuine feel-good factor for their employees would find its choice of contractors greatly restricted.

PRODUCTIVITY

It is important that office buildings are conducive to work, yet in many cases there is anecdotal evidence to the contrary. There are even accusations that some office buildings cause employees to suffer headaches, feelings of lethargy, irritability and lack of concentration – and, in some cases, can be responsible for high rates of absenteeism. Even more worrying are the suggestions that the office environment can cause irritation of the eyes, nose, throat and skin. Although some of these symptoms sound like the side effects of spending an evening in the pub, or too long in the swimming pool, they are distinguished by being prevalent among the workforce of some office buildings and not of others. In fact, the symptoms usually disappear after a few hours of leaving the 'affected' building. This type of condition is commonly referred to as **sick building syndrome** (SBS) and it clearly leads to an inefficient use of human resources. It is important to remember that ultimately buildings are 'machines for working in' and investment in green construction should also result in a more efficient working environment.

An interesting example of a highly integrated green building is the Rocky Mountain Institute in western Colorado. Here it is claimed that the staff that work in the building are productive, alert and cheerful all day – without getting sleepy or irritable. Weizsäcker (1998: 13) attributes the high rate of productivity to 'the natural light, the healthier indoor air, the low air temperature, high radiant temperature and high humidity (far healthier than hot, dry air); the sound of the waterfall (tuned approximately to the brain's alpha rhythm to be more restful); the lack of mechanical noise, because there are no mechanical systems; the virtual absence of electromagnetic fields; ...the green plants'.

Occasional days of sick leave mean that employees are being paid but not in return for any productivity. Equally worrying, and damaging to overall productivity, is the situation where employees do attend work but spend a part of each day complaining about their working environment – and ultimately they might be so fed up that they decide to look for another job. This clearly all adds up to a waste of resources.

The annual cost of absenteeism from the workplace has received little attention from economists but it was suggested in 2007 that 15 member states of the EU (with a population of 375 million) could significantly increase productivity by making indoor office environments more healthy and comfortable. The report estimated that this could produce a return of up to 240 billion euro per year (EU 2007: 8). This figure, however, does not specifically account for the effects of sick building syndrome. As the following calculation suggests, this is a significant omission.

According to a survey (Hedges and Wilson 1987) involving employees working across 46 office buildings of varied age, type and quality the incidence of sick building syndrome is quite widespread. Participants of the survey were asked how much they thought the physical conditions of the office influenced their productivity. The majority thought that their productivity was affected by at least 20 per cent. This is the equivalent of taking one day off in five. Worker self-evaluation, however, may be subject to exaggeration. But even if we take a reduced figure of 10 per cent, this would still represent a significant cost. For example, if we assume that £20,000 is the average office salary, then an organisation employing 1,000 people could be losing in the region of two million pounds each year. (The calculation is simple: a 10 per cent SBS effect on lost productivity represents £2,000 per employee per year, multiplied by 1000 gives a potential loss of £2,000,000.) Arguably, these figures are a worst-case scenario, and not everyone is equally affected by SBS – the literature suggests that 55–60 per cent of staff in problem buildings may be affected. It seems more plausible, perhaps, to accept an estimate of one million pounds per 1000 employees per year. The more worrying statistic is that the service sector employs more than 15 million people every day in offices. This calculation implies a national cost of SBS in the UK in the region of £15 billion. Again this may be a worst-case scenario, as it assumes that all buildings are affected by SBS. However, the important point is to consider how certain types of construction can result in inefficient uses of resources and set parameters for debate.

As the Egan report (1998: 22–3) stressed, construction needs to be viewed as a much more integrated process paying far more attention to the needs of the end user – even to the extent that completed projects should be assessed for customer satisfaction and the knowledge gained fed back into the industry. To a limited extent this is now happening, and **post-occupancy evaluations** have begun to identify features likely to influence and improve indoor environmental quality and productivity – some of these are listed in Table 9.6. The general message is that construction needs to change its approach. The end user requirements need to be given as much respect as construction specifications, as poorly designed buildings fail to meet the needs of the eventual occupants. Investment into good quality design can achieve a more efficient working environment and lower running costs. As an extensive report into the culture and performance of construction recently suggested the time has come for the language used in the industry to be overhauled: intelligent construction and systems integration need to become the norm, along with issues of whole life costing, quality, productivity and sustainability (IGT 2010: 155). To make sure that the interests of *all* parties involved in the construction chain – particularly the occupier – are included in the building appraisal.

Table 9.6 Internal features to improve productivity

Careful attention to a building's specifications can enhance internal environmental quality and improve productivity beyond the levels achieved in buildings which use standard practices. These characteristics are likely to assist the indoor quality.

✓	Natural systems of ventilation
✓	Building materials and furnishings that have low toxicity
✓	Use of natural daylight
✓	Energy efficient lighting with a low flicker rate to reduce headaches
✓	User control of temperature and ventilation
✓	Attention to maintenance and operation of buildings to reduce the build up of microbial agents

Source: Adapted from Heerwagen (2000: 354)

In terms of economics, the important point that emerges is that as the breadth of expertise required from a construction firm increases, the number of firms supplying the market decreases; in short, there is greater opportunity to differentiate between firms. We pointed out this type of consequence in Chapter 6 when we reviewed PFI contracts and there are interesting parallels here. Successfully completed projects, of either the green or PFI variety, have the potential to be more economically efficient and more sustainable. But both project types seem, at present, to favour the big firm and not the small firm that typifies the industry.

One possibility in the longer term is that teams of small firms will begin to work together more closely to secure a place in the green market. This was one of the ways that the Egan report hoped the industry would go forward. As Egan (1998: 32) expressed it: 'Alliances offer the co-operation and continuity needed to enable a team to learn and take a stake in improving the product. A team that does not stay together has no learning capability and no chance of making the incremental improvements that improve efficiency over the long term.'

Key Points 9.2

○ Thinking long term instead of short term makes an important contribution towards achieving greater resource efficiency in the built environment.

○ Some analysts argue that resource productivity can be increased by a factor of four, and that the barriers to achieving these gains are cultural rather than technological.

○ For traditional commercial buildings, the running costs outstrip the capital costs by a ratio of at least 10:1.

○ An important consideration of any economic activity is to consider the end user. For example, in construction, the internal design of an office building should be conducive to work.

LIFE CYCLE ANALYSIS

It should be apparent that any firm interested in producing products for the green market needs to consider a broad range of criteria. And the few firms that have begun to take their environmental performance seriously have adopted auditing procedures that go far beyond narrow financial measures. By auditing how much energy is used and how much waste is generated at each stage of a product's life, producers can increase resource efficiency and reduce the environmental impact of the product. But deciding where to start and where to stop with these environmental analyses is a contentious issue and the boundaries need to be clearly defined. For example, a construction firm could consider energy efficiency, the reuse of building materials, the energy embodied in the manufacture and transport of materials to site and the use of the building throughout its entire life span, etc. In fact, there seem to be ample opportunities to break into many new markets. In an ideal world, the complete 'cradle-to-grave' aspects of a building would be analysed, but this would take a business into making detailed assessments of first, second and third generation impacts. The important message is to identify carefully the quality and specifications of the product to be marketed, before deciding what is the 'cradle' and what is the 'grave' for specific purposes. Such an approach would take a firm on an incremental journey that would make its product differentiation clear and accountable. Figure 9.1 shows a very simplified model of the opportunities that life cycle analysis might offer to construction.

Figure 9.1 Life cycle analysis of buildings and infrastructure

In this simplified model, the environment is the source of fossil fuel and raw material inputs and a sink for waste outputs.

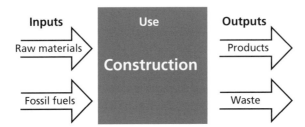

It is evident that, at each stage, the construction process burdens the environment with many costs. At the beginning of the life cycle, a large amount of natural input is needed for the construction phase and, as is well documented, across Europe the construction industry consumes more raw materials than any other industrial sector. During the operational stage, buildings are also responsible for a very significant amount (40–50 per cent) of greenhouse gas emissions, as buildings rely heavily on carbon-based fossil fuel energy for heating, lighting and ventilation. And finally, at all stages up to and including demolition, there is a large amount of associated waste. In fact, it is estimated that the construction industry accounts for 50 per cent of the total waste stream in Europe.

The life cycle analysis of a building is complicated further by fact that there may be several occupiers with different regimes of repair, maintenance and improvements throughout its life span. At all times, however, there is a flow of resources from the natural environment to the constructed product and vice versa, with varying impacts on the environment at different phases. Consequently, no matter how exemplary the initial environmental specification at the construction stage, the overall impact of a building will be dominated by the way in which it is used.

For our purposes, it is important to remember that we are not dealing here with environmental science. This text seeks to introduce economic concepts and:

- compare ideas of mainstream economists with their environmental counterparts
- understand the interrelationships between the economy and the environment.

These concerns form an important focus of the chapters comprising Part B.

Neoclassical Versus Environmental Economics

Mainstream **neoclassical economics** suggests that market forces determine the specific resources allocated to construction. We introduced these ideas in Chapter 3, where we explained how freely adjusting prices provides an efficient signalling system that determines what is made, how it is made and for whom. (Some readers may wish to review Key Point 3.1.) From this perspective, economists can easily account for why energy intensive, man-made substitutes might be used in place of more environmentally friendly products. Using neoclassical analysis, if inputs become scarce, the price rises; this, in turn, creates an incentive for an enterprising person to identify a gap in the market and produce a substitute. These substitutes often depend upon the clever use of technology and, as time goes on, more natural products are replaced (or substituted) by these man-made equivalents. So, for example, the sharp increases in oil prices during recent decades, which are outlined briefly in Chapter 14 as a cause of global inflation, have highlighted the increasing problem of demand for oil outstripping its supply. Indeed, the price of a barrel of oil reached an all time high in 2012. These significant and persistent price hikes push up the price of petrol and heating, and signal a need to substitute energy derived from oil with energy derived from other sources and to utilise developments in technology to improve energy efficiency. The reason we have presented this seemingly stark simple scenario, in which no explicit account is paid to the environment, is to stress that in traditional economic analysis the whole system is self-determining. In neoclassical terms, there is no need to resort to any form of government intervention to achieve a low-carbon economy, as given time the freely operating forces of the market will make it an economic inevitability.

In direct contrast, **environmental economics** does not accept that the ecosystem, or nature, is merely another sector of the economy that can be dealt with by market forces. Environmental economists proceed from the basic premise that there is an extensive level of interdependence between the economy and the environment; and there is no guarantee that either will prosper in the long term unless governments enforce measures that make firms acknowledge the complete life cycle costs arising from their economic activity. Daly (1999: 81) has crudely characterised the ideas of the neoclassical school: 'The economic animal has neither mouth or anus – only a close loop circular gut – the biological version of a perpetual motion machine.' The important concept that Herman Daly and his environmentally conscious contemporaries bring to economics is the greatly undervalued contribution that the environment makes to the economic system. Indeed, the environment provides all the natural resources and raw materials needed to start any process of building or infrastructure, such as land, fuel and water. The environment also provides mechanisms for absorbing the emissions and waste. In short, in this modern view, the economy is viewed as a subsystem of the environment!

In discussions of sustainability the environmental dimensions cannot be ignored, yet traditional mainstream economic textbooks do not refer to life cycle analysis or any of the equivalent auditing systems that measure environmental impacts. The sole reference point is money and the economy is presented as a linear system – similar to that portrayed by Figure 9.1. To correct this misleading picture, environmental economists usually represent the economic linear system within a larger box, or circle, to represent the environment. This approach is adopted in Chapter 11. It is used to illustrate that there is an interdependent relationship between the environment and the economy; that the environment provides resource inputs and carries away the waste outputs and cannot be taken for granted. As an example of a representation of the environmental approach see Figure 11.4 (page 184).

Unfortunately, however, the conventional mindset of those presently managing firms in the construction industry mirrors the approach taken by neoclassical economists. For this to be replaced with a genuine sustainable perspective, a commitment to understanding the ideas of environmental economics becomes most important.

It is worth closing this chapter with the observation that both neoclassical and environmental economists share a common belief that consumers and producers express preferences through their willingness to pay. This may appear ironic, but it seems that in the final analysis most economists are preoccupied with expressing everything in monetary value. This suits neoclassical economists whose main point of reference is the trade of material goods and services in markets at specified prices. It is far more problematic for environmental economists who seek to place monetary values on environmental goods and services that are commonly treated as 'free' goods. We shall elaborate on this further in the next chapter and deal specifically with valuation techniques in Chapter 11.

Key Points 9.3

○ Life cycle analysis involves a detailed study of the impacts of a product from cradle to grave. In the case of buildings and infrastructure, it emphasises the large amount of resources and waste that are involved in the construction process.

○ Neoclassical economists hold a strong belief that markets steer economies.

○ Environmental economists emphasise that the economy is dependent on the environment for several functions that cradle-to-grave analysis helps to value.

○ Environmental economics offers the construction industry a perspective that could help it to secure more sustainable outcomes.

Market Failure and Government Intervention

Throughout Part A, we emphasised that the market system allocates resources efficiently. We described how the price mechanism provides an incentive for firms to enter and exit markets in their search for profits, and how each market arrives at equilibrium. Indeed, up until the last chapter, the dominant theme has been that most economic problems can be resolved by allowing the free market to work (see Key Points: 2.1, 3.1, 5.4, 6.1, 7.1, 8.1 and 8.2). For one specific and intriguing example, see the argument put forward by traditional economists relating to the increasing price of oil as a solution to the problem of climate change, rehearsed on page 161.

The market, however, does not always work. There are some circumstances which prevent the price system from achieving productive and allocative efficiency. This seems to be particularly the case for markets involving or impacting on the environment; markets in which goods are not privately managed but commonly owned. In these cases, non-market alternatives need to be considered. One of the most important non-market forces is government, and this chapter reviews the government's role within failing markets. We shall, however, recognise the possibility that governments can also fail to achieve efficient outcomes, and this is discussed at the end of the chapter.

Market failure describes a situation where the forces of supply and demand do not allocate resources efficiently. It may be defined as:

> a marketplace where the unrestricted price system causes too few, or too many, resources to be allocated to a specific economic activity.

As we suggested in Chapter 2, the majority of environmental problems such as polluted seas, devastated forests, extinct species, acid rain and the vaporising ozone layer are examples of market failure.

Economists refer to these type of problems to justify a role for **government intervention**. As Milton Friedman, a publicist of the market system for more than 40 years, consistently emphasised, the existence of a free market does not eliminate the need for government. On the contrary government intervention is essential as a forum for determining the rules of the game, and as a provider of public goods (Friedman 1962).

WHAT CAUSES MARKET FAILURE?

Traditionally economists identify four common causes to explain why markets fail. The first set of explanations usually concerns the promotion of fair competition between firms, as monopolies and other forms of imperfect competition enable large firms to rig markets and keep prices artificially high. These problems have already been considered briefly in Chapter 8. In this part of the text, however, we are concerned with the protection and enhancement of the environment and, therefore, we restrict ourselves to the other three common causes of market failure. These are:

- externalities
- free-rider problems
- asymmetric information.

Externalities

We introduced the conceptual parameters necessary to understand externalities in Chapter 2 (see Key Points 2.4). We contrasted **private costs** and **external costs** – a distinction that helps to explain a broad set of environmental problems. The related analysis represents an important tradition in welfare economics, stretching back to the beginning of the twentieth century.

The idea that economic efficiency should describe a situation in which nobody can be made better off without making somebody else worse off dates back to around 1890 in work by Vilfredo Pareto, an Italian social scientist. According to Pareto, in a truly efficient competitive market all the exchanges that members of the economy are willing to make have to be agreed at fair prices. In such a situation, nobody can benefit unless they take advantage of someone else. There is a general equilibrium. All members of the economy face the true opportunity costs of all their market-driven actions.

In many real markets, however, the price that someone pays for a resource, good or service is frequently higher or lower than the opportunity cost that society as a whole pays for that resource, good or service. In short, it is possible that decisions made by firms and/or consumers in a transaction will affect others not involved in that particular transaction to their benefit or detriment. To put it more simply, in the competitive marketplace, a deal is struck between a buyer and seller to exchange a good or service at an agreed price; but, alongside this two-party activity, there are possible spillovers to third parties – that is, people external to the specific market activity. The spillover benefits and costs to third parties are termed **externalities**.

To clarify the concept further it might help to draw a distinction between the full economic cost and the basic cost of a good or service. The basic cost takes into account all stages of production, which could include extraction, manufacture, transportation, research, development and other business costs such as marketing. In other words, the basic cost covers all the costs that are usually added up to account for a market price. The full economic cost, however, includes all the possible basic costs plus the externalities; or, to put it another way, the full economic costs are the true burdens carried by society in monetary and non-monetary terms. In short:

full economic costs = basic costs + externalities

An example of an externality is the pollution of a river, the air or an open public space caused by a construction process. This leads to a general loss of welfare for a community. If this community is not compensated for its loss, then the cost is external to the production process. The construction firm has created a negative externality. In producing a building, the firm has paid for inputs such as land, labour, capital and entrepreneurship, and the price it charges for the finished product reflects all these costs. However, the construction firm has acquired one input – waste disposal into the river, air or open space – for free, by simply taking it. This is, indeed, taking a liberty; the construction firm is not paying for all the resources it is using. Or, looking at this another way, the construction firm is giving away a portion of environmental degradation free with every product.

Any kind of spillover that causes environmental pollution is called a *negative externality* because there are neighbourhood costs such as contaminated water and loss of habitat and associated health issues such as respiratory problems that society at large has to pay. In other words, these community costs are external to the economic transaction between the construction firm and the purchasers of the completed building. An important goal of environmental economists is to close the gap between private costs and external costs. The aim is to make the polluter pay – to make sure that those responsible for causing the pollution are made to pay the costs. This idea of making the polluter pay is discussed later in the chapter. Note, however, that if these costs are to be invoiced in some way we need to know how much to charge which, in effect, means putting a monetary value on the environment – and we shall look at ways of measuring environmental costs in more detail in Chapter 11.

Before leaving the topic here, however, we should acknowledge that not all externalities are negative. The production of a good or service can generate spillover benefits for third parties. In these instances, the market failure is not so problematic. Governments can choose to finance these goods or services that generate positive externalities through subsidies to the private sector – ensuring that companies are rewarded for production of a good or service that, if left to market forces, would be underproduced. A simpler alternative is for a government to take responsibility for the production of the good or service itself. The next section on free riders will confirm the appeal of this approach.

Free-rider Problems

Whenever positive externalities greatly exceed private benefits, the good or service concerned becomes unprofitable in the market context – in effect, some benefits associated with the good or service are allocated for free. For example, if you pay for several lampposts to light the pathway and pavement outside your house, the private benefit (to yourself) would be too small relative to the cost. And the external benefit to your neighbours from this street lighting would be significant, as they would be getting a brighter pathway for free. The problem is that the market system cannot easily supply goods or services that are jointly consumed. For the market to work efficiently a two-party agreement is preferable. If non-paying parties cannot easily be excluded from the benefits of a good of service, we have the problem of the **free rider**. Good examples of this situation are the markets for sewerage services, public open space, paving, street lighting, flood control, drainage, roads, tunnels, bridges and fire-protection services.

Asymmetric Information

Most economic texts identify the problems created by a dominant firm, or a group of colluding firms, as typical causes of market failure. As an example, reflect on the market structures that typify firms in construction and the possible opportunities for them to enter into agreements on joint profits (or at least review Key Points 8.4). In this text we have chosen to emphasise that *any* contractual agreement that is loaded in favour of one party can contribute to market failure. There is a general problem of one-sided information. In Chapter 6 (see page 86), we introduced the idea of asymmetric information.

> A situation in which some of the parties involved in an economic transaction have more information than others is defined as asymmetrical.

Markets may not achieve efficient outcomes when the consumer has to defer to a more informed producer. Let us develop this idea a little further with a simple example. When most consumers go into a music shop to buy a CD, they have enough information to make a rational decision. When they purchase services from a builder, the situation is often very different. In this situation, purchasers know roughly what they want to achieve – but they must rely on the experience and advice of the builder to specify what precisely needs to be done.

This situation – in which one party holds most of the cards – is a common cause of market failure. A new academic approach to market analysis is emerging that focuses on the contractual agreement between the 'principal' – that is, the client – and the 'agent' – the contractor. This focus on the **principal-agent** relationship questions the balance of power between the less informed client and the knowledgeable agent. The debate is around the extent to which the agent acts in the best interests of the client. This analysis of the principal-agent relationship demonstrates how the skills and experience of the agent could lead to a situation in which a trusting client may be misinformed. The initial discussions on principal-agent relationships appeared in health economics: in health contexts, it is clear that the doctor – the agent – has far more medical information than the patient – the principal. Consequently, we are very reliant on doctors to act in our best interests.

Principal-agent analysis can equally be applied in construction contexts – to project managers, engineers and architects. Many large-scale construction projects are technically complex and not easily understood by non-professionals. Although the costs of a mistaken choice may not appear as dire as in medical cases, they are equally difficult to reverse. For example, if the clients or purchasers of a major building development wish to reduce the environmental impact of the construction process, they are completely dependent on the expertise of contractors to achieve these outcomes. It is quite possible that energy usage may not be as efficient as it could be or that waste may not be minimised as requested. The hired 'agent' may not always act in the client's best interest, and they might be able to get away with it because of the 'principal's' incomplete knowledge.

Key Points 10.1

○ Market failure occurs whenever the free forces of supply and demand over-allocate or under-allocate resources to a specific economic activity. Examples seem to be widespread across the environment.

○ Three reasons for market failure are (a) externalities, (b) the free-rider problem and (c) asymmetric information.

GOVERNMENT INTERVENTION AND MARKET FAILURE

Governments intervene in various ways to correct market failures. Historically, the preference had been for correction via legislation, but increasingly correction is sought by influencing prices and knowledge. Some typical examples are outlined in Table 10.1. This indicates, in a very general way, some of the approaches that are used to tackle different types of market failure. As the press describes it, stick, carrot and tambourine approaches all play a part. To clarify what this means, in the next sections we discuss some of the approaches used by government to resolve each of the three causes of market failure, before going on to consider their effectiveness.

Table 10.1 Government policies to address market failures

Market failure	Market Based Instruments	Government spending	Publicity & information	Government regulations
Negative externalities	Landfill tax Climate change levy Aggregates levy		Cost benefit analysis	Water quality legislation
Free-rider problems		Provision of public goods Tax relief for cleaning up contamination		Habitats and species legislation
Asymmetric information	Green deal		Energy performance certificate Detailed billing	Building regulations Code for sustainable homes

Government Taxation

The UK government currently collects around £600 billion in taxes each year. The lion's share is obtained through taxes on personal incomes and business profits, and a relatively smaller amount is raised through environmental taxes such as those charged on waste, carbon usage, pollution and so on. However, the system of taxation is evolving from simply raising funds to provide essential public services to altering patterns of private expenditure. As politicians like to say, the burden of taxation is beginning to shift away from 'goods' such as employment towards 'bads' such as pollution and environmental damage. In short, modern taxation is being used to determine the allocation of resources by influencing the final market price of a good or service. Ironically, if the tax incentive is effective at changing behaviour, then the government could foreseeably collect less tax as a result. For example, it is estimated that the current revenue from taxes on motor vehicle use (vehicle excise duty and fuel tax) could drop by 35 per cent from £38 to £25 billion each year as motorists switch to electric, hybrid and fuel-efficient cars. In other words, based on the current tax incentives, the revenue from taxation on motoring is expected to decline by £13 billion a year by 2029, although the actual level of traffic on the roads is forecast to increase by 50 per cent in the same period (IFS 2012).

Taxes and subsidies that operate through the price mechanism to create an incentive for change may be described as a **market-based instrument**, as theoretically these fiscal instruments seek to internalise external costs into the price of a product or activity. Interestingly many of the recent examples of taxes introduced to reduce negative externalities relate to the construction industry.

LANDFILL TAX

The landfill tax was introduced in October 1996. It was imposed to provide an incentive to minimise waste and promote recycling – to internalise the costs to the community of waste going to landfill. Depending on the nature of the waste, the current tax can be as much as £64 per tonne for so-called 'active waste' that gives off emissions, and £2.50 per tonne for 'inert material' such as concrete bricks and excavated soil that does not biodegrade. This is potentially a significant penalty, although in the case of construction the vast majority of the waste is inert and attracts the lower rate. Currently, slightly more than 100 million tonnes of construction and demolition waste could potentially end up as landfill – of which 16 per cent apparently is material delivered and then thrown away unused. The government has indicated that it will make continuous annual increases in the standard rate of landfill tax, and it is expected to increase to £72 per tonne in 2013.

Although there is some uncertainty regarding the precise amount of construction waste, it is widely accepted that the industry is the biggest producer of waste products. Hopefully the increasing rates and targets relating to landfill will send a clear signal to those working in construction of the need to reduce the external costs associated with the large volume of waste produced, as well as provide an economic incentive to develop recycling. In fact, it is interesting to note that the proportion of waste being sent to landfill in the UK actually decreased by 11 per cent between 2004 and 2008 (DEFRA 2011).

CLIMATE CHANGE LEVY

The climate change levy commenced in April 2001. It is basically a tax on the business use of energy, and it covers the use of electricity, gas, coal and liquefied petroleum gas (LPG) used by the non-domestic sector. The levy is imposed on each business energy bill according to the amount of kilowatts used. There are differential rates for different energy sources. The levy is nearly three times higher for electricity (£0.00509 per kilowatt hour) than gas or coal (£0.00117 per kilowatt hour). This differential has been introduced because the use of each type of fuel creates different levels of greenhouse gas emissions. The levy has increased energy costs in the commercial sector by 10 to 15 per cent.

The purpose of the climate change levy is to encourage businesses to internalise – that is, pay for – the negative externalities associated with the greenhouse gas emissions that they are responsible for generating. Firms using environmentally friendly energy technologies, such as photovoltaic systems, energy crops and wind energy, or combined heat and power systems are exempt from the levy. Manufacturing, mining and utilities companies have been hit the hardest by the introduction of this levy. Ironically, the impact on the construction industry could be beneficial: the climate change levy encourages businesses to use energy more efficiently, and as all businesses occupy buildings, the expertise of the construction industry could potentially help to make offices and factories more energy efficient and, therefore, reduce energy costs.

AGGREGATES LEVY

The aggregates levy came into effect in April 2002. It is a tax applied to the commercial exploitation of rock, sand and gravel. It applies to imports of aggregate as well as to aggregate extracted in the UK. Exports of aggregate are not subject to the levy. The purpose of the levy is to give businesses operating in the UK an incentive to compare the full costs – including all negative externalities – of using virgin aggregates with alternatives or recycled materials.

To explain it another way, the levy has been established to reduce the noise and scarring of the landscape associated with quarrying. These environmental costs could not continue to be ignored, and the levy is meant to encourage the polluter to pay. The intention is that the construction industry should reduce its demand for primary materials by recycling as much as possible and by reducing waste on site. The immediate benefactors from the removal of these negative externalities would be those communities living close to the quarries. And it is interesting to note that their opinions were sought in the preparatory research that established the initial level of aggregate levy at £1.60 per tonne. By 2012 it had increased marginally to £2.00 per tonne, and in April 2013 it will see further revision to £2.10 per tonne. Finding an exact value for the environmental costs of quarrying will be examined further in the next chapter.

Government Spending

The second area that we identified as a cause of market failure relates to the free-rider problem. The basic problem here is excludability. The benefits of some goods or services – due to their very nature – cannot be excluded from non-payers. Even supporters of the free market, from Adam Smith to Milton Friedman, have recognised that there are a few goods and services that the market mechanism does not supply effectively. These are generally referred to as **public goods**.

In order to explain the precise nature of public goods, it is helpful to begin at the other end of the spectrum and clarify the definition of **private goods**. Indeed, so far in this text, private goods have been at the heart of the analysis. We have mainly discussed the activities of private construction contractors in providing private goods and services. These private goods (and services) are distinguished by two basic principles. One can be termed the **principle of rivalry**. This means that consumption by one user reduces the supply available for others. For example, when I use the services of a plumber, they cannot be working at the same time on your water and heating system. We compete for the plumber's services; we are rivals for this resource. The services of plumbers are therefore priced according to our levels of demand and the available supply of their time, and the price system enables plumbers to divide their attention between customers. The other principle that characterises a private good is the **principle of exclusion**. This simply implies that a user can be prevented from consuming a good or service unless they pay. In short, anyone who does not pay for the good or service is excluded. For example, if a road bridge is set up with a tollgate, then the communications link that this bridge offers is available only to those who pay. All others are excluded by the price mechanism.

These principles of exclusion or rivalry cannot be applied to *pure* public goods. They are non-excludable and non-rivalrous in their characteristics. National defence, street lighting and overseas representation are standard textbook examples of pure public goods. A distinction is sometimes made between *pure* public goods, which are both non-excludable and non-rivalrous, and *quasi* (near or impure) public goods, which do not have both these characteristics. The major feature of quasi public goods is that they are jointly consumed, but they might be excludable in some ways. For example, to access a gas or electricity supply, or a global positioning system requires the payment of a connection fee to benefit from a network. Hence it is possible to apply a discriminatory pricing system and provide such services as a type of private good.

There are, therefore, a number of distinguishing characteristics of public goods that set them apart from normal private goods, and these are portrayed in Figure 10.1. It shows a spectrum contrasting the characteristics of *pure* public goods against those that typify *pure* private goods. Developing Figure 10.1, we can describe public goods in more detail as follows.

- Pure public goods are usually indivisible, as these goods cannot be produced or sold in small units.
- Public goods can be used by increasing numbers of people at no additional cost; both the opportunity cost and marginal cost of one more user is normally zero.
- Additional users of public goods do not deprive others of the benefit.
- It is very difficult to charge people for a public good on the basis of how much they use, and they cannot be bought and sold in the marketplace.

Figure 10.1 A spectrum of economic goods

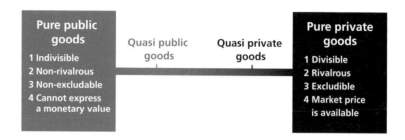

TAX RELIEF

Public goods overcome the failure of markets to supply goods or services that generate external benefits. In other words, they enable governments to intervene to provide resources that market forces would otherwise under-allocate. Equally, the government could provide tax incentives or subsidies to encourage the private sector to innovate in a way that will benefit society as a whole, both now and, more importantly, in the future. There are various tax incentives and subsidies to encourage research and development across all sectors. For example, developers are being encouraged to devise ways to clean up contaminated land through the provision of a 150 per cent tax credit for the costs incurred.

REGULATION, PUBLICITY AND INFORMATION

In each of the corrective actions described so far, businesses are being encouraged to reduce the incidence of environmental damage, either by responding to modified price signals that include environmental costs or through the government taking responsibility by providing public goods or paying subsidies through tax incentives. In contrast, another set of options is for governments to set regulatory standards or use their authority to provide information to aid decision-making. As these schemes are relatively less likely to raise business costs, they are considered rather ineffective instruments. But in some instances there are few alternative options, and these mechanisms continue to have a role to play.

GOVERNMENT REGULATIONS

In the history of government intervention in markets there are more examples of regulation than anything else. This generalisation does not only apply to regulations designed to improve markets in the construction sector, but also to the economy as a whole. In this section, therefore, we could address all types of market failure, such as regulations relating to patents and fair trading. However, we shall concentrate on those relating to problems of asymmetric information.

To begin with an example that has many implications across the property and construction sectors, it is the government's responsibility to prevent businesses denying responsibility for the environmental impact of their products and services. This seems to be particularly important as we are increasingly aware that habitat needs to be conserved, water quality needs to be maintained, carbon emissions need to be reduced, sites of scientific interest protected, and building standards observed. Legislation of this type is well established – for instance, the present system of **planning regulations** has a history of more than 50 years and the set of **building regulations** is more than 25 years old.

According to the Building Act 1984, building regulations can be made in England and Wales for the purposes of securing the health, safety, welfare and convenience of people in and around buildings, to further the conservation of fuel and power, and to prevent waste. Responsibility for complying with the regulations rests with builders and developers. The aim is to assure the public that a certain level of technical accuracy has been achieved in any construction work and environmental impacts reduced.

Building regulations (or codes as they are sometimes referred to) set a baseline of minimum standards to be expected from the industry. As such, they lag behind the standards imposed by the equivalent sets of regulations in many other north European countries. This is particularly the case in terms of energy efficiency, and even though building regulations are continually updated and revised a staggering amount of energy could be saved by going beyond the minimum standards set out in the building regulations. The new code for sustainable homes is a push in that direction as it seeks to improve on the 2006 building regulations, increasing energy efficiency by 25 per cent by 2010, 44 per cent by 2013, and ultimately reaching an energy standard of zero carbon emissions by 2016 (DCLG 2008).

PUBLICITY AND INFORMATION

Information is gathered by the government in an attempt to support and influence market decisions. This approach is particularly important when markets lack transparency due to problems of asymmetric information. Examples take many forms, from the simple gathering of data to a government campaign to raise awareness or publicise findings, through to statutory requirements to gather information to support a market transaction. Two contrasting examples of this type of initiative follow.

The Carbon Trust was established by the UK government in 2001 to raise greater awareness of the business opportunities relating to low-carbon technologies and retrofits. Its mission is 'to help governments, companies and public sector organisations to achieve the twin benefits of economic growth and carbon reduction'. It proudly boasts that in its first ten years it has helped clients to save more than £3.7 billion in energy costs and reduce carbon emissions by 38 megatonnes (Carbon Trust 2011: 2). Endorsing these achievements, a National Audit Office review confirmed that the Carbon Trust offers value for money in helping to reduce greenhouse gas emissions (NAO 2007a: 5). Despite this record, the Carbon Trust has not been exempt from government spending cuts and, going forward, this might present some new challenges. But its original purpose still represents a good example of a government-backed organisation helping society avoid inertia and ignorance in regard to its environmental performance, and encouraging more transparent and efficient energy markets.

A government can also choose to legislate about the level and quality of information required to support a market transaction. A good example of this type of intervention is the European requirement that an energy certificate must be issued prior to a building being sold or rented. For instance, in the UK it is a legal requirement that each (commercial or residential) building on the market for sale or rent is surveyed so that the person interested in occupying it will have some indication of how much it may cost to heat and light, and what cost-effective improvements could be made to achieve a better rating. These energy performance certificates (EPCs) give occupiers information on how to make their homes or offices more energy efficient and reduce the related utility costs. Although this measure is designed to raise awareness of energy efficiency issues, there is no legally binding requirement to act on the recommendations in the report that accompanies the energy certificate.

Ideally these campaigns and legislation to improve flows of information should create a greater symmetry between the expectations of consumers and the knowledge of suppliers, leading to a fairer, more efficient market allocation. It is also possible for information measures to reinforce the objectives of new fiscal incentives. For example, information on greenhouse gases can help people to respond positively to carbon taxes that differentiate between fuel sources, and the recommendations that come with an EPC can prompt people to take up the Green Deal opportunities discussed in Chapter 9 (see page 152). In this way, improved information services should raise awareness and provide more efficient markets.

Key Points 10.2

○ Governments can use several devices to correct market failure. These include taxes to internalise externalities, the provision of public goods to overcome free-rider problems, and publicity and legislation to reduce the problems associated with imperfect flows of information.

IS GOVERNMENT INTERVENTION EFFECTIVE?

The assumption that the alternative to a failing market is a brilliant government is wrong. Governments can fail, too. Several of the corrective measures we have discussed have problems. These are briefly summarised in Table 10.2 and are examined further in this section.

Measurement Problems

Government attempts to minimise negative externalities require measurement. The **polluter pays principle** is all well and good, providing that the guilty parties are easy to identify and that it is possible to determine a fair price for them to pay. Given that many externalities manifest themselves in global or national environmental issues and involve free goods, such as air, the ozone layer, habitat, flora, waterways, and peace and quiet, their measurement (and assessment) causes endless problems.

Table 10.2 Market failure and government intervention

Cause of market failure	Example of intervention	Nature of problem
Externalities	Taxes and levies	Measurement
Free-rider problem	Public goods Tax credits	Tax burden
Asymmetric information	Publicity campaigns Building regulations	Enforcement

To analyse these problems further it may help to consider Figure 10.2. Here we have the demand curve D and the supply curve S for product X. The supply curve includes only the private costs (internal to the firm). Left to its own devices, the free market will find its own equilibrium at price P and quantity Q. We shall assume, however, that the production of good X involves externalities that are not accounted for by the private business. These externalities could be air pollution, destruction of a green belt, noise pollution or any neighbourhood cost. We know, therefore, that the social costs of producing X exceed the private cost. This can be illustrated by

Figure 10.2 Internalising external costs

We show the demand and supply for X in the normal way. The supply curve S represents the summation of the private costs, internal to the firm producing X. The curve to the left, S_1, represents the total (social) costs of production. The grey arrows indicate the external costs that have been added. In the uncorrected situation, the equilibrium is Q, P. After imposing a tax (P_1 – P), the corrected equilibrium would be Q_1, P_1.

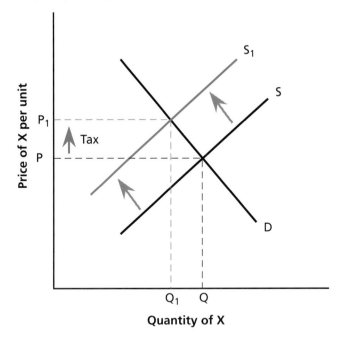

shifting the supply curve to the left, since it indicates that theoretically the costs of producing each unit are higher. (You may remember from Chapter 5 that changes in price and non-price determinants – such as a tax – are represented in different ways in graphical analysis. Review Key Points 5.3 for further clarification.)

The diagram highlights the fact that the costs of production are being paid by two groups. At the lower price P, the firm is only paying for the necessary private inputs. The difference between the lower price P and the higher price P_1 is the amount paid by the community – the external costs. For these external costs to be internalised, the government would need to introduce a tax equal to $P_1 - P$. This should result in fewer resources being allocated to this activity – with less demand and supply Q_1 – as the tax would lead to higher prices and force potential purchasers to take into consideration the costs imposed on others.

It is easy to see that in an unfettered market, external costs are not paid for and resources are over-allocated to environmentally damaging production. A tax should help to alleviate the problem, but the practical issues of precisely how much tax and who will be burdened with the expense are difficult questions to resolve.

Tax Burden

As we have explained, pure public goods would not be properly provided by a market structure because of the free-rider problem. For similar reasons, quasi public goods would also be under-allocated. The incentive to contribute to the cost of production of public goods is greatly reduced by the knowledge each individual will potentially benefit regardless of whether they pay. Consequently, most governments step in to provide goods and services such as law and order, overseas representation, infrastructure and environmental management. The concomitant demand for roads, tunnels, bridges, prisons, police and fire stations, overseas embassies, play areas, clean recreational space, flood control systems, etc. explains how governments become such important clients of the construction industry. In fact, in the UK it is reckoned that public sector spending, in value terms, accounts for nearly 40 per cent of the business done by construction firms (Cabinet Office 2011: 5).

The drawback to this level of commitment is the cost, especially as the majority of goods that governments produce are provided to the ultimate consumers without direct money charge. Obviously, this does not mean that the cost to society of those goods is zero. It only means that the price 'charged' is zero. The full opportunity cost to society is the value of the resources used in the production of goods provided by the government. For example, though nobody pays directly for each unit of consumption of defence or environmental protection, everybody pays indirectly through the taxes that finance government expenditure.

In the UK, the government collects something around £600 billion in taxes each year. Spending on law and order, defence, the environment, international co-operation and transport alone accounts for approximately 25 per cent of this expenditure. In effect, the average citizen in the UK must work from 1 January to March or April just to pay all their direct taxes.

This **tax burden** is clearly a significant proportion of any citizen's income, and it raises some of the thorniest questions that any government has to face. In the UK, and much of the developed world, public spending grew relatively unchecked until the financial crisis that commenced around 2008 brought the problem into sharp focus. Now, many governments are exercising 'austere' cut backs in attempts to meet the **golden rule** – which, in simple terms, means that governments no longer allow current spending to exceed current receipts. The golden rule forms a central plank of government policy and its function will be discussed further in Chapter 12.

Enforcement Problems

The success of any government policy cannot rely solely on a strong theoretical argument. Political support, voter appeal and luck are equally important. In other words, just because a government has carefully debated and passed through parliament a new policy, launched a publicity campaign or initiated another set of regulations does not automatically guarantee success.

Rule-based measures, such as regulations, create a whole range of associated costs. There are the compliance costs of implementing, enforcing and administering the legislation. For example, the building regulations (or codes) are devised to set minimum standards, such as how much insulation should be used, what kind of windows should be fitted and how efficient the heating boiler should be. Yet in recent history no builder has been prosecuted for non-compliance!

This rather startling fact made news during 2011 when Andrew Stunnell, the government minister responsible for building regulations, lambasted house builders for their failure to build to required standards. His comments highlighted the differences that existed between the predicted and actual energy used in new homes, drawn from a number of detailed studies. For instance one study of 15 low carbon homes reported energy losses through the fabric to be around 70 per cent higher than predicted (Bell et al. 2010). Similar studies throughout the UK suggested gaps between the measured and predicted heat loss vary considerably, from one case where the results were actually better than expected by 1 per cent to a significant number of dwelling were the heat loss was as much as 120 per cent greater than that predicted by the design specification (GHA 2011). Equivalent problems, but on a larger scale, are catalogued for commercial buildings by CarbonBuzz (see page 144). Buildings tested for compliance with ventilation standards also fail to meet the new requirements. The difference between predicted and actual performance – referred to as the performance gap – is an increasingly recognised problem. It suggests that builders cannot keep up with the stringent standards required in construction.

This raises questions about the current system of building control. The system has become over prescriptive and complex. Building regulations are subject to frequent revision. However, these regulations specify how to do things, but not what to achieve; they tend to measure inputs more than outputs. As a government review entitled *The Future of Building Control* actually acknowledged: 'the system is not broken but it has some serious failings and weaknesses that must be tackled if we are to ensure that it remains fit for purpose in today's world and in the future' (DCLG 2007: 5).

Even where building codes are enforceable, regulations provide little incentive to be innovative. In fact it has been argued that it is the heavily regulated nature of construction activity that accounts for the conservative attitudes that typify the industry. There is usually little incentive to go beyond the regulatory standard, so typically construction firms only do the minimum that is required. Few builders will exceed the standards set out in the regulations, unless of course an informed client demands something special. Similarly, if governments rely on publicity campaigns to change attitudes, rather than raise the costs of production, then all businesses, construction or otherwise, have no real incentive to refrain from using polluting products and methods.

Businesses can improve performance. Some exceptional examples of potentially resource efficient projects were given in Chapter 9 (see, for example, Table 9.4). It is suggested that in several instances it is possible to exceed traditional performance by factors of 4 and 5. But, in general terms, the construction industry suffers from inertia, and is not particularly innovative or sustainable.

Key Points 10.3

○ One method for internalising external costs is to impose a tax. But it is difficult to set tax rates so that the polluter pays the correct amount.

○ To overcome resource allocation problems, governments usually provide a range of public goods. Inevitably, these create a range of associated costs that are ultimately financed by taxpayers.

○ Just because a government has rubber stamped some regulatory procedures or launched a publicity campaign does not automatically mean that better practices will be effectively enforced or voluntarily introduced.

GOVERNMENT FAILURE

To conclude this chapter, we should recognise that market failure cannot simply be remedied by government action – that is, perfect governments do not resolve imperfect markets. Indeed, modern economic texts also acknowledge the occurrence of **government failure**.

Government failure is understandable, since the political process by its very nature is likely to be inefficient in allocating resources. When choices are expressed through the market mechanism, the price forces individuals to absorb most of the costs and benefits. Politicians, however, allocate resources more on the basis of judgement. Government judgements are often skewed by lack of financial incentives, gaps in information and pressures applied by different interest groups that need to be acknowledged for re-election.

The basis of this idea is derived from a branch of economics and political science known as **public-choice theory**. This studies the ways that government decisions are made, working from the premise that politicians are basically driven by a desire to please their supporters and be re-elected. This leads government minsters to be more concerned with what they say than what they do. Many grand ministerial announcements prove difficult to implement. As a former construction minister remarked in retrospect: 'in many cases the conclusion is that the policy is unworkable, and after a discreet period of silence, or an abortive implementation, it is quietly allowed to die' (Raynsford 2012).

There is also concern about more explicit corrupt activity in the public sector. Corruption is evidenced when elected politicians and public officials take actions for private gain by exploiting their entrusted power. Typical examples are accepting payment for information, directly embezzling public funds and taking kickbacks in relation to public procurement. Since the mid-1990s Transparency International, a non-governmental organization based in Berlin, has been monitoring and publicising bribery and corruption in the public sector across 180 countries. Its annual corruption and bribery indices suggest that there is a high level of corrupt activity in relation to construction projects throughout the world and we make some comparisons in Chapter 13 (see pages 233–7).

Even in nations that are relatively uncorrupt, the sheer scale of managing a country from the centre is problematic. There are problems of distribution, measurement, enforcement, and funding. These problems lead to inefficiency and a wasteful use of resources. Indeed, the more wide reaching and detailed an intervention becomes, the less likely it is that the benefits will justify the costs.

Consequently, there has been a tendency to believe that markets might provide a more efficient and equitable means of allocating environmental resources, particularly as government systems tend to become bureaucratic, inflexible and excessively expensive to run. Furthermore, as government intervention increases, individual liberty is reduced and the competitive spirit declines. Indeed, the idea of less intervention by government is part of the rationale for the public sector cuts (the so-called austerity packages) that were introduced across Europe as a response to the financial crisis of the late 2000s.

The present trend, therefore, is to provide incentives through the market system and wherever possible to reduce the scale of state intervention. This means that environmental taxes and other economic instruments will probably continue to be the key tools used to achieve environmental improvement. How far this trend should continue is debatable, as it is not solely a question of economic efficiency but one of politics too.

Key Points 10.4

○ Government failure is a recently acknowledged phenomenon. It suggests that intervention through policy initiatives does not necessarily improve economic efficiency.

○ Government failure is caused by a number of factors, such as poor judgment, lack of information, inadequate incentives, the scale of the problem and, in some instances, corruption.

11 Environmental Economics

Economists like to proceed from a model framework to simplify reality and create a reference point for their specific analysis. For many years, however, the models used by mainstream economists tended to overlook the interactions between the economy and the environment. The benefits of economic activity were exaggerated and accounted for, while environmental costs were ignored. We introduced this major distinction when we discussed a product's life cycle (from the 'cradle' to the 'grave') in Chapter 9. However, we have suggested that the environment cannot be ignored since it provides resources at the beginning of a product's life cycle, and absorbs waste at the end of the cycle. This important strand of economic thinking began to gain credence in the 1960s, drawing an important dividing line between environmental and neoclassical economics (see Key Points 9.3).

The significance of the environment for businesses of all types is succinctly represented by the flow diagram in Figure 11.1. The environment is the resource base that provides renewable and non-renewable resources that enable production to begin. At the other end of a product's life, the environment is also expected to provide a sink facility to assimilate the waste matter. (The environment can also provide opportunities, as an amenity, for various leisure pursuits.) The process is easy to exemplify within the construction sector, as the industry consumes resources and generates waste on a scale that completely dwarfs other sectors of the economy.

Figure 11.1 The environment: beginning and end

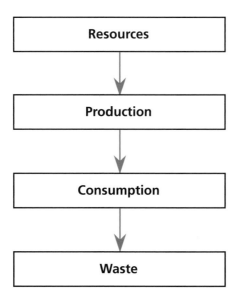

In the first place, it is the environment that provides the land on which buildings and infrastructure are located. Second, it is the environment that provides many of the resources that are used to make building material products. As acknowledged in the *State of the World Report*, the construction industry consumes more than one third of global resources, over 10 per cent of all fresh water, and is responsible for 30 to 40 per cent of solid waste. Buildings also account for 25 to 40 per cent of the total energy used and approximately 30 to 40 per cent of global carbon dioxide emissions (Taipale 2012). Finally, but by no means least important, it is also the environment that is ultimately responsible for assimilating and processing the waste that arises from the various phases of construction, from building through to demolition. In other words, without the environment there would be no resources for construction and no way of managing some of the waste and outputs arising from the processes involved in maintaining the building stock and associated infrastructure.

In this chapter, we concentrate specifically on the ways in which the environment and economy interact and explore three important conceptual areas that characterise environmental economics. These are the mass balance model, private costs versus social costs, and environmental valuation.

THE MASS BALANCE MODEL

The mass balance model has become a standard introductory reference for students new to environmental economics. It focuses on the energy used and wasted by economic activity. The first law of thermodynamics provides the starting point. This law explains the 'physics' of economic activity, since it states that we cannot destroy matter, we can only change it. Any resource that is extracted from the environment must be returned to it in some form or other – what goes in must come out.

The model conveys the process of production as the transformation of a certain number of inputs into certain a number of outputs. Some of the outputs may be positively valued, but other outputs will be undesired and of negative value. In other words, all economic activity produces a desired output (the conventional good or service) plus undesired output (such as pollution). For example, when clay and sand are combined and heated at high temperatures for several days there are several outputs: the conventional or desired output of bricks, and the undesired output of carbon particles and sulphur dioxide. In short, we have introduced two related notions: the principle of **joint production** that highlights how any economic activity inevitably produces several types of output; and the **mass balance model** that works from the premise that matter cannot be created or destroyed. The concept of joint production is derived from classical economics and the mass balance model from physical science. Their general significance to environmental economists is to emphasise that everything is connected to everything else.

Figure 11.2 presents the basic idea behind the mass balance model, that the mass of material inputs into an economy is balanced by the mass of products and waste outputs leaving the system. In other words, there is equivalence between the mass of resource inputs (raw materials) flowing into a system and the mass of outputs (product and waste) flowing out of it – hence the term mass balance.

Figure 11.2 The mass balance model

The mass balance for the whole economy is represented by the flows: *A = B + C*

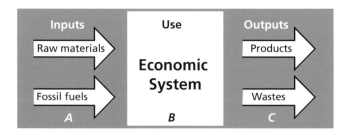

In the model, the environment is portrayed as having a similar relationship to the economy as a mother to an unborn child in so far as it provides sustenance and carries away wastes. An alternative perspective is to view the environment as a large protective shell surrounding the economic system. Unfortunately, this shell is often treated as a 'free' good and was for many years ignored by traditional economics.

The relevance of the mass balance model is referred to in the *Strategy for Sustainable Construction* (HM Government 2008). This report noted that the industry absorbs nearly a quarter of all raw materials used in the economy (an estimated 420 megatonnes annually from a total of 1508 megatonnes). This makes the construction industry the biggest user of materials in the UK, and its resource efficiency is therefore central to the sustainability agenda.

Data derived from the first trials to audit resource flows in the construction industry should clarify the significance of the mass balance theory. As stated, the industry uses about 420 megatonnes of material each year, of which the additions to stock (buildings, roads and other infrastructure) account for about 270 megatonnes. This leaves a balance of 150 megatonnes of waste (nearly 3 tonnes of construction waste per capita in the UK). It was estimated that approximately 46 megatonnes of materials were recycled and, although this is clearly not an adequate amount, it should be acknowledged that it is not possible to quantify resources that are reused on site without processing, such as excavated soils or hardcore that may be 'reused' as part of 'cut and fill' works on the same site (Smith et al. 2003; Forum for the Future 2006).

Although the precision of the data is questionable, the magnitude of the figures highlights the important role that the construction industry plays within the economy and its potential environmental impact. Governments across Europe are now beginning to account for resource flows; regular updates on material flows through the economy can now be found in the UK environmental accounts, which are updated annually. This data monitoring is part of a government drive to improve resource productivity and it is clearly related to achieving sustainable development.

The data and concepts arising from the mass balance reports should encourage businesses to think beyond traditional resource intensive methods of production. Resource inefficiency reduces competitiveness, eats up primary resources, and creates excessive waste. In theory, construction waste could potentially be reduced to zero if the materials thrown away by one contractor could be used by another, but this would require a process of putting the two together. As it is, contractors tend to throw away valuable materials. Anecdotal evidence would suggest that as much as £10 million could be saved each year by the larger contractors.

Maintaining the stock of natural capital is an essential prerequisite for understanding and managing sustainable development. And according to economists, there are two different approaches to achieving this goal. As outlined in Chapter 9, the basic premise of neoclassical economic analysis is that, left to its own devices, the market mechanism will provide the necessary incentives to encourage technological solutions to resource problems. In short, technological innovation should provide substitutes for any 'shortages' in the environment. As a result, the economy could grow forever. As Robert Solow (1974: 11), a Nobel laureate economist, wrote more than 30 years ago: 'It is very easy to substitute other factors for natural resources...The world can, in effect, get along without natural resources.' An old phrase used to highlight the supremacy of the economic system was that the economy would 'grow around' particular shortages such as resource problems. Technology would invent an easy way out! This traditional technocentric view is based on the **constant capital approach**, in which any decreases in natural resources are substituted by increases in man-made assets.

Environmental economists present things differently. According to their analysis, the economy and the environment are inextricably integrated. The ecosystem and the economic system are viewed as complements rather than substitutes. In fact, most environmental economists proceed from the mass balance model to argue that the economy is a subsystem constrained by the ecosystem, as it depends on the latter as a source of raw materials inputs and a sink for waste outputs. This ecocentric view adopts the **natural capital approach**, which is based on the premise that it is impossible to substitute natural capital with physical capital. According to Lovins et al. (1999: 158), advocates of this type of approach, 'production is increasingly constrained by fish rather than by boats and nets, by forests rather than by chain saws, by fertile topsoil rather than by ploughs'. In other words, the material factors comprising the ecosystem are unique and cannot be substituted at any price.

Empty World Versus Full World

To draw a distinction between these contrasting approaches, it is common for environmental economists to present two models. One, an open system, which we alluded to in Figure 9.1 when we portrayed resources flowing in a linear fashion, that pays no respect to the limits imposed by the environment. And the second, a more constrained closed system, in which the environment physically contains and sustains the economy. Some economists refer to the traditional view of the open system as an 'empty world' model to contrast with the 'full world' model that is the modern closed system.

Figure 11.3 is another way of illustrating the empty world model – the economic subsystem is small relative to the size of the ecosystem. It depicts the past, when the world was empty of people and man-made capital but full of natural capital.

In contrast Figure 11.4 (see page 184) represents the full world model, describing a situation nearer to today, in which the economic subsystem is very large relative to the ecosystem. This highlights the fact that unless qualitative changes occur the ecosystem is going to be pushed beyond its limits. In fact, there are signs that this point is imminent with, for example, global warming, ozone depletion, soil erosion, biodiversity loss, population explosions and resource depletion.

Figure 11.3 Empty world

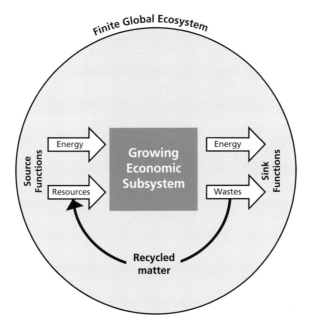

Source: Adapted from Goodland, Daly and El Serafy (1992)

Professor Kenneth Boulding presented this contrast in a colourful way in his 1966 paper *The economics of the coming spaceship earth*. In this paper he drew an analogy between the empty world and the full world. He referred to the empty world as a cowboy economy, as this effectively characterised the traditional economists' view of the earth's resources – abundant, limitless and capable of sustaining reckless, exploitative and violent behaviour. In stark contrast, he referred to the full world of the future as the 'spaceman economy' – a single spaceship, without unlimited reservoirs of anything, either for extraction or for pollution. The gist of his argument was that the planet should not be pushed beyond its limits. Taking this contrast to its logical conclusion we can envisage two types of economy: an inefficient economy in which all ecosystem services are treated as free goods and used abundantly; and an efficient economic system in which all resources are allocated according to price. This analysis clearly highlights the problems of pursuing unlimited technological growth, and makes a strong case in favour of sustainable growth that does not take the world's natural support functions for granted.

Figure 11.4 Full world

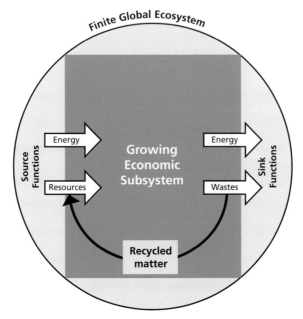

Source: Adapted from Goodland, Daly and El Serafy (1992)

Key Points 11.1

○ The economic system and the environmental system are inextricably integrated, since goods and services can not be produced or consumed without the environment providing resources at the beginning of a product's life cycle and absorbing waste at the end of the cycle.

○ The mass balance model focuses on energy used and wasted by economic activity. It highlights the resource intensive nature of construction and prompts a contrast between the 'empty world' of the past and the 'full world' of today.

○ There are two contrasting approaches to achieving sustainable development: one based on technology, in which decreases in natural capital are substituted by man-made assets; and another which places far greater emphasis on the critical nature of natural capital.

PRIVATE COSTS VERSUS SOCIAL COSTS

When environmental economists talk about costs they fall into three categories: private, external and social. First, there are the costs of an individual's actions that are known and paid for directly. For example, when a business has to pay wages to workers, it knows exactly what its labour costs are. When it has to buy resources to commence production, it knows what these will cost. Similarly, when tenants have to pay rent for their flat, they know exactly what the cost will be. These are the normal everyday costs associated with most traditional economic activity; they are called **private costs** and formed the focus of Part A. They were formally introduced in Chapter 2 (see Key Points 2.4). Private costs are those borne solely by the individuals who incur them. They are *internal* in the sense that the firm or household must explicitly take account of them.

Second, there are **external costs**, created by the actions of other people. (These have also been discussed previously in Chapters 2 and 10, so it might help to review Key Points 2.4 and 10.1.) We have considered situations in which a business dumps the waste products from its production process into a nearby river, and individuals drop litter on a beach. Obviously, there is a cost involved in each of these actions. When the firm pollutes the water, people downstream suffer the consequences. They may not want to swim in the river or drink the polluted water. In the case of fly tipping or simply common litter, the people who come along after the waste has been dumped are the ones who bear the costs. The costs of these actions are borne by people other than those who commit them. The polluter has not paid, the costs have not been internalised; they are, by result, referred to as external costs.

When we add the external costs to the internal or private costs, we arrive at the third category – total or **social costs**. Pollution – as all problems pertaining to the environment – may be viewed as a situation in which social costs exceed private costs. Because some economic participants do not pay the full costs of their actions but only their (lower) private costs, their actions are 'socially unacceptable'. If there is a divergence between the social and private costs of a specific activity, we may see 'too many' resources allocated to that activity. To take just one of the many examples, when drivers step into their cars, they certainly don't pay the full social costs of driving. They pay for petrol (which includes a significant element of fuel duty), maintenance, depreciation, road tax and insurance on their cars. However, they cause an additional cost – that of air pollution – which they are not forced to take fully into account when they make the decision to drive. The air pollution created by exhausts is a cost that drivers do not bear directly, but it causes harm to other individuals who suffer the inconvenience of respiratory ailments, and dirtier clothes and buildings. The social cost of driving includes all the private costs plus the external costs such as the costs of air pollution, which society bears. Decisions made only on the basis of private cost, therefore, lead to 'too much' driving!

Externalities

As we know from Chapter 10, when a private market cost differs from a social cost it is a problem of **externalities** – individual decision-makers are not internalising all the costs. In other words, some of the costs remain outside of the market and are external to the decision-making process. The individual decision-maker is the firm or the customer, and external costs (and benefits) will not enter into that individual's or firm's decision-making processes. The important point for environmental economists to highlight is that the full cost of using a scarce resource is borne one way or another by society. That is, society must pay the full opportunity cost of any activity that uses scarce resources.

It may help to view the problem as it is was presented in Figure 10.2. There we used a market demand and supply diagram to emphasise that the conventional supply curve includes only internal, or private, costs. Consequently, the market-determined equilibrium price and quantity does not incorporate externalities, as these are not taken into account by the individual producers or purchasers. The quantity produced is 'excessive', and the price is too low because it does not reflect all the costs. In Chapter 10, we considered the possibility of internalising the external cost through some form of taxation to correct the market failure and charge the full social costs of production. This theme is repeated in Figure 11.5.

Figure 11.5 The economic effect of a pollution tax

The supply curve S is based on the costs to the firm producing the good, without any consideration of its environmental costs. If a tax is introduced, the firm's total costs costs increase and the supply curve shifts upwards, by the amount of the tax, to S_1. In the uncorrected situation, in which pollution was being taken for granted, the equilibrium price is P_e and the equilibrium quantity is Q_e. After the tax is introduced, the equilibrium price rises to P_t and the quantity sold falls to Q_t.

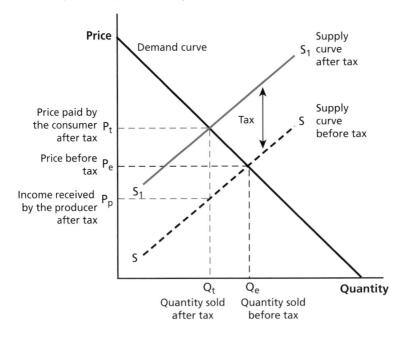

THE POLLUTER PAYS PRINCIPLE

A firm facing a charge or tax on pollution could respond in one of three ways.

• It could install pollution abatement equipment or change production techniques to reduce the amount of pollution.
• It could reduce pollution-causing activity.
• It could simply choose to pay the price to pollute.

The relative costs and benefits of each option for each polluter will determine which one or which combination will be chosen. Allowing the choice is the efficient way to decide who pollutes and who does not. In principle, each polluter is given the incentive to meet the full social cost of their actions and adjust production accordingly. In theory, there should be less environmental pollution and the price paid by the consumer should increase. This is explained in Figure 11.5.

The pollution tax causes the supply curve S to rise to S_1. The effect of the tax is to increase the price of the good. As a result, the quantity consumed decreases as the cost is higher for the purchaser. But the income received by the producer after tax is paid to the government P_p is lower than it was before the tax was introduced P_e. The environmental cost is therefore shared between the producer and the consumer of the good. Their relative contributions depend on the slope of the supply and demand curves. The more competition there is in the market, the less the consumer will pay.

IS A UNIFORM TAX APPROPRIATE?

Though taxation offers a way of forcing producers to take account of social costs, it may not be appropriate to levy a *uniform* tax according to the physical quantity of pollution. The same activities do not necessarily have an identical impact everywhere – we must establish the amount of *economic* damage rather than the amount of the physical pollution. A major motorway in London causes much more economic damage than a motorway routed through a less populated area. There are already innumerable demands on the air in London. Millions of people could breathe polluted air and thereby incur health problems, such as sore throats and asthma, which may even lead to premature death. Buildings would become dirtier. A given quantity of pollution, therefore, causes more harm in concentrated urban environments than in less densely populated rural areas. If we are to establish some form of taxation to align private costs with social costs – to force people to internalise externalities and to make the polluter pay – we somehow have to come up with a measure of *economic* costs instead of *physical* quantities. Because the economic cost for the same physical quantity of pollution varies according to factors such as population density, the natural formation of mountains and rivers, so-called optimal taxes on pollution should vary from location to location. (Nonetheless, a uniform tax might make sense given that the costs of administering a variable tax, particularly of ascertaining the actual economic costs, are relatively high.) Either way, the dilemma a government faces is how much should it charge for a 'permissible' amount of pollution and some light will be shone on this difficult question in the next section.

PLANNING GAIN

Before closing this section, it should be noted that not all externalities represent a cost, some are benefits. The principle of **planning gain** provides a classic example. These are agreements in which a developer provides community benefits that extend beyond the marketed output in return for gaining consent for the development to take place. In recent years, these agreements have become a recognised feature of planning negotiations. They are justified on the basis that when planning permission is granted, two things happen. First, the development imposes externalities on the local community, and second, the private developers typically benefit from the fact that their land or property is now of a greater economic value. In some sense, from a developer's perspective, planning gain may be viewed as a sort of 'bribe' or 'sweetener' – offered to accelerate the proposal through the planning process. From the local authority perspective, planning gain, if properly negotiated, can open up opportunities to deliver the right infrastructure to support wider community benefits and enable local plans to take shape. Examples of facilities provided through these arrangements include roads, parks, social housing, schools and health centres.

Because of the protracted nature of the negotiations, these 'deals' often remain low profile. Both parties endeavour to secure their own broad interests and there is a wide range of outcomes. The exchange that is agreed as a result may, from some perspectives, seem underhand or unfair, but it has been encouraged by legislation such as the Town and Country Planning Act 1971.

In an attempt to make the negotiations more equitable, a scheme has been introduced that allows planning authorities to impose a set charge for developments in their local areas. This optional scheme came into force in April 2010. In effect, the **community infrastructure levy** is a type of development tax that local authorities can choose to enforce. The levy could be calculated as so much per dwelling (such as the £18,000 'roof tax' that is currently being trialled on some new housing estates in Milton Keynes) or as so much per area (such as the £50 per square foot charged on new office space in Westminster or the astronomical £575 per square foot charge proposed for new office space on prime waterfront sites at Nine Elms in the London Borough of Wandsworth). The actual rate of the levy is left up to local authorities to decide (and at the time of writing only five authorities have formally established a charging schedule) so the London Borough of Wandsworth might find that the Nine Elms levy proves to be too high, as it could cause some development to become unviable. Each local authority needs to take account of market values in its particular area and the local infrastructure required, but if an authority gets the balance right, a planning gain system based on a levy, or tariff, could provide the way to a clearer and quicker basis for securing both private and public benefits and developments.

External benefits cause less of a problem than external costs. Consequently, environmental economists tend to give them less coverage. Conceptually, however, they should not prove too difficult to comprehend, as the symmetry of the first two key points suggests.

Key Points 11.2

○ Internal (private) costs + external costs = total (social) costs.

○ Internal (private) benefits + external benefits = total (social) benefits.

○ Private costs are those explicit costs that are borne directly by consumers and producers when they engage in any resource-using activity.

○ Social costs include private costs plus any other costs that are external to the decision-maker.

○ When social costs exceed private costs, externalities exist.

○ One way to make private costs equal social costs is to internalise the externality by imposing taxes or regulations. In theory, these taxes should be equivalent to the economic damage caused by the activity.

ENVIRONMENTAL VALUATION

As we discussed in this chapter and at several points in Chapter 10, to overcome some of the failings of resource allocation created by the market mechanism it is important to attribute monetary values to externalities. This can be attempted through various methods. Three of the most commonly employed approaches are contingent valuation, the travel cost technique and hedonic pricing. Each of these is briefly outlined below.

CONTINGENT VALUATION

The **contingent valuation method** relies on survey data. People are asked, through a questionnaire or interview, to express the value they attach to a specific environmental asset. For example, in the research leading up to the introduction of the aggregates levy in 2002, some 10,000 people living in communities surrounding quarries were asked how much they would be willing to pay, in the form of increased taxes over a five-year period, for the local quarry to be shut down. The question was based on the assumption that the quarry site would be restored to blend in with the natural landscape, and that all the quarry workers would be found suitable alternative employment. A further 1,000 people, who did not live in close proximity to a quarry, were asked what they would be willing to pay to shut down a quarry in a National Park such as the Peak District or Yorkshire Dales (HM Treasury 2002: 19).

TRAVEL COST

The **travel cost method** places a monetary value on environmental assets by calculating the cost of getting to recreational facilities such as wildlife reserves, forests and canals. The relevant costs include public transport fares and/or petrol, and travelling time.

HEDONIC PRICING

The **hedonic pricing method** attempts to identify an implicit market value for environmental services by analysing the bundle of characteristics that makes up a product in order to attach a specific value to the environmental element. This process has been described as establishing a 'surrogate market'. The method is easily exemplified in the property market: two properties may be identical except for their location, so the price that people are willing to pay for a house close to a river, woodland, beautiful view or conservation area may be quantified.

The Value of Human Life

The underlying principle employed in all valuation exercises is opportunity cost and this is most clearly illustrated when economists attempt to identify the value of a human life. The conventional method of calculation is to base the value of life on estimates of lost (forgone) contributions to the economy; the stream of lost career earnings. In other words, the opportunity cost of the average victim involved in an accident is based on a wage risk approach or, as some economists have preferred to call it, the human capital approach. In contrast, a more recent method follows the principles of contingent valuation: estimates are based upon an individual's willingness to pay to reduce risks and improve safety or, conversely, on the amount that people are willing to accept to compensate for a loss of life. Not surprisingly, the values based on the willingness to pay approach tend to be twice as high as those estimated by the human capital approach as people tend to value their own lives well above their earnings potential. And those living in poorer countries tend to place lower monetary values on their lives than people in richer ones.

Government economists throughout the world have adopted these seemingly crude methods to evaluate major infrastructure and transport projects. In resource terms, this enables them to consider a dimension that traditional investment appraisal normally overlooks – namely, the value of lives, lost or saved, due to accidents. The estimated net benefits to society begin to look very different once the appraisal incorporates a monetary value imputed for life.

Table 11.1 Statistical values of human life

Country	€ 000
United States	3,189
United Kingdom	2,107
Sweden	1,954
Germany	1,266
France	589
Greece	206
Portugal	56

Source: Adapted from SafetyNet (2009: 9)
Note: In 2002 Prices

Table 11.1 shows the values used in different countries to assess the monetary value of a fatality in a road accident. Interestingly, due to different cultures, different stages of economic development and different methods of calculation, governments seem unable to agree on the monetary price that should be allocated. Indeed the figures suggest that an American's life is worth far more than any other.

The Precautionary Principle

Ascribing monetary values to any externality is an inexact process; the figures produced are no more than estimates, based on value judgements. It is often argued, however, that it is better to have a rough estimate than completely ignore externalities. In fact, this is the thinking that underpins the precautionary principle that most governments have followed since the Earth Summit of 1992. The Rio declaration defined the **precautionary principle** as follows:

> if there are threats of serious or irreversible damage, the lack of scientific certainty should not be used as a reason for postponing cost-effective measures to prevent environmental degradation.

In short, care should be taken today to alleviate the possibility of problems tomorrow. As a consequence, several externalities have been brought into the financial frame. These range from global issues such as atmospheric pollution to local issues such as the loss of an archaeological site. A comprehensive set of examples is listed in Table 11.2. It is based on a seminal piece of research that was

Table 11.2 Monetary value of global ecosystem services

Ecosystem service	Examples of ecosystem function	Monetary value in 1994 prices (trillion dollars per annum)
Coastal	Processing of nutrients Controls pollution Treats waste Provides for tourism	12.6
Oceans	Generates oxygen and absorbs carbon dioxide Provides habitat for fish	8.4
Wetlands	Stores and retains water Storm protection and flood control	4.9
Forest	Regulates greenhouse gases Prevents soil erosion Provides timber and medicines Opportunities for recreation	4.7
Lakes and rivers	Stores and retains water Flood control Opportunities for fishing	1.7
Cropland and grassland	Provides pollination for plants Pollution control Regulates atmospheric conditions	1.0

Source: Adapted from Costanza *et al.* (1997: 254, 256)

published in 1997 and subsequently debated in many journals – in fact, according to Peterson 2010, it is the most cited paper on ecosystem services that has ever been published. The aim of the research (Costanza et al. 1997) was to make the case that precautionary action is justified: it showed that given the relative importance of ecosystem 'services', we can no longer risk squandering them. The authors wanted to highlight the enormous economic contribution of ecosystem services and stress that it made no sense to ignore externalities that damage or threaten the global ecosystem. The research findings suggest that the total economic value of ecological services was nearly twice the annual value of the global output of all goods and services; $33 trillion per year (in 1994 prices) compared to an annual global output of $18 trillion (in 1994 prices).

Clearly the vast economic systems that have evolved over the centuries have degraded the ecosystem, and this is largely due to the ecosystem services being taken for granted. In the terms introduced in Chapter 10, the problem with ecosystem services (such as fertile soils, clean air, a stable climate and a balanced carbon cycle) is that they have the characteristics of a pure public good. No individual can be excluded from the benefits of an improved ecosystem (they are non-excludable) and an individual benefiting from an improved ecosystem is not going to affect other individuals (the services are non-rivalrous). In short, although there is a real value to these services, there are no incentives to pay to maintain them so everyone in effect gets free rides. This creates an imbalance in the way that economic and environmental decisions are made and how the natural environment is used, and the tendency is to focus on short-term material gains at the expense of a declining ecosystem and stock of natural resources.

The work of Costanza highlighted this challenge and began a broader and more accepted focus on placing value on environmental services. This can be seen in the United Nations *Millennium Ecosystem Assessment Project* that ran from 2001 to 2005 and *The Economics of Ecosystems and Biodiversity* (TEEB) programme that was initiated by the European Commission in 2007 and continued by the United Nations Environment Programme (UNEP) from 2011 onwards. What these studies had made abundantly clear is that ecosystem services lie outside of the market and are too often ignored or undervalued.

Interestingly at Rio+20 – in 2012 – the precautionary principle was still high on the agenda. However, the debate had broadened to reflect greater concern over biodiversity and the quality of environmental assets, and a deeper assessment of natural wealth. The European Union's strategy now includes a focus on more efficient use of raw materials and on pollution reduction, and it also demands a greater protection of natural capital assets, such as land, water and biodiversity. Furthermore, it has been accepted that ecosystems need to be restored, residual waste reduced to close to zero and systemic risks to the economy from the environment avoided. A vision of the future can be sensed from a suite of indicators that measure resource productivity by referring to material flows, carbon emissions and other environmental impacts (EU 2011). The United Nations has compiled a wealth index by carrying out a comprehensive analysis of a country's capital assets, including manufactured, human and natural capital, and their corresponding values. The index reflected the state of natural resources, or ecological conditions, from 1990 to 2008 for an initial group of 20 countries, to determine whether national policies are sustainable in the long term (UNEP 2012). Significant gaps in the valuation process remain, but the work of Costanza et al. (1997) has clearly triggered an international movement that will influence national accounting systems for many decades to come. The only problem is that the environmental valuation debate is in some ways seen as a distraction from the real problem of pursuing the goal of continual and sustained economic growth – this will make more sense once you have studied the conventional accounting systems used to measure economic activity outlined in Chapter 13.

Key Points 11.3

○ Placing monetary value on externalities is problematic. There are three basic methods: contingent valuation, travel cost and hedonic pricing.

○ Following the publication of Costanza's (1997) influential paper, a controversial catalogue of indicators have been produced to capture the value of the world's ecosystem services.

○ At Rio+20 it was confirmed that nations need to move beyond measuring economic success purely on the basis of the amount of goods and services their citizens consume each year.

COST-BENEFIT ANALYSIS

When the total costs and benefits of a project cannot be adequately represented by market prices, there is a case to use an alternative method of investment appraisal. Resources that are not allocated through markets, and which therefore do not have a market price, need to be brought into the picture. This is where the technique of **cost-benefit analysis** may be useful.

This method of resource allocation includes externalities as part of the process. It therefore manages to include aspects that most methods of resource allocation omit. In these analyses, external costs and benefits are considered together with internal (private) costs and benefits to provide a fuller picture. Obviously, for this technique to work, all issues need to be expressed in a common denominator for a 'total price' to be calculated. As a result, what may not seem viable in conventional financial accounting terms may appear viable in the broader cost-benefit appraisal.

Cost-benefit analysis is easier in theory than in practice. There are the problems of valuation (as we have discussed above) and there are issues about selecting the externalities. Inevitably, the selection of what is – and what is not – relevant relies on normative decisions based on value judgements. For example, when identifying the external costs of a new underground line in London, the economists conducting the cost-benefit study did not include anything about terrorist attacks on passengers.

Despite the difficulties, economists have carried out many cost-benefit analysis studies to evaluate specific public sector investments. Examples include the construction of the Channel Tunnel, the relocation of Covent Garden and the proposed third runway at London Heathrow airport. In most of these cost-benefit studies, a recurring aspect has been the measurement of time saved expressed in monetary terms. The pricing of an indirect benefit (such as time saved) is as difficult as the pricing of an indirect cost, because market forces fail to determine these prices. In fact, a good way to distinguish between direct (internal) and indirect (external) prices is that the private (direct, internal) cost or benefit will always have a definite monetary value. In practice, some external costs and benefits will be expressed in monetary values in a cost-benefit analysis, and others will not. For example, when appraising new road schemes, government economists usually choose not to put monetary values on traffic noise, visual obstruction, community severance, impact on pedestrians and cyclists, disruption during construction and impacts on climate change. Their excuse is that these costs are uncertain and difficult to pin down.

However, the influential Stern Report (2007) claimed that national governments were committing a serious error in ignoring the future costs of climate change. Nicholas Stern urged them to identify and use a social cost of carbon. The UK government responded by establishing a shadow price for carbon of £25 per tonne. This represents what we are willing to pay now to account for the cost of future damage caused by carbon emissions (Defra 2007: 2).

As Stern (2007: 33) explained, consumption is not just about goods that we might buy from the supermarket, as people also value their health and the environment; and although expressing mortality and environmental quality in monetary terms raises profound difficulties it is essential to avoid global disaster. Slightly further on in the report Stern (2007: 54) adds the dimension of 'time', arguing for the need to assess the flow of costs and benefits over a number of years to effectively compare the welfare and value of future generations with those of today. He strongly supports the case of adopting the convention of expressing calculations in present value terms, using the process of 'discounting' (which is described in the next section). In other words, Stern endorsed the environmental economists approach that the best way to tackle the complex problem of carbon emissions is to ensure that each activity which produces carbon is priced in a way that reflects its true cost to society over time.

The recommendations made by Nicholas Stern were taken on board, to some degree, when the Department for Transport (DfT 2007) conducted a cost-benefit analysis to support its enquiry into the proposal to build a third runway at London Heathrow airport. The DfT's calculations assumed a project with a 70 year life span, and it concluded that the total benefits were an estimated £17.1 billion (at net present values) while the total costs were between £11.9 and £12.7 billion (depending on infrastructure connections). These costs include £4.8 billion to allow for the damage caused by climate change (that is, as much as 40 per cent of the total costs). In short, the total economic benefits of a third runway at London Heathrow would outweigh the total economic costs. To put it bluntly, the project generates a net benefit of between £4.4 and £5.2 billion – or, to put it cynically, around the same amount as the cost allowed for climate change. In interrogating the basis of the calculations (such as the amount of carbon emissions projected and the costs allowed to mitigate their damage), critics have accused the government of producing Alice-in-Wonderland economics (Elkins 2008).

Cost-benefit analysis, therefore, does not provide clear-cut answers to public sector investments. It involves value judgements and (often dubious) estimates which inevitably raise concerns about the reliability of the process. The cost-benefit appraisal offers no more than a framework for government departments to build upon when considering various options. It should be emphasised that the technique is merely a method of *identifying* externalities; it does not automatically control them. However, in the UK, the Government Economic Service strongly endorses the use of cost-benefit analysis on the understanding that the public sector – concerned with wider economic, environmental and welfare issues – needs to consider social costs and benefits when making decisions relating to policy (Price 2010).

DISCOUNTING

Another difficulty associated with the pricing of environmental externalities – and with cost-benefit analysis generally – is that governments have to assess the costs and benefits of investment today against the cost of environmental damage that accrues over a number of years in the future. Private entrepreneurs face a similar type of problem in making decisions about capital investments that will secure a stream of revenue and profits in the future. The process used to assist with these decisions is known as **discounting.** This involves calculating the **present value** of flows of money that are expected to arise at some time in the future.

Table 11.3 Present values of a future pound (sterling)

	Compounded annual interest rate				
Year	3%	5%	8%	10%	20%
1	0.971	0.952	0.926	0.909	0.833
2	0.943	0.907	0.857	0.826	0.694
3	0.915	0.864	0.794	0.751	0.578
4	0.889	0.823	0.735	0.683	0.482
5	0.863	0.784	0.681	0.620	0.402
6	0.838	0.746	0.630	0.564	0.335
7	0.813	0.711	0.583	0.513	0.279
8	0.789	0.677	0.540	0.466	0.233
9	0.766	0.645	0.500	0.424	0.194
10	0.744	0.614	0.463	0.385	0.162
15	0.642	0.481	0.315	0.239	0.0649
20	0.554	0.377	0.215	0.148	0.0261
25	0.478	0.295	0.146	0.0923	0.0105
30	0.412	0.231	0.0994	0.0573	0.00421
40	0.307	0.142	0.0460	0.0221	0.000680
50	0.228	0.087	0.0213	0.00852	0.000109

The main reason for making this calculation is that any capital outlay has an opportunity cost: this may be the cost of borrowing or the returns that can be made on an alternative investment. In either case, interest rates can be used to link the present with the future. For example, if you have to pay £110 at the end of a year when you borrow £100, the 10 per cent interest charged gives you a measure of the cost of borrowing. So, the present value (the value today) of £110 to be received in one year is £100 if the market rate of interest is 10 per cent.

The point is that £1 in the pocket today is worth more than £1 in the future because, if you have the £1 today, you could be putting it to work to earn interest. From this, it follows that £1 receivable in the future has to be discounted to find what it is worth today. For instance, at a 10 per cent rate of interest, £1 receivable in three years is worth about £0.75 today. To express it another way, at 10 per cent a year compounded annually, £0.75 will grow to £1 in three years. It is not necessary to work these figures out for yourself since there are plenty of programs available for computers, as well as compound interest tables such as Table 11.3.

In terms of cost-benefit analysis, or any other investment appraisal covering a number of years, it is possible to use discounting techniques to compare the present value of the costs with the present value of the benefits. The problem with this technique is that the present value of any sum of money is dependent on correctly judging three factors: the timing and extent of future costs (and income) and the interest rate. The further into the future a benefit is accrued (or a cost is incurred) or the higher the interest rate used, the lower will be the present value of that benefit (or cost). To take an extreme example, if a nuclear accident occurs in 500 years time, imposing a £10 billion economic cost on the unlucky future generation, it will be reduced to a present value of 25 pence assuming a discount rate of 5 per cent. Likewise with benefits: a tree that needs many years to mature and costs £100 in current terms would, assuming a discount rate of 5 per cent, have a future value after 50 years of £8.70. Discounting, therefore, provides a system for comparing current costs with future benefits; however, it necessarily downplays the cost of environmental damage for future generations

Key Points 11.4

○ Cost-benefit analysis accounts for the monetary values of all internal and external costs and benefits to identify a total (social) price.

○ Government departments undertaking investment appraisal are encouraged to use cost-benefit analysis.

○ The problems with cost-benefit analysis are: (a) what to include as 'relevant' externalities; (b) how to quantify the externalities in monetary terms; and, (c) how to effectively 'discount' the criteria to judge all values in today's terms.

Part B has emphasised that the construction industry is significantly resource intensive. It has also demonstrated how economists have broadened their analysis to integrate the material world with the natural world. As a consequence, addressing issues relating to the environment and taking opportunities to enhance the image of a firm should be of increasing importance to those involved in modern business.

The following extract reports on the different attitudes to sustainability and corporate social responsibility in the construction sector and beyond. The data is based on a detailed content analysis of the annual reports and mission statements of 300 construction-related companies listed on the New York Stock Exchange. The analysis covers 75 contractors, 75 (engineering and architectural) design firms and 150 client companies from the energy, utility and transport sectors. In each category, the sample was drawn from the leading companies based on size of turnover in 2006. The findings suggest that relatively few construction-related companies at the time positively embraced the sustainability agenda. This research raises a number of interesting problems that might be worth thinking about. As Milton Friedman (1962) emphatically stated, 'there is one and only one responsibility of business – to use its resources and engage in activities designed to increase its profits'. This raises several significant questions. First, why should a business concern itself with social and environmental responsibilities? So long as it engages in free and fair competition, what's the problem? Second, what kind of examples could a researcher look for as evidence of a firm genuinely demonstrating corporate social responsibility? Finally, what kind of measures might a government adopt to promote sustainable development across the built environment sector as a whole?

Timothy Jones, Yongwei Shan and Paul Goodrum (2010) 'An investigation of corporate approaches to sustainability in the US engineering and construction industry', *Construction Management and Economics* 28: 971–83

The global population has exploded over the course of the last century. The Earth's population is expected to exceed 8 billion within the next 20 years (US Census Bureau, 2008). However, the available resources have remained finite, and many natural resources are now consumed at rates far exceeding their replenishment (Woodruff, 2006). The challenge to provide an improved standard of living to an exponentially growing population, with finite available resources, is becoming increasingly difficult. To address this issue, the concept of sustainability has emerged in the past decade.

In 1987, the World Commission on Environment and Development (or Brundtland Commission) issued the Brundtland Report (United Nations, 1987). In this report, sustainability refers to 'development that meets the needs of the present without compromising the ability of future generations to meet their needs' (Beheiry et al., 2006). Since the Brundtland Report, the concept of sustainability has been further developed and is now commonly thought to consist of three aspects, or pillars: the environmental, economic and social pillars (Beheiry et al., 2006). Sustainable development is now classically portrayed as the interface between environmental, economic and social sustainability (Goodland and Daly, 1993).

The environmental aspect of sustainability involves reducing ecological impacts today in order to preserve the environment for future generations. The economic aspect refers to providing for positive economic growth. The social aspect is concerned with the responsibility of corporations to conduct business ethically, especially while pursuing developmental opportunities abroad. The overlap of all three key aspects, social, economic and ecological, is referred to as sustainable development. Other researchers have suggested that sustainability efforts have appeared to focus more on the environment in comparison to the other two aspects, but this aspect of sustainability has not been examined in great detail and is therefore one of the primary motivations for the research described herein.

This paper examines construction-related companies' perception on sustainability. Companies' perception with regard to sustainability is examined through analyses of their company documents, such as mission, vision and values statements and annual reports. Mission statements provide a guide to behaviours and decisions (Falsey, 1989; Ledford et al., 1995), develop a unity of purpose within the organization (Campbell et al., 1990), motivate staff (Collins and Porras, 1991) and enhance performance (Pearce and David, 1987). The future is the domain of the vision statement. A vision statement describes the desired future state of the organization within the arena of competition defined in the mission (Raynor, 1998). Values statements depict the principles or concepts of intrinsic worth with which to align one's action as an end in itself (Raynor, 1998).

Hopwood (1996) suggested that annual reports have become a highly sophisticated product of the corporate design environment, the main purpose of which is to proactively construct a particular visibility and meaning rather than to reveal 'what was there'. To some extent, these documents reflect the practices and principles about how a company conducts its business.

Research objectives

This paper presents a study of the current engineering and construction industry's strategic approach towards sustainable construction. Three broad groups of companies were included in the study to be representative of the industry: client, contractor and design firms. Typically, contractors perform construction-related activities; thus, contractors and construction firms are used interchangeably throughout the paper. This study was undertaken to achieve three primary objectives:

(1) Identify which sustainability related concepts are most frequently used by companies to describe their sustainability programmes. This objective includes understanding whether or not fundamental concepts of sustainability are being evenly emphasized.

(2) Determine whether any disconnects currently exist between company groups, for instance, whether or not there are significant differences in the focus on sustainability between contractors, designers and their potential clients.

(3) Determine whether any disconnects in emphasizing the concepts of sustainability currently exist between commercial and industrial subgroups for owners, contractors and design firms...

Discussion of results

The results...tend to support the hypothesis that client companies and contractor companies use the concepts of sustainability in different ways. From the data, it can be seen that 20.0% of client companies published a dedicated Corporate Sustainability Report for public viewing, while only 2.7% of contractor and 4.0% of the design firms like-wise published such a report. Clients, more than construction and design firms, may attempt to use sustainability efforts to gain a positive public image for public relations purposes. This is reasonable, since most companies typically market a product to the general public, while construction and design firms typically do not. Client companies more often create a report focused on sustainability efforts and then display this report prominently to the public on their websites. Since green concepts and environmentalism are very popular in the public forum, publishing a sustainability report for public viewing may be a way of improving the company's public image; therefore, they are well motivated to ensure the most favourable public opinion of their companies. Publishing a Corporate Sustainability Report may also be viewed favourably by government agencies. Both construction and design firms, on the other hand, rarely market a product to the public. They are focused on marketing services to various owner entities (owner companies, government agencies, etc.). They were less likely to offer a Corporate Sustainability Report, choosing instead to highlight safety and quality achievements on their websites and within their documents...

References

Beheiry, S.M.A., Chong, W.K. and Haas, C.T. (2006) Examining the business impact of owner commitment to sustainability. *Journal of Construction Engineering and Management,* 132(4), 384–92.

Campbell, A., Devine, M. and Young, D. (1990) *A Sense of Mission,* Hutchinson: London.

Collins, J. and Porras, J. (1991) Organizational vision and visionary organization. *California Management Review,* 34, 30–52.

Falsey, T. (1989) *Corporate Philosophies and Mission Statements,* Quorum Books: New York

Goodland, R. and Daly, H. (1993) *Poverty alleviation is essential for environmental sustainability,* Divisional Working Paper 1993-42, Environmental Economics and Pollution Division, World Bank, Washington, DC.

Hopwood, A.G. (1996) Introduction. *Accounting, Organization and Society,* 21(1), 55–6.

Ledford, J., Wendenhof, J. and Strahley, J. (1995) Realizing a corporate philosophy. *Organizational Dynamics,* 23, 5–19

Pearce, J. and David, F. (1987) Corporate mission statements: the bottom line. *Academy of Management Executive,* 1(2), 109–16.

Raynor, M. (1998) That vision thing: do we need it? *Long Range Planning,* 31(3), 368–76

United Nations (1987) *Development and International Economic Cooperation: Environment Report of the World Commission on Environment and Development: 'Our Common Future',* Official Records of the General Assembly, 42nd session, Supplement No. 25, Brussels, Belgium.

US Census Bureau (2008) North American Industry Classification System 2007 Version, available at www.census.gov/epcd/www/naics.html (accessed 4 August 2008).

Woodruff, P.H. (2006) Educating engineers to create a sustainable future. *Journal of Environmental Engineering,* 132(4), 434–44.

Extract information: From pages 971–2 plus relevant references from page 983.

Part C

Economic Growth that Meets the Needs of Everyone

WEB REVIEWS: Economic Growth that Meets the Needs of Everyone

On working through Part C, the following websites should prove useful.

www.hm-treasury.gov.uk/

This was one of the first government websites, and it is still one of the best. The site opens with opportunities to search for information on Budgets, the economy (including forecasts and the latest indicators) and the euro. There is a comprehensive index, which offers an alphabetical listing of around 100 subjects.

www.bankofengland.co.uk

The role of the Bank of England's monetary policy committee and the significance of a central bank is discussed in Chapter 12. For both of these topics the Bank of England website is a useful reference point. The home page has a comprehensive index and its own search engine. You can easily find the latest monetary policy committee minutes, up-to-date monetary and financial statistics, and summaries of the Bank's most recent Quarterly Bulletin and Inflation Report – both of which can be downloaded in full (in pdf format). There is also a link to some museum pages that provide a historical description of the UK's role as a financial centre.

www.worldbank.org

The World Bank is not a bank in the common sense of the word. It is owned by 185 member countries with the primary objective of providing financial and technical assistance to developing countries around the world. As a consequence, it collates a broad range of development indicators ranging from GDP to levels of literacy for low-income as well as high-income countries. This is a good website to find comparative macroeconomic data that is updated on an annual cycle and is supported by commentary and methodological notes. For example, the statistics in Table 13.1 could be updated from this site.

www.nao.org.uk

The National Audit Office investigates the services provided by government funding to assure that taxpayers are getting value for money. It produces around 60 reports per year, and makes electronic versions freely available for download from this website. Many are relevant to this text as they review specific public sector projects as well as exploring broad themes such as PFI and modernising construction. Some examples are listed in the references and the website can be searched for more resources by title or text subject.

www.euroconstruct.org

This website specialises in analysing and forecasting construction markets across Europe. It provides up-to-date information on 19 countries in western and central Europe. The data is provided by research organisations based within each country and there are links to equivalent organisations in the Baltic States and Japan. It is a good source to access comparative data on GDP and construction as well as specific country reports.

Managing the Macroeconomy

The management of macroeconomic issues appears to dominate the pages of the national press. In the 1930s economists and politicians of the industrial world wrestled with the **great depression**, which led to millions being made unemployed. Forty years later, during the 1970s, the big problem was **stagflation**, a combination of stagnation (slow or zero growth) and inflation (rising prices). Now in 2012, another forty years later, the macroeconomic issue is the crisis in the global financial system, resulting from the **credit crunch** and sovereign debt problems. This has disturbed the long-standing equilibrium in the financial markets and caused rising inflation and a deep recession. Against this background, governments strive to deliver economic stability or, as it is sometimes expressed by politicians in the UK, to achieve high levels of growth that meets the needs of everyone.

In this chapter, we review the policies that underpin this vision of economic growth that meets the needs of everyone. We shall also consider economic forecasting. All this has direct relevance for anyone who needs to understand, manage and/or plan construction capacity over a medium-to-long time period, as the demand for construction products is always derived from activity in other sectors. Inevitably, then, construction economists need the ability and confidence to interpret economic statistics relating to the wider economy. Economic statistics are considered throughout Chapters 12 and 13, so the chapters should be read as a related pair.

FIVE MACROECONOMIC OBJECTIVES

All governments, regardless of their political persuasion, seek to achieve economic goals. In economics, these goals are referred to as **macroeconomic objectives**. There seems to be some political and economic consensus about the five dominant macroeconomics objectives: price stability, full employment, a sustained rate of economic growth, a positive trade balance with overseas partners and effective protection of the environment. Each of these objectives is considered in turn below. Recent macroeconomic statistics for the UK economy, which show the extent to which these objectives are being achieved, are presented in Table 12.1 (see page 204).

Stable Prices

Stable prices are crucial for business confidence, facilitating contracts and enabling the exchange rate system to function smoothly; in contrast, persistently rising prices cause problems for most sectors of an economy. Price stability has become the primary objective of most governments that wish to secure long-term growth and full employment. Economists no longer believe that tolerating higher rates of inflation can lead to higher employment or output over the long term. In the UK, the government's target is to keep inflation within a range of 1–3 per cent and the

consumer price index (CPI) is monitored on a monthly basis. A sample of annual CPI statistics for the UK economy is presented in Table 12.1. As the table shows, up until recently, the *general* trend has been around 2 per cent, and this is regarded as encouraging, not as an end in itself but due to its significance in meeting all other government objectives. Price stability is so central to understanding modern macroeconomic management that we present a full explanation of its measurement and its impacts upon business expectations in Chapter 14.

Table 12.1 UK macroeconomic statistics

There are various sources for this data, we have used the Office for National Statistics website (see the ONS web review on page 30). Three points need to be emphasised, economic data is subject to revision, it is calculated in different ways by individual nations, and proper comprehension relies on footnotes.

	1995	2000	2005	2007	2011
Inflation [1]	2.6	0.8	2.1	2.3	4.5
Unemployment [2]	2.3	1.0	0.9	0.9	1.5
Economic growth [3]	2.8	3.1	2.7	2.5	0.7
Balance of payments [4]	−8.5	−24.8	−31.0	−57.8	−27.8

Notes: 1 Consumer prices (percentage increase on previous year)
2 Claimant unemployment (annual average, in millions)
3 Annual percentage increase in real GDP
4 Current account (total for whole year, £ billions)

Full Employment

Full employment does not mean that everybody of working age is employed, as in any dynamic economy some unemployment is unavoidable. For example, there will always be individuals moving in and out of employment, as they change from one job to another, and there will always be seasonal, technological and overseas factors that cause fluctuations in the jobs available in different sectors. Problems arise, however, when there are large numbers of unemployed for long periods of time; as a large pool of unemployed labour represents wasted resources. It also has many costs, not just in terms of loss of output but also in terms of human suffering and loss of dignity. All governments record the number of workers without a job, although the precise way this is measured changes from time to time. At present 'official' unemployment in the United Kingdom is estimated by the number of people registering for Jobseeker's Allowance – known as **claimant unemployment**. Unemployment is either expressed as a percentage rate – the number of claimants as a percentage of the total workforce of 31 million – or as an absolute number. The unemployment rate has been below 10 per cent for over a decade, with unemployment reaching a high of approximately 2.3 million in the mid 1990s.

Sustained Economic Growth

A long-term objective of all governments is to achieve steady increases in productive capacity. Governments measure **economic growth** by the annual change in the rate of output, and the commonly used measure of economic output is GDP – **gross** (total) **domestic** (home) **product** (output). GDP figures are used worldwide as a proxy for a country's progress towards prosperity: since the more money a country makes, the higher its GDP growth, the assumption is that increases in GDP mean that the citizens of that country are enjoying a higher standard of living. The way GDP measures output can be seen as a giant till ringing up all the transactions taking place inside a country. To accurately portray the rate of change of actual output, GDP must be corrected for price changes from one time period to another. When this is done, we get what is called 'real' GDP. So a more formal measure of economic growth can be defined as the rate of change in real GDP over time (usually one year). As the footnotes indicate, the growth data in Table 12.1 has been corrected accordingly. It is, therefore, a clear indicator of **boom** or **recession**. A fuller coverage of GDP and how it is calculated is given in Chapter 13.

External Balance

All international economic transactions are recorded in a country's **balance of payments** statistics. The ideal situation represents a position in which, over a number of years, a nation spends and invests abroad no more than other nations spend or invest in it. Economic transactions with other nations can occur on many levels and, for accounting purposes, these transactions are often grouped into three categories: current account, capital account and financial account. Of these three, the most widely quoted is the current account. This involves all transactions relating to the exchange of visible goods (such as manufactured items, which would include building materials), the exchange of invisible services (such as overseas work undertaken by consultants) and investment earnings (such as profits from abroad). Clearly, in any one year, one nation's balance of payments deficit is another nation's balance of payments surplus – ultimately, however, this is not sustainable and, in the long run, debts must be paid. The data in Table 12.1 show a worrying trend, in so far as the UK current account figures are all negative amounts. However, in addition to buying and selling goods and services in the world market, it is also possible to buy and sell financial assets and these are recorded separately in the financial account. The UK's annual position on its financial account is usually in balance. A further qualifying remark regarding foreign trade is to recognise that balances of payment figures are notoriously difficult to record accurately. (In fact, of all the statistics shown in Table 12.1, the balance of payments estimates are subject, by far, to the biggest amendments.) In practice, therefore, statistics relating to the external balance need to be considered in a broader historical context.

Environmental Protection

The environment has gained a high political profile in recent years, and there is a recognition that a healthy economy depends upon a healthy resource base. We have already reiterated several times that the protection of the environment, and indeed its enhancement, forms an important strand of the sustainable development agenda. We have also examined (particularly in Part B) the problem that – left to their own devices – markets cannot deal effectively with environmental impacts. At present, there is a broad-ranging debate arising from concerns that some economic activity damages the environment, and politicians are beginning to consider environmental protection together with other macroeconomic goals. As a result, governments are increasingly choosing to intervene to influence resource allocation, preserve biodiversity and reduce pollution. However, there is no agreed way of measuring the impact of environmental protection initiatives and so there is no indicator for this objective in Table 12.1. We have already discussed the problems of integrating environmental costs into the marketplace and we will briefly assess the implications again when considering the measurement of national output in Chapter 13 and when introducing the possibility of measuring a nation's ecological footprint.

PRIORITIES: A HISTORICAL PERSPECTIVE

The order of priority that these five macroeconomic objectives are given varies from government to government. But all governments, in all nations, ultimately seek these objectives in their quest for macroeconomic stability. Indeed, since the end of the Second World War, there has been a consensus that governments should take action to stabilise economic activity. For example, the White Paper on Employment published in May 1944 stated that the government accepts responsibility for the maintenance of high and stable levels of growth and employment, and these themes dominated government agendas in the twentieth century.

Since the 1980s, however, the order of priority has changed and governments have become more concerned with curbing inflation. Budget statements made during the 1990s typically emphasised that price stability was considered a precondition to secure high and stable levels of growth and employment. As the political and economic scene evolves, it is foreseeable that at some time in the not too distant future there will be another shift in emphasis, as the objective of protecting the environment becomes sufficiently important to raise its profile and demote inflation, employment, trade and growth from their current positions as governments' top four macroeconomic objectives. In other words, while there may be some doubts at present about the priority to attach to environmental issues, the interest expressed in climate change and sustainable development is gathering momentum throughout the world and a new order is imminent.

Key Points 12.1

- ○ To achieve economic stability, five main macroeconomic objectives are pursued: (a) full employment, (b) stable prices, (c) external balance, (d) steady growth and (e) environmental protection.

- ○ The order of priority accorded to these macroeconomic objectives varies from government to government.

GOVERNMENT POLICY INSTRUMENTS

In their attempts to achieve their macroeconomic objectives, all governments, regardless of political persuasion, employ the same types of policy instrument. Again, it is only the emphasis that seems to change. These instruments can be grouped into three broad policy categories:

- fiscal policy
- monetary policy
- direct policy.

Fiscal Policy

In the UK, fiscal policy emanates, on the government's behalf, from HM Treasury. **Fiscal policy** consists largely of taxation (of all forms) and government spending (of all forms). The word fiscal is derived from the Latin for 'state purse' – and this is most appropriate as taxation is the main source of income from which governments finance public spending. In short, fiscal policy is concerned with the flow of government money in and out of the exchequer.

Important elements of the current fiscal framework are to make sure that both sides of the government balance sheet are managed efficiently. Any public sector debt must be held at a prudent and stable level in relation to GDP, and borrowing is only acceptable to cover capital expenditure. In technical terms, these two Treasury rules are:

- the **golden rule**, which states that over the economic cycle the government can only borrow to invest and not to fund current spending
- the **sustainable investment rule**, which states that over the economic cycle public sector debt expressed as a proportion of GDP must be held at a stable and prudent level.

Spending that provides goods and services that are to be consumed in the same year as purchase (such as expenditure on drugs in the NHS) is classed as current spending. Spending that produces a stream of goods and services for use over several years (such as funding a new hospital) is classed as capital spending and can be funded within reasonable parameters by public sector borrowing.

The two Treasury rules provide a benchmark against which the government can judge its fiscal performance and establish a stable framework for the broader economy. They provide a framework by which the UK seeks to manage its fiscal affairs in ways that are transparent, stable, fair and efficient. Excessive government borrowing creates instability and needs to be kept to an absolute minimum. The problem is that much of current public sector spending is demand determined (for example, the expenditure on drugs in the NHS is determined by the levels of illness that are diagnosed), and much public sector capital spending is politically determined in the sense that it is committed years in advance (for example, the hospitals built under PFI schemes incur annual payments that were determined many years ago). So meeting the government's objective to balance the current public sector accounts is challenging.

Monetary Policy

Monetary policy is implemented in most countries by a **central bank**, such as the Bundesbank in Germany, the Federal Reserve in the USA and the Bank of England. In the UK, prior to 1997, monetary policy was set by the government – in other words, the Bank of England simply followed government instructions. In May 1997, the government established a new monetary policy framework, transferring operational responsibility to an independent **monetary policy committee** (MPC). The committee is responsible for monetary and financial stability. In this role it sets interest rates each month and provides support for economic growth via a system of quantitative easing. The overall measure of the MPC's effectiveness is judged by its ability to maintain the government's overall inflation target. At present, the target is for a 2 per cent increase in the annual consumer price index. (A fuller account of the work of the MPC and the measurement of inflation is explained in Chapter 14.)

The monetary policy committee consists of ten experts drawn from outside and inside government circles. At the monthly meetings, the panel of experts carry out in-depth analysis of a wide-ranging set of data. Published monetary policy committee minutes (extracted from the Bank of England's monthly *Inflation Report*) suggest that the analysis includes the general state of the world economy, trends in domestic demand, the labour market, the housing market and the financial markets, and last, but by no means least, various measures of inflation and costs in specific sectors of the economy. The committee works from the premise that interest rates represent a general cost of activity and, therefore, after allowing a period of time to transmit through the economy, interest rates ultimately control the level of prices, funding liquidity, and aggregate demand.

To enable the Bank of England to concentrate on issues relating to monetary stability, responsibility for supervising individual financial institutions was handed over to a newly created Financial Services Authority in 1997. This separating of functions survived a decade of economic stability (from 1998 to 2007), however serious problems in the banking sector caused by the credit crunch led to global calls for a greater regulation of the financial system. In particular, high street banks needed to be safeguarded from the riskier activities of their investment arms. As a consequence, regulators around the world have strengthened their rules and raised the level of capital required by banks to back up their assets. The UK plans legislation to separate the high street (retail) and investment (casino) functions of a bank's activity and these requirements are expected to be fully implemented by 2019 (see Vickers (2011) and Chapter 14 for more details).

CO-ORDINATION OF FISCAL AND MONETARY POLICY

An important point to note at this juncture is that fiscal and monetary policies are equally important in managing the macroeconomy. A change to either policy has broad effects on many of the core macroeconomic objectives. Consequently, all governments employ both fiscal and monetary instruments, although the emphasis alters from government to government. Until 1997, the Chancellor of the Exchequer directed the operation of both UK fiscal and monetary policy. Although this theoretically meant that there could be a high degree of co-ordination between both arms of macro policy, in practice this was often not the case. The 1997–2010 Labour government took pride in the levels of stability that followed the introduction of the new monetary regime in 1997. The coalition government that came to power following the May 2010 election set up the **Office for Budget Responsibility** (OBR) to make independent assessments of public finances. These two changes have resulted in a greater clarity of roles and responsibilities between the Bank of England and the Treasury.

Direct Policy

Many other government economic policies tend to be more 'objective specific' compared with the broad macro fiscal and monetary policy options we have considered so far. We refer to these instruments as **direct policy**, but it is also known as direct control or direct intervention. A feature of this type of policy is that it tends to have less impact on overall market prices than the broad macro changes to tax or interest rates.

Direct policy tends to be of a legislative nature. Conventional economic textbook examples include legislation designed to control prices, wages or imports to assist with the stabilisation of prices and trade; legislation to support research and development, education and training to influence long-run growth; and general support to encourage small businesses. Good examples of direct policy within the area of construction economics include building regulations and codes established to raise the quality and performance of finished products, and the Strategy for Sustainable Construction (HM Government 2008) introduced to change cultural attitudes towards productivity, safety and the environment. These initiatives are aimed at stimulating growth, stability and environmental performance within the construction sector.

MACROECONOMIC OBJECTIVES AND POLICY

Effective macroeconomic management is not an easy task. The basic objectives and policies are summarised in Figure 12.1 (see page 210). A finely tuned macroeconomy is elusive. There are trade-offs to be made between one objective and another, and one policy and another. As a consequence, there are no magic formulas or miracle cures that economists can agree on. There is instead vigorous political debate that highlights the difficulties that lie behind the management of the macroeconomy.

Let us briefly consider one example of macroeconomic instability. The financial crisis that troubled economies across the world from 2008 to 2012 was triggered by instability in the US financial markets in mid 2007. However, the crisis soon spread across the world, with serious implications for the UK and beyond. The international financial system came close to collapse in the autumn of 2008, following the failure of Lehman Brothers, and the subsequent panic across global markets prompted unprecedented action by central banks. For example, the Bank of England cut the official rates of interest several times and supplied billions of extra funds as it tried to stabilise the financial system.

Despite these monetary policy initiatives, banks and building societies (the lenders) raised the cost of borrowing and restricted the supply of loans in their

Figure 12.1 Government objectives and government policy

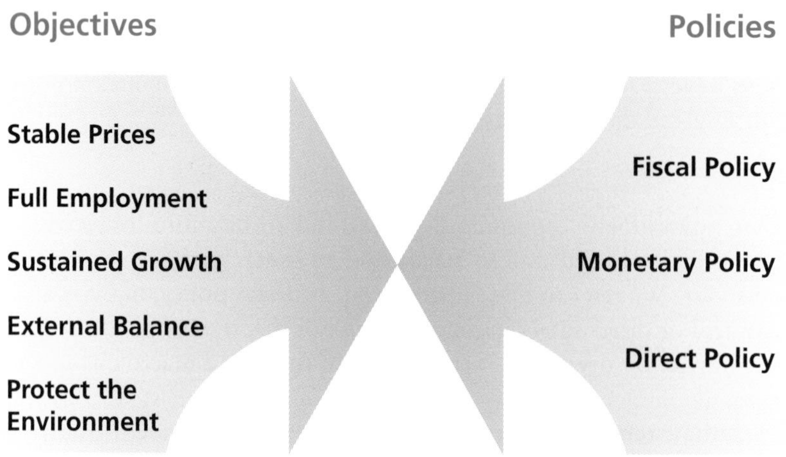

Objectives   Policies

Stable Prices

                                                      Fiscal Policy

Full Employment

Sustained Growth   **Monetary Policy**

External Balance

                                                Direct Policy

**Protect the
Environment**

efforts to remain solvent. In other words, interest rates available to borrowers increased across the financial sector (despite all efforts by the central banks to lower them). This action was unprecedented as the usual practice (outlined in Chapter 14) was for changes in interest rates on mortgages and loans offered by retail banks to move in exactly the same direction as changes made to the official rate – and usually on the same day. Post 2008, however, institutions in the financial market were less quick to change their rates and less inclined to mirror changes in the official rate. They began to revise their margins and increase the costs of borrowing to move to a new equilibrium.

As a result, consumer spending and business investment was cut back – and this, in turn, reduced demand for goods and services. This cut into business profits leading to job losses. Increases in unemployment typically put a strain on the fiscal stance as the unemployed no longer pay income tax (reducing government revenues) and receive benefits from the state (increasing government spending). Lower business profits also reduce government revenues received through corporation tax. The **public sector net borrowing** – the government's annual deficit – consequently increased. In turn, this provoked a debate about the need to cut back government spending in other areas, leading to further unemployment. As output falls, then obviously economic growth slows down. Yet, as economic growth declines, there is the possibility of relatively less environmental damage.

This period of economic instability emphasises the complex nature of the macro economy, the incompatibility of some government objectives and policies, and the global nature of macroeconomic interaction. A further point of complication is the fact that few policies have an immediate effect. Monetary decisions are taken monthly and fiscal decisions are effectively only made annually in the government's Budget, and the policies that are executed can take years to play out.

Key Points 12.2

○ All governments utilise a combination of fiscal policy, monetary policy and direct policy.

○ Fiscal policy is concerned with government expenditure and taxation.

○ Monetary policy is concerned with achieving monetary and financial stability by manipulating interest rates and quantitative easing.

○ Effective management of the macro economy is difficult due to the incompatible nature of the policy objectives and time lags.

MACROECONOMIC MANAGEMENT

Most economies aim to increase their rate of output each year. This has been achieved in the United Kingdom – on average, during the last 60 years, there has been a 2.6 per cent annual increase in economic activity. Unfortunately, however, long-term growth is not achieved at a steady rate, and there are always periodic fluctuations above and below the general upward trend – this concept is portrayed in Figure 12.2. These fluctuations are related to activity in the broader economy.

At times, the overall business climate is buoyant: few workers are unemployed, productivity is increasing and not many firms are going bust. At other times, however, business is not so good: there are many unemployed workers, cutbacks in production are occurring and a significant number of firms are in receivership. These ups and downs in economy-wide activity used to be called 'business cycles', but the term no longer seems appropriate because *cycles* implies predetermined or automatic recurrence and, today, we are not experiencing automatic recurrent cycles. The subtle changes that characterise the general nature of economic activity in the twenty-first century are now more usually referred to as **business fluctuations**.

In fact, empirical evidence highlights how the nature of contractions and expansions has changed. A review of data from 28 advanced economies covering the past century shows that the recession phase has tended to shorten in duration, and rarely exceed 15 months. Periods of expansion have increased, and in some instances in the newly industrialised Asian economies sustained growth has been achieved for more than 20 years. In recent years, periods of expansion have become the norm while recessions have been relatively rare (International Monetary Fund 2007: Chapter 5). Neither the United States nor the United Kingdom experienced a recession (that is, two consecutive quarters of negative growth) during the 12 year period from 1995 to 2007; instead, they had more than a decade of uninterrupted growth. Some economists and journalists began the twenty-first century celebrating the 'end of boom and bust' economics, yet ten years later, as the years from 2008 to 2012 witnessed an atypical downturn, commentators spoke of depressions, double-dip recessions and 'crisis'.

When there are successive real declines in GDP economists and politicians alike are fazed. These declines in output are usually classified as a short period of

Figure 12.2 Business fluctuations

The solid black line represents the long-term growth trend around which economic activity fluctuates.

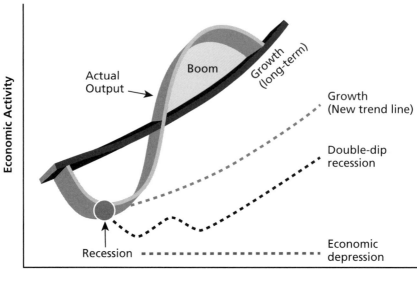

recession. However, when the sub-trend growth continues for several consecutive quarters there becomes a worrying line between a recession and a depression. For example, in April 2012 the UK economy still remained more than 4 percentage points below its 2008 peak. This flat period of inactivity could well be categorised as a depression. When an economy emerges from a recession with a short period of growth but quickly falls back into recession, there is talk of a 'W-shaped' or 'double dip' recession (as shown in Figure 12.2). This occurred in the UK when the economy experienced a trough between April 2008 and July 2009 and subsequently went through a further period of negative growth from October 2011 to March 2012.

The language, terms and character of macroeconomic fluctuations are constantly under review. It is interesting to note that the National Bureau of Economic Research in the United States has formally moved to a position where a recession is any period of diminishing economic activity. It identifies a month when the economy reaches a peak of activity and a later month when the economy reaches a trough – the time in between is a recession. For example, in the late 2000s recession, NBER identified December 2007 as the peak of the previous economic cycle in the US economy and June 2009 as the trough when the recessionary cycle was formally declared complete (NBER 2010). In terms of Figure 12.2 a new upward trend had commenced, all be it far below the longer term growth path. In NBER's terms, the business cycle is characterised by peaks and troughs in economic activity. The period between a peak and a trough is described as a contraction or a recession, and the period between the trough and the peak as an expansion. NBER does not separately define depressions nor does it define a double-dip recession. Two periods of contraction will be either two separate recessions or parts of the same recession. Time will tell whether the NBER's formulation will become more widely accepted.

Interpreting Business Fluctuations

There is a collection of economic statistics that can indicate where we are, where we've been and, most importantly, where we seem to be going. Their origin date back to the 1960s and they are still referred to as **cyclical indicators**. They enable governments to predict changes that are happening in an economy. These predictions are based on a composite set of statistics that are regarded as running ahead of the general economic trend. This is because things do not happen simultaneously – some indicators may point in an upward direction while others portray a downward trend especially at the 'peaks' and 'troughs' (the turning point) of a cycle.

Statistics that are assumed to precede the general trend of the economy by changing six-to-twelve months ahead of the main trend are referred to as **leading indicators**. This group is broken down into two subgroups: a longer leading index (which looks for turning points about one year ahead) and a shorter leading index (which indicates turning points approximately six months ahead). Examples of leading indicators are housing starts, new car sales, business optimism and the amount of consumer credit. **Lagging indicators**, by contrast, alter in retrospect, usually about one year after a change in the economic cycle – they confirm what we already know and, in forecasting terms, are not so important. Examples of lagging indicators include unemployment, investment in building, plant and machinery, levels of stock, and orders for engineering output. Economic statistics that are thought to trace the actual cycle are called **coincident indicators** and the obvious example is GDP figures.

Economic Forecasting

The interpretation of economic events is a complex process, especially as macroeconomic policy instruments can affect several variables at once. To take the simplest example, some cynical forecasters are quick to suggest that the sight of an increasing number of cranes on the skyline, visible from their office window, means that we are about to witness the start of the next recession. But this would be a ridiculous suggestion as there are many other variables that need to be used as a basis for forecasting. Indeed the art of forecasting involves completing a picture using as much existing data as possible and combining this analysis with anecdotal evidence to arrive at an overall view. To avoid too many subjective judgements, models tend to be computer based, using mathematical equations to link a number of economic variables. For our purposes, these variables can be categorised into two main types. **Exogenous variables** are external to the economy in so far as they are determined by world events and policy (examples include oil prices and exchange rates). **Endogenous variables** are dependent on what goes on within an economy (examples include employment and inflation). There are more than 120 exogenous variables, and hundreds of endogenous ones – and the larger traditional models of the macroeconomy contain upwards of a thousand relationships. The present trend, however, is for models to be smaller and rarely exceed twenty core equations.

Treasury forecasts are informed by 36 independent models managed by institutions in the city and beyond. Table 12.2 shows the May 2012 forecast for the UK economy up until 2016. The table summarises the averages, it does not show the range of estimates making up the forecast. For example, the highest and lowest forecasts for the percentage change of GDP in 2013 were 3.0 per cent and 0.5 per cent respectively; the respective figures for CPI were 3.2 per cent and 1.5 per cent.

Table 12.2 Economy forecasts 2012 to 2016

Average projections	2012	2013	2014	2015	2016
GDP growth (%)	0.3	1.8	2.2	2.5	2.4
CPI (%)	2.8	2.1	2.3	2.3	2.4
RPI (%)	3.2	2.6	3.4	3.7	3.9
Unemployment (mn)	1.69	1.77	1.71	1.65	1.57
Current account (£bn)	-28.4	-24.4	-22.4	-18.4	-17.7

Source: Adapted from HM Treasury 2012

Understandably, since it is difficult to predict accurately the behaviour of millions of consumers and businesses to the last detail, economic forecasts are often wrong. Furthermore, forecasts are limited, since they rely on assumptions about policies that may need to change owing to sudden events or revised statistics. There are also problems relating to time lags, since it often takes years for a specific monetary or fiscal instrument to fully work through an economic system.

The important point, however, is the message conveyed by the forecast; the trend does not have to be 100 per cent accurate. Forecasting models are no different from any other economic model in that they attempt to simplify reality. In the case of the economy, this is a complex reality and a forecasting model only identifies, measures and monitors the key variables. Understanding half of the picture, however, is better than not seeing any of it at all.

Managing the Construction Industry

As you might expect, fluctuations in construction output share a similar pattern to the broader economy and they are referred to as **building cycles**. Interestingly, economists have a long history of studying building cycles, though not because they are particularly interested in the construction industry but because there is a strong possibility that it may contribute to a better understanding of business fluctuations. The symmetry between business fluctuations and building cycles is, however, complicated by the fact that economic development is usually associated with a shift from investment in new construction to spending on repair and maintenance. In other words, as GDP increases the proportion of new construction work decreases.

Bon and Crosthwaite (2000) confirmed this pattern. Their international review also demonstrated that there is an inevitable decline in the share of construction in GDP as economies mature. In fact, it has been observed that newly developing countries experience up to double the rate of expenditure on construction as their more developed counterparts. We shall explore this further in the Chapter 13.

Another notable comparison between the construction sector and the rest of the economy is that building cycles tend to display far greater amplitude than the equivalent cycles in general business activity. In other words, periods of decline are far more severe in construction than in the general economy, and conversely periods of expansion are more buoyant. For example, when the UK entered recession during the first quarter of 2012, a relatively small 0.3 per cent decline in GDP was accompanied by a significant 4.8 per cent decrease in construction output (percentage changes based on previous quarter). There is, therefore, a unique and distinctive situation in construction – the long-term trend is for growth in construction output to slow as economies mature, yet the variation from year to year can be quite volatile.

This line of analysis is not meant to portray a depressed industry, as construction is, without doubt, a permanent and important sector of any economy. Indeed, in Chapter 13, we will argue that it is possibly the most important sector. It is certainly not like other sectors that may expand and contract and then disappear. The construction sector always has work – to maintain existing stock and to replace demolished stock and construct new stock – even if it may be at a declining rate.

From the perspective of managing the economy, the construction industry is important and most developed economies have a government department or commission to co-ordinate activities in the sector. In the United Kingdom, there are many government interactions with the industry. The departments concerned with roads, housing, health, education, energy, defence and the environment procure products from the industry – in fact, in one way or another, as much as 40 per cent of construction activity, is derived directly or indirectly, from the public sector. There is also a Construction Sector Unit (CSU) based in the Department for Business Innovation and Skills (BIS). This is headed by a senior civil servant (the government chief construction adviser) reporting to a politician (the Minister for Business). The main concerns of the CSU are to deliver a marked improvement in business performance in construction, in terms of productivity, profitability and competitiveness. The day-to-day functions of the unit are summarised in Table 12.3. In effect, this means that the industry has representatives at government level to speak on its behalf and the government has a mechanism to deliver its messages from the top down. Current government initiatives relate to encouraging the adoption of **building information modelling** (and this is covered further in Chapter 15), cutting the costs of public procurement, promoting innovation across the sector, and increasing the uptake of green construction (Cabinet Office 2011).

The Green Construction Board (GCB) provides one example of how these various issues are processed. Seven working groups consisting of more than 150 representatives drawn from government and industry meet regularly to try and develop a long-term strategic framework. This group is gathering evidence to create a 'knowledge database' of low carbon research and development to support the work of the Sustainable Construction Task Force. Together these groups intend to create a route map to set out the industry's direction to 2050. The only problem then will be implementing the line of travel. In short, the GCB is a consultative forum that provides an opportunity for government and the broadly defined construction industry to promote innovation and sustainable growth. Formally launched in November 2011, it is too early to judge its impact, or what policies it might adopt to enforce its intentions (BIS 2012).

Table 12.3 Functions of the Construction Sector Unit

✓ Promoting R&D to increase innovation and the adoption of new practices

✓ Improving the construction process, technologies and techniques

✓ Tackling people issues, such as recruitment, health & safety, training & diversity

✓ Improving payment practices in the construction industry

✓ Improving awareness of the benefits of information technology

✓ Leading on sustainability in construction

✓ Promoting overseas activities by the construction industry

✓ Engaging the industry in regulation and policy development

Source: Adapted from BIS website (2012)

Key Points 12.3

○ Most modern economies achieve long-term growth, but the pattern is not steady as there are characteristic periods of fluctuation above and below the general upward trend (see Figure 12.2).

○ Leading (cyclical) indicators are of particular significance in macroeconomic forecasting as they change six-to-twelve months ahead of the main business trend.

○ Models used for forecasting and policy evaluation are often based on computer programs linking endogenous and exogenous variables.

○ The symmetry between business fluctuations and building cycles is complicated. Construction output declines as a percentage of GDP as economies mature, yet the variation from one year to the next can be quite volatile.

○ Most developed economies have a government department or commission to co-ordinate activities in the construction sector. The Construction Sector Unit carries out this function in the UK

13 The Economy and Construction: Measurement and Manipulation

The construction industry is an important focus of government policy. This is largely due to the recognition of the importance of construction to national economies. Broadly defined, the construction industry – including manufacturers of building products, equipment and components, and the various professional services provided by architects, surveyors, engineers and property managers – typically accounts for about 15–16 per cent of total annual economic activity. (You may remember that we compared broad and narrow definitions of construction activity in Chapter 1; and it may be useful to review Key Points 1.3.)

Official statistics, however, generally tend to restrict the construction sector to the narrower definition of the industry, estimating the activity of firms that construct and maintain buildings and infrastructure – that is, just those businesses that undertake on-site activities. Consequently, the share of total annual economic activity attributed to construction by the official statistics in the fully industrialised countries is now rarely larger than 12 per cent and usually in the 7–10 per cent range. In the 27 countries of the European Union, the construction sector accounted for 9.6 per cent of economic activity on average in 2011. Note though, as we explained in Chapter 12 (see Key Points 12.3), that construction output tends to decline as a proportion of a country's GDP as its economy matures.

Apart from the industry's contribution to the total economic flow, it also has a significant impact on living standards and on the capability of society to produce other goods and services. In other words, construction is important to the economy because it produces investment goods. These are products that are not wanted for their own sake, but on account of the goods and services that they can create. Across the 27 member states of the European Union, construction counts for 51.5 per cent of all investment goods. Construction also can have extremely significant effects on the level of employment as it tends to be more labour intensive than other sectors. In fact, in Europe construction is the largest industrial employer, representing nearly 15 million jobs. (All statistics are taken from the European Construction Industry Federation, FIEC 2012.)

In this chapter we explore the relationships between the construction industry, other sectors and the national economy. As the contrast between the broad and narrow definitions of the industry illustrates, there are many sectors closely associated with construction activity. A case can also be made that construction indirectly affects and supports activities in the financial, manufacturing, wholesale, retail, residential and service sectors. Consequently, data for construction-related activities is frequently muddled into manufacturing and service industry surveys. These direct and indirect relationships have important implications for management of the macroeconomy and their analysis is facilitated by the annual publication of national income accounts, often referred to simply as the **national accounts**.

MEASURING ECONOMIC ACTIVITY

The national accounting framework provides a systematic and detailed description of the UK economy and, by following agreed international accounting conventions, it enables comparisons to be made with other countries. It is not our intention to delve into the minutiae of this system, but just to establish the general measurement concepts necessary to discuss the broader role of construction.

To begin the analysis, we consider a simple economy without a government sector, a financial sector or an overseas sector – that is, our starting point is a simple two-sector model economy and we analyse only the relationship between households and businesses. The complications of the real world will be considered later. We have already portrayed economic activity using this type of a model in Chapter 1 (see Figure 1.4), and for convenience a modified version is presented in Figure 13.1.

To make our starting model effective, we make these assumptions:

- households receive income by selling whatever factors of production they own
- businesses sell their entire output immediately to households without building up any stocks
- households spend their entire income on the output of the businesses.

These three assumptions seem realistic. Businesses will only make what they can sell. Production does involve paying for land, labour, capital and enterprise, and these services generate income payments – rent, wages, interest and profit – which, in turn, are spent. The model of the circular flow outlined in this way suggests that there is a close relationship between income, output and expenditure. These relationships are presented in a traditional format in Figure 13.1.

From Figure 13.1, it is clear that businesses reward the owners of factors of production (land, labour, capital and enterprise) by paying them rent, wages, interest and profit and, in turn, these factor rewards (incomes) form the basis of consumer expenditure. This model shows that it is possible to measure the amount of economic activity during a specified time period by adding up the value of total output, or total income, or total expenditure. In effect, it is only necessary to adopt one of these three approaches since conceptually they are identical – even in the actual national accounts they rarely differ by more than 0.5 per cent. The small discrepancy is due to each being calculated using different statistical methods.

To get a better idea of the magnitude of the numbers involved, readers are advised to look at a copy of the *UK National Accounts: The Blue Book* (ONS 2011c). Tables 1.1 and 1.2 summarise all three methods of measurement – namely the **output approach**, the **expenditure approach** and the **income approach**. By analysing these statistics, it is possible to gain a good insight into the economy, especially as the data usually covers the last 10 years. Although very few people actually study the detailed breakdown of the accounts from cover to cover, they are an essential data source for anyone concerned with macroeconomics. Indeed, the national accounts are far more important than just indicating changes in GDP; they form a central reference for those who wish to broaden their understanding of the economy and its measurement.

Figure 13.1 The circular flow of income, output and expenditure

The diagram highlights two flows: a monetary flow and a real flow of goods and services. The two lower flows indicate the factor market – households exchange their factors of production with businesses in return for payment. The two upper flows show the product market – businesses provide a flow of goods and services in return for monetary expenditure.

GDP and Growth

Before considering any figures, however, we must fully understand what they convey and the significance of any changes in their size. In simple terms, **gross domestic product** (GDP) can be regarded as the annual domestic turnover; or, to employ the analogy used in Chapter 12, the result of a giant till ringing up all the transactions that occur within a specific territory. In formal terms:

> GDP represents the total money value of all the production that has taken place inside a specific territory during one year.

An alternative measure is **gross national income** (GNI). This is very similar to GDP, but includes a net figure for employment, property and entrepreneurial income flowing in and out of a nation's economy from overseas – in other words, GNI aggregates all the activity that generates income to a specific nationality. In practice, GDP and GNI represent very similar amounts. For example, in 2010 GDP in the UK totalled £1,458 billion and GNI was £21 billion more at £1,479 billion. In European states, GDP and GNI rarely differ by more than 1 or 2 per cent, but the difference may be substantially larger in less developed economies.

When GDP figures are adjusted from current prices to constant prices, to allow for inflation, it is possible to calculate the **real value** of any change in economic activity between one year and the next. Effectively, economic growth can only be declared if 'real' GDP has increased. If real GDP has declined, this is described as a **recession**. In the majority of years during the last half-century the recorded figures have been positive.

Since the UK shares a common set of accounting conventions with other countries, we can make international comparisons of GDP and GNI. Some figures are shown for six selected countries in Table 13.1. The final column, which shows GDP growth, is obviously expressed relative to economic activity in the previous years. The term 'real' emphasises that inflation has been removed from the calculation, with each year's GDP values being expressed at an agreed base year (to convert current prices to constant prices). Worldwide economic activity tends to be on an upward path and, as Table 13.1 shows, all the selected economies grew strongly in the period 2000–2006. In particular, there were big increases in India and China, although both countries were growing from a relatively low economic base. To facilitate international comparison it is necessary to take into account the size of the country. This is achieved by expressing GDP or GNI on a **per capita** basis, by dividing total GDP or GNI by the total population to arrive at an amount per head (see the dollars per capita column in Table 13.1).

Table 13.1 Macroeconomic statistics for selected economies

| Countries in rank order | Gross National Income (GNI) | | GDP |
	Billions of dollars 2010	Dollars per capita 2010	% real growth rate 2000 – 2010
Norway	427	87,350	1.7
United States	14,646	47,340	1.8
Japan	5,334	41,850	0.9
United Kingdom	2,377	38,200	1.8
China	5,721	4,270	10.8
India	1,554	1,270	8.0

Source: Adapted from World Bank (2012: Tables 1.1 and 4.1)

The statistics in Table 13.1 are taken from World Development Indicators, a comprehensive set of data produced by the World Bank each year. The first table in the series always addresses 'The Size of the Economy'. The current publication lists relevant data for most major national economies in alphabetical order. In previous years, however, this data was presented in rank order according to GNI per capita. The concept of 'rank order' demonstrates the importance of these figures, as they are used to create a type of league table, in which (in 2010) Norway, the United States, and Japan come ahead of the United Kingdom, since the gross national income (GNI) divided by the population is higher in each of these countries than in the UK.

If 'total' GNI were the reference point, then Norway would move to the bottom of our list and the richest nation in the world would be the United States with a GNI in 2010 of $14,646 billion. As you can see from Table 13.1, there is a huge difference between the largest and second largest economy (measured in terms of GNI per capita). There are 309 million Americans and 5 million Norwegians and, once population size is taken into consideration, Norway can offer a higher material standard of living than the United States – $87,350 compared to $47,340 respectively. At the other extreme, China and India with the highest populations in the world are relatively poor, with an annual per capita income of $4,270 and $1,270 respectively. According to the World Bank (2012) low-income countries are those with less than $1,005 per capita (which is possibly less than some students in the Western world earn each month – even while they are studying) and this gives some indication of the poor average standard of living in India.

It should be stressed that figures derived from national accounts do not effectively measure the distribution of wealth or levels of contentment within society. Recent research in the UK found that every £100 increase in national income since 1977 has benefited the rich disproportionately (particularly when bonus payments are included) and since 2008 the poorest half of the population (defined as 11 million working-age adults earning low-to-middle incomes) had not seen any increase in their wealth at all. In the authors' words, average wages in the UK were stagnant from 2003 to 2008 despite GDP growth of 11 per cent in the period. Similar trends are evident in other advanced economies from the US to Germany. For some time, the pay of those in the bottom half of the earnings distribution has failed to track the path of headline economic growth (Whittaker and Savage 2011: 2). Similarly the per capita income figures shown in Table 13.1 hide the stark reality of unequal income distribution. For instance, 70 per cent of those who live in India and 30 per cent of those who live in China survive on less than $2 per day, which is far less than the respective annual per capita figures would suggest (World Bank 2012).

There is a debate opening up about 'happiness' in an economic context, as there are those who argue that although people have got richer in Western society they have not become happier. There are things that money cannot buy, such as good health, lasting relationships, strong communities, interesting employment and a beautiful environment. So the happiness of a society does not necessarily equate to its level of income. The problem is that no accepted measure for the quality of life, or wellbeing, has been established to date. Richard Layard's work at the Centre for Economic Performance at the London School of Economics may eventually fill this void. He has certainly had encouragement from politicians of all persuasions to identify what is sometimes referred to as a 'happiness index'. It is Layard's (2011) contention that as the consumer society has become dominant over the past 50 years, happiness has declined. As he points out we are richer, healthier, have better homes, cars, food and holidays than we did half a century ago, but despite economic growth increasing by leaps and bounds, happiness in the West has stagnated. This raises some serious questions about the pursuit of economic growth, and some argue that this political objective is neither sustainable nor desirable.

Key Points 13.1

○ National accounts measure the annual level of economic activity, and economic growth is identified by changes in 'real' GDP.

○ GDP represents the total money value of all production created within a country during a year. GNI includes the income generated for the nationals of that country by overseas activities.

○ Per capita GDP is a broad measure of national income but it is less useful as an indicator of the wellbeing of a society.

GDP AND CONSTRUCTION

Construction is a significant part of the total economy. In 2010, construction in the narrowest sense of the definition produced about 7 per cent of UK GDP. In comparison, manufacturing produced 16 per cent of GDP, while agriculture accounted for just 1 per cent. The lion's share of economic activity fell into the service category, which broadly defined includes the wholesale and retail sectors. In 2010, the service sector accounted for 76 per cent of GDP (ONS 2011c).

In newly developing countries, the construction sector can contribute as much as 15 to 20 per cent of GDP because it accounts for a significant amount of investment during a country's development. As industrialisation proceeds, factories, offices, infrastructure and houses are required and construction output as a percentage of GDP reaches a peak. In other words, construction is responsible for the output of buildings and infrastructure upon which most other economic activities depend.

Once an economy has developed, the demand for construction products, in relative terms, declines and construction output as a percentage of GDP tapers off. In an industrialised nation the building stock is well developed, so the need to add to it is less. Much of the infrastructure and many of the buildings may be ageing, but the requirement for new build work is generally smaller – however, there is likely to be a far greater need for repair and maintenance work. Generally, therefore, the higher the GDP per capita, the higher the proportion of repair and maintenance work in the construction sector. This relationship is described by the Bon curve (see Figure 1 of Reading 5, page 266).

Bon's analysis (1992) studied the entire path of development from the least developed countries to those of an advanced industrialised status. The picture that emerges is one in which the share of construction in gross domestic product tends to increase with the level of per capita income during the early stages of economic development. However, once a country reaches a certain level of development, construction output, in relative terms, declines in relation to national output – that is, it decreases relatively but not absolutely. Bon's analysis suggests that when a certain level of development is achieved the share of construction activity will settle at around 6 to 8 per cent of a country's GDP. Take China as an example, during the first two decades of its reform era (1978–1999) as it emerged from being a less developed country, construction as a percentage of its GDP grew from 3.8 to 6.6 per cent. During the next decade as roadways, ports, rail networks, airports, power stations, factories and houses were added to the capital stock, construction output in China increased to 13 per cent of GDP. Once China has fully matured into a high income economy, this phenomenal rate of construction growth would be expected to subside, but given the country's scale and population this will not be the case for at least another decade.

Following the logic of Bon's analysis (1992) and the subsequent body of research that it triggered – see Reading 5 and Choy (2011) for references – development agencies and international bodies such as the World Bank and the United Nations became more confident of the key role played by infrastructure in the development of a poor country. They argued that poor countries find themselves in a poverty trap and escaping that trap requires large investments on building basic infrastructure (roads, electricity grids, ports, water and sanitation, accessible land for affordable housing) and on environmental management (Foster and Briceno-Garmendia 2009). In short, there is a high correlation between economic growth in developing economies and construction activity.

The relationship between GDP and construction output is discussed further in the three case studies in this chapter (see pages 233–7). However, we end this section by entering some general caveats concerning the reliability of construction data. It is difficult to obtain accurate data on the output of the construction sector for three reasons. First, the construction industry comprises a very large number of small geographically dispersed firms that undertake mainly small projects. This makes compiling comprehensive data sets difficult for government agencies monitoring the industry. Second, much of the work in construction is subcontracted and, consequently, there is a risk of double counting; the construction statistics division at the ONS does not seek output data from property developers as they are regarded as the clients of the reporting contractors. Third, alongside the official activities 'put through the books' and recorded in company accounts, there is a significant informal economy – unofficial work carried out for cash in hand. (There is also the associated conundrum of DIY.)

The most extreme examples of unreliable data seem to come from countries that are the least developed or those in a stage of transition. Meikle (2011) goes as far as to suggest that 'informal' construction could possibly represent the majority of construction activity in some developing countries. He claims that in Africa typically 90 per cent or more of rural residential building and more than 50 per cent of urban building is informal output. In the higher income OECD countries, informal construction is also an issue, with some estimates valuing this activity at around 18 per cent of the officially recorded construction output (Jewell et al. 2005). Construction output is difficult to measure as many small building firms are unlikely to be able to (or want to) meet the bureaucratic demands of national accounting systems. Data can be unreliable. Surveys might be completed on the basis of the value of quotations for new orders, and these may be resubmitted later, possibly at a lower price, creating a credibility gap in the data. There are also construction businesses that fail to declare some or all of their work to avoid tax. Tax evasion reduces government income, and also can result in the true value of construction output being underestimated. Some countries have placed a value on informal construction output, but as far as the UK is concerned, the last set of revisions to the method for compiling construction output data, which came into effect in 2010, removed any estimate for unrecorded output. It could be argued that there is no guarantee that any estimate for informal activity is fair, so it is more transparent to leave it out completely. However, this change in approach does introduce a discontinuity in the UK figures, as data published before 2010 was compiled on a different basis and included a small allowance for informal activity.

Key Points 13.2

○ The construction industry can be defined in many ways. Broad definitions suggest that the construction industry typically accounts for 14–16 per cent of total annual economic activity; on the narrower definition, it only accounts for about 6–8 per cent.

○ There is a relationship between the level of construction activity and a country's stage of development. Construction can be regarded as the engine of economic growth.

○ In most countries, construction activity is difficult to accurately assess and, in many cases, it is under-reported in the national accounts.

FROM CIRCULAR FLOW MODE TO REALITY

The two-sector circular flow model presented in Figure 13.1 suggests that the amount of money flowing around an economy is always constant – the GDP figures never change; economic growth is always equal to zero. This is because the model is based on the assumption that expenditure levels precisely determine income levels, and expenditure is in turn determined by income, and so on. In this theoretical model of a two-sector economy, income and expenditure levels are permanently static: there is no growth and no decline. This economy could be classed as being in neutral equilibrium.

In reality, however, every economy experiences leakages (withdrawals) from the circular flow. These occur through the sectors we initially specifically excluded: overseas, financial institutions and government. Simultaneously, there may be injections of funds into the economy through these sectors, for example from exports (that is, earnings flowing in from abroad).

Leakages and Injections

Figure 13.2 extends the circular flow model to include leakages and injections. Three leakages from expenditure are shown: savings, imports and tax. Counterbalancing these leakages, there are three injections: investments, exports and government spending. The decisions that determine the overall size of these leakages and injections of funds are carried out by different groups of individuals with different motivations. It is most unlikely, therefore, that leakages and injections will be equal and cancel one another out.

Figure 13.2 The circular flow model with injections and leakages

To complete the circular flow diagram, leakages need to be subtracted as households save, spend money on imports, and pay taxes to the government. And injections need to be added as businesses benefit from investment funds, export earnings and government spending.

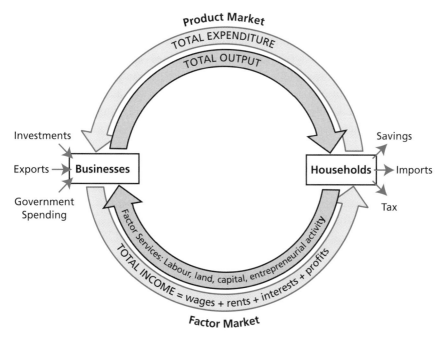

If the total level of leakages is greater than the level of injections, the economy will become run down, raising unemployment and reducing standards of living. To take an extreme example, if every household decides to spend its money on imports from abroad, then this would represent a major leakage of funds from the domestic circular flow and a significant boost to other countries. Conversely, if the total level of injections is greater than the level of leakages, the economy will be boosted, increasing employment opportunities and raising the amount of national income.

EQUILIBRIUM OF THE MACROECONOMY

Equilibrium means a 'balanced state'. In the macroeconomic context, this means that income, expenditure and output levels continually adjust upwards and downwards to keep in line with one another. For example, when leakages exceed injections, expenditure on domestic output will be less than factor incomes. Consequently, firms will not receive sufficient revenue to cover their output costs. Stocks will accumulate and firms will cut back output and incomes until they equal expenditure again. A new level of equilibrium will have been reached. It is the nature of imbalances between leakages and injections that prompt changes in output from year to year. These changes lead to different levels of income circulating within the economy – representing different levels of economic activity. The dilemma for economists, however, is that although all economies tend towards equilibrium, the associated level of activity is not always sufficient to support full employment.

It should be clear by now that national income analysis is one-dimensional, in so far as it focuses entirely on the monetary value of material goods and services. Whenever we discuss leakages and injections, the environmental dimensions (as covered in Part B) are ignored. The interaction between the economy and the environment is not effectively measured in conventional national accounts. In fact, any money spent on cleaning up the environment from pollution simply contributes to total expenditure in just the same way as money spent on dumping waste, dealing with crime or preparing military attacks. At present, all output is regarded as contributing to economic wellbeing. In historical terms, an explanation for this positive attitude to all output is due to its association with employment. This will become clearer in the next section.

Key Points 13.3

- ○ Exports, investments and government spending represent injections to the circular flow. Imports, savings and tax represent leakages. The size of the injections set against the leakages determines the annual level of economic activity.

- ○ All economies tend towards equilibrium but that does not guarantee full employment.

- ○ Conventional national accounts do not systematically measure the interactions between the economy and the environment.

MANIPULATING THE LEVEL OF ECONOMIC ACTIVITY

In the last 60 years, there has been some debate between economists about the policies that should be adopted by governments to manage the economy. The discussion has proceeded in two main phases. During the years following the Second World War, the consensus of opinion seemed to be for an interventionist strategy. This approach was informed by Keynesian **demand management** theory. The major objective was to keep the economy running near full capacity without incurring wide fluctuations in output. During the last twenty or thirty years, the Keynesian consensus has given way to **supply-side economics**, with unfettered free markets providing the theoretical basis of the approach. The dominant economic objective of supply-side economics is to control inflation in order to achieve high and stable levels of growth and employment. These two contrasting approaches to managing the level of economic activity are considered below.

Demand Management

When a government is faced with a situation in which resources are unemployed and the economy is generally running below full capacity, it can intervene in various ways to reflate the economy. The easiest option is to increase its own spending, and thereby inject funds into the circular flow. This idea, known as demand management, was fashionable throughout Europe from 1945–1975. It was an attractive option because injections of government funds were seen to have a multiplier effect on national income level and employment.

THE MULTIPLIER THEORY

The theory of the **multiplier** builds on the circular flow concept – on the idea that expenditure determines the level of output and its associated income. In other words, when people are employed they spend their wages on goods and services produced in other sectors of the economy which, in turn, generate employment and spending elsewhere, creating an upward spiral. Keynes argued that if the current amount of expenditure is insufficient to maintain full employment it becomes advisable for the government to intervene – or, to express it in journalistic terms, to 'kick start' or 'pump prime' the economy.

Consider this scenario: assume a government invests £40 million for a new road. This will cause expenditure and output to raise by the same amount. To increase output, more labour will be taken on. New firms may be started. The newly employed resources will be rewarded with incomes to the value of the initial injection. However, as this money circulates around the economy, some of the £40 million will leak out of the flow in the form of savings, imports or taxes. Economists refer to this as the **marginal propensity to leak** (MPL). The concept of the margin – as we discussed in Chapter 7 – focuses on additional or incremental amounts. The marginal propensity to leak, therefore, represents the proportion of the 'additional' income that does not get used on consumption. If we assume an MPL of 25 per cent, we can quickly calculate that households will spend £30 million of their increased income on **consumer goods**. (Certainly if the recipients of income injected by government spending were previously unemployed, we would expect these households to spend any additional income coming their way on consumption rather than saving.) This additional spending will add a further boost to total expenditure. In turn, firms producing consumer goods will increase output, and they will take on more resources and have to pay out more in interest, wages and rent in order to earn more profit. Again incomes will increase. This will lead to successive rounds of further expenditure. If we continue to calculate the increase in additional expenditure occurring as a result of the initial additional government investment of £40 million, we find that national income is 'pumped up' by a significantly larger amount. In this example, it would actually be £160 million. The determining factor is the size of the leakage, since the multiplier is equal to the reciprocal of the MPL. In developed European economies the leakages are quite large and, accordingly, the multiplier effect is significantly smaller than our example suggests.

This scenario is not too far removed from the immediate post-war reality. The government policy was to invest in the construction industry to increase the general volume of economic activity. Construction was specifically chosen because in most countries it was, and still is, a labour intensive activity, and it plays an important role in the development of the productive capacity of the economy. In fact, many post-war economists regarded the construction industry as a 'regulator' of the economy. On this basis a large group of contractors put forward a case for continued public sector investment during the recession of 2009, claiming that for every £1 invested on construction, output would increase GDP by £2.84. They argued that in relative terms construction relies less on imports than other sectors (such as motor manufacturing), so the additional activity would significantly benefit the UK (UKCG 2009). In other words, supporting jobs in construction should promote further employment in the economy as a whole.

Aggregate Demand

A central part of Keynesian analysis is **aggregate demand** (AD).

> Aggregate demand can be defined as the total spending on goods and services produced in a whole economy.

At the beginning of this chapter we considered total expenditures on a theoretical level in a two-sector economy. In such a model, aggregate demand would be equal to consumption expenditures (for example, on beer and chocolate) by households and investment expenditures (for example, on buildings and machinery) by businesses. In reality, however, we need to add government expenditures (such as on road construction) and export revenue from UK output (such as US purchases of Jaguar cars). Aggregate demand (AD), therefore, consists of four elements: consumer spending (C), investment spending (I), government spending (G) and expenditure on exports (X). Aggregate demand is the total of these four elements once one further adjustment is made: to be technically correct, spending on imports (M) from abroad needs to be subtracted as this is not money spent on UK products. It is traditional for the shorthand notation to be used to express aggregate demand using the formula:

$$AD = C + I + G + X - M$$

At this juncture you could be excused the feeling of déjà vu. Earlier in this chapter we discussed national income accounting and derived a monetary value for economic activity. Aggregate demand is in fact analogous to GDP. Table 13.2 shows the components of UK aggregate demand in 2010.

Demand management techniques proved to be a difficult tool to use. One of the difficulties was the timing of the action. It becomes particularly difficult when the sector used for delivery is construction, as the time lags tend to be long and variable. A second and more obvious problem was overshooting – adding a too large injection which causes the economy to overheat. The subsequent excess demand achieves nothing except continually increasing prices – resulting in higher inflation. In short, it proved very difficult to use demand management techniques to shift aggregate demand to the precise level to secure full employment at the right time.

Table 13.2 Measuring aggregate demand in 2010 market prices

Components of aggregate demand	C + I + G + X − M = AD
Related amount in £bn during 2010	1151 + 224 + 125 + 437 − 476 = £1458 billion

Source: United Kingdom National Accounts (ONS 2011c)
Note: AD total may not sum as figures rounded to nearest billion

These problems were neatly analysed by Professor Frank Paish during the 1960s. He suggested that the problem with Keynesian demand management techniques lay with the concepts of productive capacity and actual output. He argued that by adding to productive capacity, you generate income before you generate actual output. Therefore, to achieve stability, Paish recommended maintaining a margin of unused productive capacity.

The focus has now shifted to concerns about potential output and actual output. Once spare capacity has been used up – and full employment of resources has been achieved – actual output will be restricted in the short-term. The answer to the problem, therefore, seems to involve ensuring an amount of unused productive capacity and/or increasing the potential output of an economy. To increase the level of output, at the full employment point, requires more capacity and this is determined by resources being used efficiently. Technically speaking, economists had become concerned about the supply side. Patricia Hillebrandt (2000: 66) has always regarded the capacity of the construction sector as a constraint and she identified three episodes – in 1964, 1973 and 1989 – when the industry or its material suppliers were stretched to capacity. Interestingly each episode resulted in periods of sharp inflation, and the latter two also led to periods of recession in 1974 and 1991.

Key Points 13.4

○ During the years immediately following the Second World War, the strategy for manipulating economic activity was informed by Keynesian demand management theory.

○ An increase in government expenditure causes a multiplier effect on the level of national income. The larger the marginal propensity to leak, the smaller the multiplier effect.

○ Aggregate demand is the sum of all expenditures in an economy. It is usually considered in four categories, and using the standard notation, defined by the formula C + I + G + X − M.

○ Today, demand management techniques are closely associated with the problem of inflation.

SUPPLY-SIDE ECONOMICS

One of the main hallmarks of economic policy since 1979 has been concern over the supply side of the economy. The focus has shifted from government spending and aggregate demand to production and aggregate supply. **Aggregate supply** can be regarded as total production, and clearly many factors influence its size such as the level of profits, ease of movement into and out of markets, the level of wages, the efficiency of capital and labour, the level of fixed costs, etc. As a result, policy oriented to the supply side has given rise to measures to increase incentives within the economy. Indeed, economists concerned with this perspective prefer market forces to government intervention. The Statement of Intent on Environmental Taxation (HM Treasury 1997) was an early example of this approach. This set out the UK government's intention to use the tax system to promote sustainable growth, by shifting the tax burden away from 'goods', such as employment, towards 'bads', such as pollution. The policy sought to achieve a 'double-dividend' – of increasing capacity and reducing environmental damage.

Supply-side policy, therefore, is geared to making markets work more efficiently. This has been achieved by reducing the structural rigidities that clutter many markets. For example, **wages councils** were stripped of many of their powers in the 1980s and formally abolished in 1993; trade union activities have been restricted and union membership significantly reduced as a result; and market competition has been opened up through privatisation and deregulation. Similarly, governments have made broad reductions in rates of income tax – from 33 per cent to 20 per cent at the basic rate and from 83 per cent to 50 per cent for the top rate – to provide more incentives for people to work harder. Indeed, there seems to be no market that has escaped from the plethora of supply-side measures.

Aggregate Supply

By devising supply-side methods to promote competition in as many markets as possible, governments believe they can reduce the money they spend on direct intervention and encourage the entrepreneurial spirit that drives production. In technical terms, this policy is designed to shift the aggregate supply curve to the right. This is demonstrated in Figure 13.3. An aggregate supply curve represents the relationship between the output firms would be willing to supply and the general price level.

It would be nice to simply conclude that the aggregate supply curve slopes up because, the higher the price level, the greater the incentive for producers to produce more. But it must be emphasised that we are not talking about changes to individual specific prices – the vertical axis represents changes to the general price level; it is an index of the weighted average of all prices. In order to understand the true purpose of the aggregate supply curve, we examine four situations:

1   large amounts of unused capacity
2   full capacity
3   an intermediate range between the two
4   increasing capacity in the long run.

Figure 13.3 The aggregate supply curve

At a price level of P_1, the AS curve is a horizontal line up to the point where output is Y_1. Then, there is an intermediate stage – some sectors of the economy are experiencing excess capacity, but others are not. At Y_2 there is no immediate capacity in any sector of the economy and prices rise. In the long run, however, it should be possible to increase productive capacity and shift the AS curve towards the right.

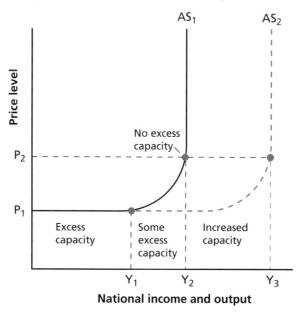

UNUSED CAPACITY

When an economy has many factories operating at less than capacity, there is a general under-utilisation of the productive capabilities and it is possible to increase output without any pressure on prices. If there is unused capacity, producers can increase supply without having to pay higher prices for factors of production. If they need more labour, they can hire someone who is unemployed – they do not need to pay higher wages to attract people. Providing there are significant levels of unemployment and unused capacity, per-unit costs of output remain the same, no matter what the volume of output. In these circumstances, we would expect the aggregate supply curve to be a horizontal line at the current price level. Consider a current price level of P_1, as given on the vertical axis of Figure 13.3. The horizontal line labelled 'excess capacity' represents that part of the aggregate supply curve AS_1 where an increase in output causes no pressure on prices. Within this range, supply is perfectly elastic.

NO EXCESS CAPACITY

Now consider the other extreme, in which the economy is running at full capacity and there is no unemployment. In such a situation it is impossible for any additional output to be produced. There is only one thing that can happen – the price level can rise, but no further increases in output are physically possible. The aggregate supply curve has to be a vertical line, as shown at output rates Y_2 and Y_3 in Figure 13.3. At this point supply can be said to be perfectly inelastic; any price rises above level P_2 produce absolutely no changes to the quantity supplied.

INTERMEDIATE RANGE

When there is some excess capacity in parts of the economy but no excess capacity in other parts of the economy then as production increases so-called bottlenecks or **supply constraints** may develop. As firms try to increase output they may experience shortages of some inputs, most frequently in certain kinds of skilled labour. When this happens, firms can try to attract more of the scarce input by offering a higher price. They compete with each other for the limited supply of people with scarce skills, thus driving wage rates up. This raises their costs of production, and they then react by raising their prices whenever possible.

The shape of the aggregate supply curve in the intermediate range is explained by these bottlenecks. As the aggregate supply curve starts to slope up, it becomes steeper as full capacity output is approached because more and more supply constraints appear. These constraints force some prices to increase and, as output nears full capacity, sellers can put prices up without losing customers. Since the price level is an index of all prices, if some prices stay constant and some go up, the general price level will rise too. This means that if we increase production from output Y_1 in Figure 13.3 – at the end of the excess capacity output rate – price levels will rise along with national income. In this range, there is a positive relationship between national income and price level. As supply constraints become greater, supply becomes less and less elastic. Successive increases in spending lead to smaller and smaller increases in output and income, up until we reach full capacity.

INCREASING CAPACITY IN THE LONG RUN

In the long run, as technology advances and the stock of capital increases, it will become possible to increase capacity and produce more – so the vertical line showing full capacity output will shift gradually to the right. This is shown by AS_2 in Figure 13.3. The increases in national income and output from Y_2 to Y_3 are achieved by new capacity. This represents how, over a fairly long period, output can begin to increase and prices fall.

As far as construction is concerned, manipulating the supply side has led to the government's role becoming less transparent. Today, governments increasingly rely on subtle messages to improve the market efficiency of the construction industry by encouraging firms to adopt innovative ideas such as factory-built modular technology and partnering. The aim is to encourage construction to embrace innovation and vertical integration, and to capture some of the advantages achieved by manufacturing at the site level. If the capacity of the construction industry could effectively increase, the overall aggregate supply curve would certainly shift to the right, enabling total output to increase without putting pressure on resources. This is because construction influences the capacity of so many other sectors of economic activity. The best evidence of increasing capacity and growth through hi-tech construction methods can be seen in Japan, which is why it is one of the three national case studies presented in this chapter.

Obviously, governments want the aggregate supply curve continually shifting to the right to achieve sustainable growth and stable prices. Unfortunately, however, this is frequently complicated by a construction industry that seems wedded to low-tech approaches.

Key Points 13.5

○ Since 1979, economic policy has been concerned with the supply side of the economy. The focus of government activity has shifted to production and aggregate supply.

○ Economists have become increasingly concerned with the principles of microeconomics that underpin aggregate supply.

○ Supply-side economics revolves around the freeing up of markets and the reduction of direct intervention by the government.

○ The aggregate supply curve describes several stages based on different levels of capacity. The nearer the economy gets towards full capacity, the more likely inflation becomes (see Figure 13.3).

THREE CASE STUDIES

There is a close relationship between a country's construction sector and its level of GDP. This is partly because the construction industry involves the assembly of many different products from a large number of industrial sectors. As we suggested in the introduction to this chapter, construction activities have far broader impacts than the main official statistics imply. Of greater importance, however, is the fact that construction provides the investment that underpins development, as it provides houses to live in, buildings to work in and infrastructure to support communication and transport. In fact, several studies have shown that the construction industry has significant linkages with other sectors of the economy. Consequently, it has become common practice to use the construction industry as some kind of regulator for the overall economy. The following three case studies demonstrate how this may be achieved in different cultures.

CASE STUDY 1: CHINA

The People's Republic of China is a twenty-first century miracle, as its economy has been rapidly transformed from a rural and somewhat backward centrally planned state to a modern urban market-oriented system. As a result, China is experiencing a building boom. The figures are astonishing: more than 260 new cities to be built within 20 years, 75,000 miles of rail track to be laid, a further 100 airports to be added to its existing network of 140 facilities, and over $120 billion worth of upgrade to the national power grid, and water and sanitation infrastructure systems. Beyond doubt, China has become the world's largest construction market, with an annual spend equivalent to £500 billion (Richardson, 2012).

So even though there is a global financial crisis and economies of many countries are wilting, China has managed to continue to expand. Annual economic growth has averaged 8 per cent since 1990, and despite the ongoing European economic crisis China achieved 9 per cent growth in 2011 and was forecast to expand by another 7.5 per cent in 2012. China has consistently been the fastest growing economy in the world for several years. The proportion of value added by construction to GDP is currently in the region of 10 per cent, and employment in the industry is around 40 million. To put this in context, note that China has a total population of over 1.3 billion people and a labour force of 798 million.

Clearly the labour intensive construction industry has played an important part in this rapid transition. At the Chinese Communist Party meeting in 1994, construction was explicitly identified as one of the four pillar industries that could help China to begin a new cycle of economic growth. (The other three pillar industries were automobile manufacturing, oil and chemical refinery, and mechanics and electronics.) The establishment of a new ministry of construction gave impetus to the sector. The ministry took a lead role in implementing new strategies for the industry. These strategies include opening up construction markets, establishing a competitive bidding system, allowing autonomy in state-owned enterprises and trying to eliminate bribery as a means of obtaining contracts, loans and materials. These policies resulted in a substantial expansion in the number of Chinese construction firms, and they have been regarded as an important motor in driving economic growth. The Chinese government has been keen to avoid foreign domination, and although foreign investment is allowed, it is not permitted in contracts for the supply of water, gas or subway systems.

The construction industry has clearly made a significant contribution to the development of the Chinese economy, but academics recognise that the process of rapid urbanisation and modernisation means that it still has a long way to go to reach world-class standards, and it should be noted that it started from a low base. Essential reforms are required in areas such as private investment, use of the price mechanism, quality of materials and equipment and, most significantly, reducing resource consumption and severe environmental impacts (Yang and Kohler 2008). Interestingly, in 2009 the Chinese State Council bowed to international pressure and committed to reducing its carbon intensity – a measure of a country's carbon dioxide emissions per unit of GDP – by at least 40 per cent by 2020. And its current five-year plan to 2015 identifies the built environment as key to achieving this target, and includes measures such as introducing a retrofit programme, committing to a series of major renewable energy projects, the creation of 200 eco cities and the development of eight low carbon projects across five regions (Richardson 2012).

Finally it should be noted that the streets are not entirely paved with gold. In some areas corruption is rife and China is a difficult market to penetrate. *Guanxi* (roughly translated as relationships) is extremely important, and getting this right is key to doing business. The need to invest time and effort in developing connections and networks cannot be overstated, and rather worryingly a survey of commercial companies across 28 countries indicated that the Chinese are quite likely to engage in bribery when business networks are formed (Hardoon and Heinrich 2011).

CASE STUDY 2: JAPAN

During the last couple of generations, the Japanese have rapidly developed their country into one of the richest nations in the world and much of this success is attributed to Japan's active construction sector. Indeed, Japan is worthy of a case study not only because the Japanese construction industry is extremely efficient in terms of output, but also because the sector has proved to be 'unusually profitable' (Constructing Excellence 2010).

In advanced industrialised economies, construction rarely accounts for more than 10 per cent of GDP. However, during the 1990s, Japanese construction – in the narrowest sense – accounted for about 12 per cent of GDP. The Japanese construction industry is in fact characterised by many exceptional features. Here we just give three of the most significant examples.

- The majority of commercial buildings are constructed for owner-occupiers rather than for speculative developers.
- The relationship between general contractors, specialist subcontractors, labourers and clients is frequently characterised in terms of collaboration and integration rather than conflict and fragmentation. In an analysis of the Japanese construction sector, Reeves (2002: 421) even goes as far as suggesting that the whole mechanism in the sector 'operates to provide mutual benefit to all of the players involved'.
- Firms compete on the basis of technology, in contrast to the usual price-based competition that drives the construction process.

It is this third point that makes Japan's construction industry unique. In most countries, construction is regarded as traditional, conservative, labour intensive and not particularly interested in innovation; in Japan, it is quite the opposite. The Japanese government regards 'hi-tech intelligent' buildings as central both to the nation's infrastructure and to the development of a knowledge-based economy. It therefore promotes technology use in the management of a building.

Technology is also embraced at the site level, where the construction of all buildings utilises prefabrication, robotics, automation and information technology. Indeed, 'computer integrated construction' is already a key part of the corporate policy of several major Japanese contractors. For example, Toyota the car manufacturer is also responsible for the building of about 5,000 houses each year. The metal-framed prefabricated modules are 85 per cent completed before they leave the factory, are more or less defect free and guaranteed for 60 years. The total time from taking an order to the final completion on site is typically around 25 days, making for a very fast process that uses the minimum of on-site time and labour; in fact once the foundations have been laid construction of the house on-site takes around 12 hours. A sharp contrast to the traditional timber- or block-framed house that would take 12 weeks to build on site.

As Professor David Gann (2000: 113) explains in his analysis of innovation and change in the global construction market: 'Rich Japanese companies invested heavily in electronic technologies for their new buildings. Moreover, government played a bigger role than elsewhere in sponsoring the concept of intelligent buildings. Japanese construction firms vied with electronics, telecommunications and office equipment manufacturers to produce prestigious buildings.' In other words, the level of technological innovation in construction is associated with a higher than usual commitment to research and development (R&D). In most OECD countries, R&D in construction rarely exceeds 0.4 per cent of the industry's total annual output. In the UK, it rarely exceeds 0.1 per cent. In Japan, it is usually above 1 per cent – and has exceeded 2 per cent. Compared to manufacturing, this is still very low, as generally manufacturing firms invest 3–4 per cent of their total annual output in R&D. The important point for our purposes, however, is that Japan again stands out as the exception to the general rule. It has developed five, very large, vertically integrated construction companies (Kajima, Obayashi, Shimizu, Taisei and Takenaka) that believe in using technology to obtain competitive advantage. Again in contrast with other countries, construction research is concentrated in the private sector and there are more than 20 companies that possess their own technological research institutes. To finish with one of Gann's (2000: 198) historic examples, in one financial year (1992–93) Kajima – one of Japan's 'big five' – invested over £125 million in R&D. In contrast, in the same period, all the UK private contractors only managed a combined R&D spend of £17.5 million.

Ten years later, official statistics from UK private construction companies suggest that they continue to invest comparatively little in formal R&D, preferring to adopt new ideas and technology from other sectors to improve their operations (Ozorhon et al. 2010: iv). Total UK private sector construction R&D annual expenditure is far too low given the size and significance of the sector. Japanese construction, however, is clearly at the other extreme, demonstrating a commitment to R&D and reaping the profits of being the most innovative construction sector in the world.

CASE STUDY 3: SOUTH AFRICA

South Africa has been chosen as a case study because it is a classic developing economy. The average resident survives on less than $6,000 a year and about one third of South Africans actually survive on less than $2 per day (World Bank 2012). It has a population of 50 million (compared to 62 million in the UK) but only a residential stock of 12.5 million units and the quality of these gives some idea of the standard of living. Eight and a half million of the homes are formal units (typically constructed from clay or cement bricks) and about one third (4 million units) of the residences are categorised as informal dwellings in squatter settlements and shacks in backyards, often constructed from earth with corrugated iron roofs and limited services. Cooking in these informal dwellings requires biomass such as wood, dung or charcoal. This sector is the focus of the South African government's housing programme, which makes a stark comparison with the affordable housing agenda in the UK (cited in Milford, 2009: 14–16).

Blacks and whites in South Africa were segregated under apartheid, but this ended almost 20 years ago, with the first truly democratic elections taking place in 1994. However, a number of social and economic distortions still exist from the apartheid era; and much of the labour force has low skill levels. Significant improvements have been made to infrastructure networks providing electricity, running water and roads but much remains to be done, particularly in the areas of housing, health and education. The government's usual line is that resources are limited, but this has limited credibility when ministers are accused of spending exorbitant sums from the public purse on cars and entertainment (Drew 2010).

The construction market in South Africa is currently worth an estimated £16 billion, which represents about 9 per cent of GDP (within the expected range for a developing economy). with infrastructure being the dominant sector. The key areas of development include energy, telecommunications, transport, water and sanitation, and these accounted for about half of the construction output generated during the last decade. The South African government's 2011 budget has prioritised expenditure on social infrastructure, and this means a greater focus on education and health over the forthcoming decade.

Clearly public sector infrastructure projects are a major contributor to economic growth and poverty reduction. However mismanagement and corruption during planning, implementation and monitoring of construction projects undermines the expected social and economic benefits. As a government report recently acknowledged, there are strong indications that corruption in the South African construction sector has increased rapidly, and is reaching a tipping point beyond which it may be very difficult to reverse (CIDB 2012).

Fraud and corruption is a major problem, evidenced by bungs, bribes and backhanders, particularly during poorly managed tender processes. This problem is not unique to South Africa, and in many ways construction has acquired the dubious honour of being the most corrupt industrial sector across the globe. As Transparency International (2010: 76) acknowledged, in both developed and developing countries, construction is consistently ranked as one of the most corrupt sectors. In terms of the three countries profiled in these case studies, China is ranked as more corrupt than Japan or South Africa. In fact according to the corruption perceptions index (produced annually by Transparency International) of 178 countries around the world only Japan appears relatively clean; it is ranked at position 17, three places above the United Kingdom.

A role for government

Governments have a central role to play in promoting change in the construction industry. In the period dominated by Keynesian policy (1950–1975) governments opted for direct intervention. Indeed, until the 1970s, 50 per cent of all construction work was purchased by the public sector in the UK. From 1975 to 2010, however, government policy has placed far greater reliance on laissez-faire market forces. This has eroded the capacity of governments to directly control construction output. Now following the near collapse of the banking system in Europe and the United States, there is some reluctance to allow free markets to continue unchecked (see Chapter 14 for details). So many governments today find themselves facing a dilemma as they have policies to encourage environmental protection, promote innovation, and set legal standards to support transparency and competition, but they are not sure how best to implement these strategies. Initiatives to encourage sustainable construction are an interesting example of how governments have attempted to manipulate building activity and these are reviewed in the concluding chapter of this book, where we also take a further review of the role that governments play in the management of the construction industry.

Key Points 13.6

○ Regardless of the level of economic development, and a nation's culture and tradition, the construction sector plays an important role in any economy.

○ In China, Japan and South Africa, the construction industry has provided an engine for economic growth.

○ Japan's construction investment in R&D is the highest of all OECD countries. In most countries, however, it is left to governments to initiate policies to promote long-term goals.

MEASURING ENVIRONMENTAL IMPACTS

As indicated in Chapter 12, for the past 50 years or more academics have debated how to establish a standard international measure that effectively monitors the broad usage of environmental resources. This is now becoming increasingly urgent. Various ad hoc measures have been trialled, and many of these defer to GDP as a reference point. For example, by expressing changes in real GDP in relation to energy used (measured in million of tons of oil equivalent), it is possible to calculate the amount of energy required to support a certain level of GDP or, more simply, the energy consumed per unit of output. With efficiency gains it is possible for energy, and emissions, per unit of output to decline. The formal way to refer to this phenomenon is to say it is possible to 'decouple' growth from energy use, or in plain English to accept that it is not inevitable that economic growth always produces an equivalent deterioration in environmental quality.

Research into resource usage has been progressing since the 1960s, and the final part of the *UK National Accounts* is now dedicated to environmental issues by detailing oil and gas reserves, atmospheric emissions, energy use, etc. Much of the data in the section is detailed in units of physical measurement or volume and this leads to debates about its use or value. While there is some agreement that governments need a yardstick to monitor and measure the total environmental impacts placed by humanity on to the ecosystem, there is no accepted consensus about the best way forward. In short, we need an authorised metric designed to capture environmental activity, which functions in a similar way to GDP in capturing economic activity.

The metric that currently perhaps has the best potential to meet this role is the **ecological footprint**. Accounts based on this metric already exist for 150 countries for the years 1961 to 2008. They are produced by an organisation called the *Global Footprint Network*, which uses data provided by the United Nations. A major source of interest and funding for this work comes from the EU. The Global Footprint Network's vision is to make the ecological footprint as prominent a metric as GDP by 2015.

The basis of the system is to provide a way of measuring how much productive land is required to sustain life in terms of producing food, energy, water and timber and carrying away the associated wastes. The calculation of the demands we place upon the planet can be made at the individual, national and global levels. This is done by taking resource use and comparing it to what is actually available. The convention is to state the data with reference to the global hectares available per person. This is explained in the next section.

Measuring the Ecological Footprint

The calculation begins with the assumption that productive land upon which life depends could be anywhere in the world. This seems reasonable, as people of one nation usually consume resources and ecological services from all over the world. The ecological footprint calculations are, therefore, stated in global hectares (gha). A global hectare is one hectare of biologically productive space. In 2008, the globe had 12 billion hectares of biologically productive area corresponding to roughly one-quarter of the planet's surface. The 12 billion hectares include 2.4 billion hectares of water and 9.6 billion hectares of land. The land area includes 1.8 billion hectares of cropland, 3.4 billion hectares of grazing land, 4 billion hectares of forest land, and 0.2 billion hectares of built-up land. Non-productive marginal areas, such as deserts, ice caps, and deep oceans, are not included in the 12 billion global hectares of biologically productive space (GFN 2012).

To express the capacity available to each person, the 12 billion global hectares of biologically productive area are divided by the number of people on earth. The world's population was approximately 6.7 billion in 2008, giving 1.8 global hectares per capita. This means that, in principle, the average amount of ecological productive capacity (the biocapacity) that exists on the planet per person is 1.8 global hectares. (This simple calculation of available biocapacity assumes that no capacity is set aside for the demands of wild species.)

If each person could survive on 1.8 hectares, or less, the world would be sustainable. However, the latest ecological footprint accounts published by the Global Footprint Network indicate that we are far away from achieving sustainable development. The figures show that, on average, every person on earth is taking 50 per cent more from the planet than it can naturally regenerate. In other words, humanity's ecological footprint is 50 per cent larger than the planet's capacity to produce these resources. This means that there is an ecological overshoot – it now takes about one year and six months for the planet to regenerate what we use in a single year. You might well ask how can this be possible? Well drawing on the financial analogy used by WWF (2012: 40) just as it is possible to withdraw money from a bank account faster than the interest that this money generates accrues, renewable resources can be harvested faster than they can be regrown. But like drawing on the capital from a bank account, eventually the resource will be depleted. At present, people are often able to shift to another source when a resource is depleted. However at current consumption rates, these sources will eventually run out of resources too – and some ecosystems will collapse.

In the language of sustainable development, to meet today's needs we are raiding our children's inheritance, leaving them with less for the future. If all human beings adopted the lifestyles of people living in the high-income countries of Western Europe and the United States, the world would be living way beyond its means. To sustain the entire world's population with a lifestyle currently enjoyed by the average resident of the United States would require a productive area equivalent to five planets.

A brief summary of ecological footprint data is presented in Table 13.3. For comparative purposes, this table shows results for the same countries as those selected in Table 13.1 (on page 220). Analysing the two tables, therefore, indicates the resources that are being consumed to maintain the present levels of economic activity (GNI). The figures must be read with considerable care. It is evident that on a per capita basis the low and middle-income countries, represented here by India and China, have smaller ecological footprints than high-income countries. However, because of their extremely large populations, both India and China have a far bigger total ecological footprint than many high-income countries, bigger than countries such as Japan, Norway and the UK. Table 13.3 ranks the countries according to their total ecological footprint. In fact, two countries alone – the United States and China – are responsible for over a quarter of the world's ecological footprint; 28 per cent on the basis of 2008 figures and possibly higher today.

Table 13.3 The ecological footprint of selected economies

2008 Data	Population (millions)	Total ecological footprint (million gha)	Per capita ecological footprint (gha/person)	Per capita biocapacity (gha/person)	Ecological surplus (deficit) (gha/person)
World	6739.6	18192	2.7	1.8	-0.9
China	1358.8	2895	2.1	0.9	-1.3
United States	305.0	2193	7.2	3.9	-3.3
India	1190.9	1036	0.9	0.5	-0.4
Japan	126.5	528	4.2	0.6	-3.6
United Kingdom	61.5	290	4.7	1.3	-3.4
Norway	4.8	23	4.8	5.4	0.6

Source: Adapted from National Ecological Footprint Accounts, 2011 edition.

Another way of assessing sustainability at a national level is to compare a country's consumption against its national biocapacity – that is, the resources it can produce sustainably (including managing wastes) within its own borders. A national biocapacity can be expressed in terms of global hectares, or global hectares per person in the country being considered, and national per capita biocapacities for our featured six countries are shown in the fifth column of Table 13.3. Apart from Norway, all the countries in Table 13.3 exceed their own biocapacity. As the final column shows, most countries have a negative ecological surplus – in other words, they have an ecological deficit. This ecological deficit measures the amount by which a country's footprint exceeds its biocapacity. Only about a third of the 150 countries for which a footprint has been calculated show a ecological surplus – they have some ecological reserve as not all their national (biological) capacity is used for consumption, or absorption of waste, by their own populations. This ecological reserve, however, does not necessarily remain untouched, as it may be used by other countries through production for export.

Most countries' biocapacity (determined by the resources available within their borders) is smaller than their ecological footprint (which is based on the resources that they actually consume). You might think that this is particularly the case in high-income countries, and certainly the United States and most nations in Western Europe have significant ecological deficits. However, some developed countries, such as Norway (as Table 13.3 shows), Canada and Australia, have ecological surpluses because they have relatively abundant resources. Conversely, many countries in the developing world have ecological deficits, because they are resource poor or because they have large populations relative to their resource base.

This paints rather a daunting picture, as the only way to reduce a nation's ecological footprint, or more importantly the global ecological footprint, is if populations become smaller, or if we can find a way to reduce the average consumption per person and increase resource efficiency through changes in technology. The latter could be achieved in part by using fewer fossil fuels to reduce carbon emissions and becoming more reliant on energy from renewable sources.

THE ECOLOGICAL FOOTPRINT AND CONSTRUCTION

The scale of the demands made on the natural environment is easily exemplified by construction, as the sector (on the broad definition) consumes resources and generates waste on a scale that completely dwarfs other sectors of the economy. To begin with, it is the environment that provides the land on which buildings and infrastructure are located. Subsequently, it is the environment that supplies many of the resources that are used to make building material products. Finally, it is also the environment that is ultimately responsible for assimilating and processing the waste that arises from the various phases of a property's life, from construction through to demolition. The scale of the problem was discussed in Chapter 11. There we considered the resource intensive nature of construction inputs and outputs, and discussed the problems of an industry that produces 3 tonnes of construction waste for each man, woman and child in the UK

Changes in the ways that property is designed and used could have a significant impact on the scale of the ecological footprint. At present, a large part of the ecological footprint is generated through energy usage, as carbon emissions make up about half of the total footprint. These could be reduced to zero by designing buildings to harness renewable sources, so-called carbon-neutral solutions that rely on wind, sun or geothermal energy sources. Similarly, the **embodied energy** that accounts for a significant amount of the energy used in the construction phase could be reduced by recycling materials, conserving buildings to reduce the need for new build, and sourcing materials locally to avoid unnecessary transportation.

Three Scenarios

Scenarios are often used by economists to explore logical consequences of different sets of conditions or choices. The World Wide Fund for Nature (WWF) has developed a tradition of reviewing ecological footprint accounts in terms of three scenarios: business as usual, a slow shift and a rapid shift. The business-as-usual scenario is based on steady growth of economies and population and it suggests that by 2050 humanity's demand on nature would be twice the biosphere's productive capacity. At this level, the collapse of the ecosystem is highly likely. The other two scenarios however suggest the possibility of achieving sustainability, either as a slow shift from the current development path or a more rapid transition towards sustainability.

The slow-shift scenario suggests that humanity's ecological footprint could be reduced by 40 per cent from 2.7 global hectares in 2008 to 1.5 global hectares in 2100, providing that respective cuts can be achieved in global CO_2 emissions, etc. The rapid-shift scenario, however, could result in an elimination of overshoot by 2080 and a modest 10 per cent biocapacity buffer, which could be used by wild species. This scenario assumes that the demands made on the ecological footprint could be reduced by 50 per cent from 2.7 global hectares in 2008 to 1.3 global hectares in 2100. Achieving this would require reductions in CO_2 emissions, improved technology and significant growth in the capacity of cropland, fisheries and forest yields (WWF 2012).

Figure 13.4 Three ecological footprint scenarios, 1960 to 2100

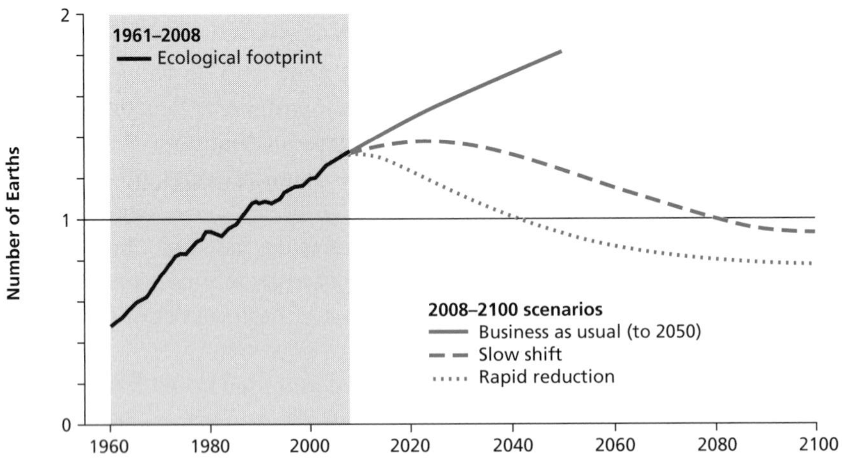

Source: Adapted from WWF, 2006: 3 & 2012

The three WWF scenarios are distinct from previous economic exercises due to the focus on ecosystem services and their ability to support humanity. The scenarios begin in 1960 and run until 2100. Figure 13.4 suggests the outcome of the alternative approaches.

A similar line of analysis was presented by Nicholas Stern (2007: xv) in his international report on *The Economics of Climate Change*. He concluded that the cost of reducing carbon emissions was likely to be about 1 per cent of global gross domestic product, while inaction would lead to a cost of between 5 and 20 per cent of global GDP. In short, Stern found that the costs of a business-as-usual scenario are extortionate and far outweigh the savings that could be achieved by immediate government action.

CONCLUSION

The ecological footprint metric is still being developed and refined. Adopting this measure – or something similar – will be controversial. It forces governments (and people) to compare the current ecological demands placed on the planet with the capacity of the earth's life-supporting ecosystems. It strongly suggests that we are not living within the sustainable limits of the planet. Ecosystems are suffering, and the further we continue down this path of unsustainable consumption, the more difficult it will become to protect (and restore) the planet's biodiversity. Governments will be unable to address this question properly until they have an agreed way to monitor and measure the problem. The challenges are huge. They are addressable, but it will require concerted action at local and international levels and significant changes in resource intensive sectors such as construction.

Key Points 13.7

○ The ecological footprint measures the natural resource consumption required to support the existing standard of living. Calculations have been carried out for 150 nations.

○ The Global Footprint Network's vision is to make the ecological footprint as prominent a metric as the gross domestic product (GDP) by 2015.

○ The global ecological footprint changes with population size, average consumption per person, resource efficiency, changes to technology, recycling, and attitudes to conservation.

14

The Business Case:
Inflation and Expectations

In any government literature that aims to encourage sustainable construction, the key argument is the business case. This makes sense, as the private firms – the primary target for these government appeals – are first and foremost in business. The bottom line means that companies have to earn enough to pay their bills. Consequently, although the ultimate aim of the policy on sustainability is to create a more socially and environmentally responsible construction industry, the business case is seen as the way to drive this forward. The UK government Strategy for Sustainable Construction made the point explicitly in its opening pages when it stated that increasing profitability by using resources efficiently lies at the heart of the case for sustainable construction (HM Government 2008: 5). This requires a stable and competitive economy, run in line with the macroeconomic objectives set out in Chapter 12 (see Key Points 12.1).

At the heart of a stable and competitive economy lies a low inflation rate, and an explicit aim of monetary policy is the control of inflation. If prices are continually changing, entrepreneurs are hesitant to enter into contracts as they cannot work out the long-run results of their investments. This is compounded by the problems caused by changing interest rates and fluctuating foreign exchange rates, which often accompany inflation. It is simply much easier to work within a stable economic environment. Stability means that the costs and prices of any project or investment can be estimated with greater certainty and transparency, allowing businesses to plan with increasing confidence. Indeed, economists define the associated impacts of inflation on business as **menu costs**. These costs arise due to the need to revise existing contracts, new tenders and bids for work as inflation sets in. Obviously, as inflation rates become higher and more volatile, menu costs get more demanding and, in extreme circumstances, they might include costs associated with alterations to vending machines, the costs of printing revised price lists, the time spent renegotiating, and so on.

Inflation also affects those that are not economically active. If prices increase, then those on fixed incomes such as pensioners, students and people on state benefits suffer. In consequence, many aspects of economic activity are index-linked to allow for inflation. For example, savings, business contracts and pensions can all be adjusted in the light of inflation. All that is needed is a reliable price index.

INFLATION AND HOW IT IS MEASURED

Inflation is a *persistent* increase in the general price level.

The italicised word in the definition is important, as any increase in the price level must be sustained to be categorised as an inflationary situation. Continuous annual price rises, however, such as those experienced in the UK, are definitely categorised as **inflation**.

Table 14.1 shows UK annual average inflation rates for the period 1951–2010. It highlights that the 1950s and 1960s were periods of low inflation, with prices rising by less than 6 per cent in most years. In the 1970s inflation rates were much higher and very volatile – in fact, during 1975 alone prices rose by 24 per cent. The 1980s saw the second highest rises in consumer prices in the post-war period, and much of this reflected the increased costs in mortgage payments and council tax. The most recent period shown in the table, from 1990 to 2010, could be characterised relative to the previous two decades as a period of **low inflation** with rates rarely exceeding 3.5 per cent.

Table 14.1 UK inflation rates

Period	Annual average change in price
1951 – 1960	4.1%
1961 – 1970	4.1%
1971 – 1980	13.8%
1981 – 1990	6.6%
1991 – 2000	3.1%
2001 – 2010	2.8%

Source: Office for National Statistics

Deflation, which is defined as a sustained fall in the general price level, is the complete reverse of inflation, even to the extent that it typically results from a fall in aggregate demand, in response to which firms reduce prices in order to sell their products. Inevitably, as a form of price instability, deflation also has adverse economic consequences and its poses particular difficulties for economic management – one being that as interest rates cannot fall below zero, the Bank of England monetary policy committee would not be able to reduce interest rates to stimulate demand. The macroeconomic aim, therefore, is to steer a stable course between inflation and deflation. To achieve price stability, the current target is to maintain the UK inflation rate around 2 per cent – and it certainly should not be allowed to go below zero.

Measures of inflation involve representing changes in price over a period of time. The statistical device best suited for this purpose is index numbers. Index numbers are a means of expressing data relative to a given **base year**. They enable the cost of a particular range of products to be expressed as a percentage of the cost of the same group of products in a given base year. The basic principle is shown in Figure 14.1. The same range of goods must be put into the basket for the base year and comparative year. That is, the system is dependent on like-for-like price comparisons, or as near as possible, from one time period to the next.

This approach is used to produce construction related indices. For example, the Investment Property Databank compiles property information by comparing

Figure 14.1 Calculating a price index

In the example two baskets of goods are compared and a base year of 2000 is selected. 2010, expressed in relation to 2000, gives a price index of 117; in short an increase of 17%.

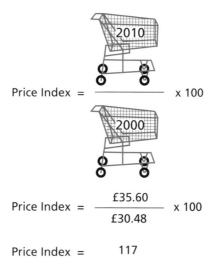

$$\text{Price Index} = \frac{2010}{2000} \times 100$$

$$\text{Price Index} = \frac{£35.60}{£30.48} \times 100$$

$$\text{Price Index} = 117$$

capital and rental values for more than 50,000 commercial properties on a regular basis. The data surveys 23 countries, with some markets going back to a base year of 1981. Similarly, the largest two mortgage lenders – Nationwide and Halifax – publish a house price index each month, based on comparisons of the prices of four different house types across 13 regions of the UK. These two financial institutions base their indices on the mortgage offers they have made in the previous month. Houses costing over one million pounds are excluded, as it is assumed that this may distort the average picture. A more comprehensive but less up-to-date measure comes from the Land Registry. This index is based on stamp duty transactions and, therefore, covers every property deal in England and Wales. However, it is slow to reveal what is happening in the market since the stamp duty is paid at the end of the property transaction and that can be four or five months after the sale price has been agreed – and house prices can rise or fall by as much as 20 per cent within three months. Recently, two websites – Hometrack and Rightmove – have added further indices to this set. They collect information from 4,000 estate agents on asking prices and agreed prices and cover more properties than Nationwide and Halifax. All these indices are monitored and collated by the Department for Communities and Local Government (DCLG) to form a further measure of house prices.

Each house price index produces a different estimate of house price inflation. This is because the indices are dependent both on the sample data that has been used to make up each specific basket and the way that each index is subsequently calculated. For interest, we contrast the DCLG data with the Nationwide, Halifax and Land Registry indices in Figure 14.6 (see page 261).

House price indices are important because they are used by the government and the monetary policy committee as an indicator to assess the economy. For a discussion comparing their strengths and weaknesses, see the Office for National Statistics methodological commentary by Dey-Chowdhury (2007).

Many other institutions and professional bodies also produce specialised indices. Some are particularly relevant for the construction industry. For example, the Royal Institution of Chartered Surveyors produces the **Building Cost and Information Service** (BCIS) which is published quarterly. The two main BCIS indices are briefly considered on page 250, but first we need to explain how the most common measures of inflation are calculated.

Cost of Living Indices

In the UK, the most commonly used price indices are the **retail price index** (RPI) and the **consumer price index** (CPI). The two indices are very similar. They both provide an assessment of the prices of goods and services purchased by the typical household. In fact, a good way to think about the RPI and the CPI is to imagine a shopping basket full of goods and services on which people typically spend their money; but in this case the basket includes everything from food and housing to entertainment. Movements in the RPI and the CPI, therefore reflect changes in the cost of living, and they are the accepted measures used to calculate the percentage increase in inflation.

To be precise, the CPI and RPI measures are based on the regular comparison of the prices of about 650 items. Obviously, it is only possible to calculate a meaningful average price for standard products – goods and services that do not vary in quality from month to month or from one retail outlet to another. The prices of these 650 standard items are recorded at various retail outlets in 150 locations throughout the UK. In comprising this monthly index, therefore, nearly 100,000 (150 times 650) prices are recorded; however, only the national averages are published.

An important objective is to make sure that the basket is representative of the consumer goods and services typically purchased by households. Hence, the basket is reviewed each year to keep it as up to date as possible and to reflect changes in buying patterns. Items are dropped from the basket when they become more difficult to find in the shops or no longer typical of what most people spend their money on. For example, purchases of vinyl records were common during the 1960s and 1970s. However, with the advent of compact discs and music downloads, records now form a niche market and have not been included as a representative item since the early 1990s. Similarly, spending on conventional CRT (cathode ray tube) televisions no longer warrants inclusion, and spending on flat-panel televisions has taken their place. It is equally important that new items are added to represent emerging markets, consequently satellite navigation systems, DAB radios and tablet computers were all added in the last few years.

The various items in the RPI and CPI basket are given a statistical weighting to take account of their importance to the typical household. The items that take more of people's incomes are given a higher weighting. For example, the statistical weight for food is higher than that for tobacco and alcohol, as changes in food prices affect everybody, whereas tobacco and alcohol prices only affect smokers and drinkers.

The contents of the baskets used for the RPI and CPI are very similar but not identical. The RPI basket includes several items chosen to represent housing costs, including mortgage interest payments, council tax and rent, all of which are excluded from the CPI. In contrast, the CPI basket includes the fees paid by people living in communal accommodation such as nursing homes, retirement homes and university halls of residence. These so-called institutional households are excluded from the RPI, as its focus is exclusively on private UK households. In addition, the CPI also includes some items to represent costs faced by foreign visitors to the UK. These subtleties form the main technical differences between the two indices.

The more distinct difference between the indices is in their use and origin. The RPI has been established as a means of measuring price inflation in the UK since 1947. The CPI has its origin in a system that is common to the European Union (EU). It was previously called the **harmonised index of consumer prices**, and it allows direct comparisons to be made with the inflation rates in other member states of the EU. It became the official index used by the UK government to target inflation in January 2004. The CPI tends to produce a lower rate of inflation than the RPI. For example, if CPI is used to compile a version of Table 14.1 showing UK inflation rates, the average rate for the last two decades would be 2.7 and 2.1 respectively; on average, a difference of approximately 0.6 per cent.

In order to distinguish between the two it might be useful to think of a headline rate and an official rate. The **headline inflation rate** is the one that appears in the press and usually refers to the RPI. The **official rate of inflation** forms the target for government purposes and, at present, is the CPI. Finally a passing reference should be made to the retail price index excluding mortgage interest payments (RPIX), as this had been the official rate for government purposes from January 1998 to December 2003. The three measures are compared in Table 14.2.

Table 14.2 Summary of inflation indices

CPI	Inflation measured by the consumer price index, the UK government measure since 2004.
RPI	Inflation measured by the retail price index, in use since 1947.
RPIX	RPI *excluding* mortgage interest payments, was the UK government measure from 1998–2003.

Building Cost Index and Tender Price Index

None of the cost of living indices – CPI, RPI, and RPIX – reflect the prices facing builders and their clients. These indices only indicate the movement in prices of goods and services purchased by the average household, they do not take into account the movement of prices affecting construction. The property and construction consultants Davis Langdon have tracked movements in the input costs of construction work in various sectors since 2000 and these highlight how the rate of increase in construction costs has considerably outstripped the consumer

Figure 14.2 Construction cost indices

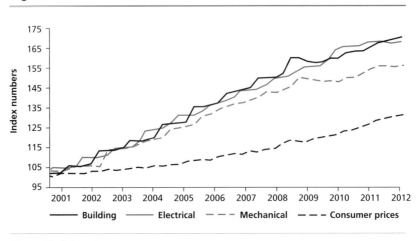

Source: Davis Langdon

price index. As Figure 14.2 indicates, the construction indices show year-on-year inflation for the period 2000 to 2012 but at a far faster pace than that recorded by the consumer price index. The trend shown in Figure 14.2 has been typical for the past hundred years, as construction costs (especially building materials) have always increased faster than the general price index – in other words, the real cost of building has risen. This can prove potentially difficult for companies in the construction industry because the long duration of large contracts means that work is often let at fixed prices before a project commences.

To assist and support the bidding process, the Building Cost and Information Service publishes a building cost index and tender price index that may be used to monitor changes in construction prices. The building cost index measures changes in costs of labour, materials and plant – that is, it covers the basic costs faced by contractors. The basket is compiled from nationally agreed labour rates and material prices. The index also includes forecasts to help predict any changes in prices that may occur in the period between submitting a tender and project completion.

The tender price index involves an analysis of successful tenders for contracts worth more than £250,000. It includes movement in wage rates, discounts, plant costs, overheads and profit – that is, it indicates the basic cost of construction work to the client. In effect, the tender price index is a measure of the confidence in the industry about its current and future workload. When demand for the industry's services is high, not only do contractors' margins increase but so do the margins of their suppliers and wage rates, and you would expect to see rises in the tender price index. And, conversely, when demand for the industry's services decline, all these factors decrease and thereby exert downward pressure on the price index.

The size of the gap between these two indices suggests something about market conditions. During a recession you would expect to see the indices converge, with tenders certainly at a lower level than during a boom. The explanation is simple: when there is less work available, contractors will be satisfied just to get a contract and cover their costs; when the market is buoyant, tender prices will increase as contractors take advantage of the opportunity to more than cover their costs.

As with retail price indices, it is important to make sure that in compiling construction indices the prices being recorded are for comparable items. Are apples being compared with apples, not with oranges? In other words are the baskets being used for comparisons consistent? The physical and functional qualities of the structures being priced must be broadly similar to make a valid index.

Key Points 14.1

○ Inflation is a persistent increase in prices, and controlling it is a major priority for all governments.

○ Many modern contracts are index linked to protect the economically active and those on fixed incomes from the effects of inflation.

○ Price indices compare the current cost (the cost today) with the cost of the same item(s) in a base year.

○ The RPI and the CPI are both used as measures of the general inflation rate in the UK.

○ There are several specialised price indices, and examples that apply to the construction sector include the Davis Langdon construction cost indices, the building cost index and the tender price index.

THE CAUSES OF INFLATION

Although there are many different explanations for inflation, we shall just consider the two main causes. According to these two explanations, inflation either occurs because an increase in demand pulls up prices or because an increase in the cost of production pushes up the price of final products.

Demand-pull Inflation

As we have already explained in Chapter 13 (and reviewed in Key Points 13.4 and 13.5), when aggregate demand in an economy is rising inflation may occur. The severity of the problem depends on how near the economy is to its full capacity level (that is, to full employment). As Figure 13.3 (see page 231) illustrates, there is during any specific time period a fixed output rate in the economy and if demand increases beyond that point the only way businesses can respond is by increasing their prices. Indeed, beyond a point of full capacity no more output is physically possible and the pressures of excess demand can only be countered by price increases. To put it in other words, when total demand in the economy is rising and the capacity available to produce goods is limited, **demand-pull inflation** may occur.

Cost-push Inflation

Prices also rise when the economy is nowhere near full employment. The UK economy experienced inflation during a recessionary period in the mid-1970s and again during 2011. Consequently, other explanations of inflation developed. A common feature of these explanations is the focus on changes in business costs caused by factors such as wage rises, widening profit margins or increased import prices for raw materials.

Let's consider oil prices, as they have a history of rising rapidly. For example, oil prices trebled in 1973–1974 following action by **OPEC** to limit oil production. They more than doubled in 1978–1979. Thirty years later a similar saga played out, with crude oil prices rising from around $30 a barrel at the beginning of the 2000s to more than $100 a barrel in July 2008 and again in 2011. In each instance, most oil importing countries experienced periods of inflation. This is because petrol, propane (LPG), diesel and kerosene (jet fuel) are all derived from crude oil, and oil is also used in the manufacture of chemicals, plastics, fertilisers and textiles. In short, the lifeblood of any developed economy is oil. As the price of oil increases, the costs of production are pushed up, which, in turn, are passed on to the consumer in the form of higher prices. This phenomenon is known as **cost-push inflation**.

Two closing reflections on these periods of oil price instability put the problem of cost-push inflation into a sharper focus. First, one response has been to make energy efficiency an important criterion when designing buildings, and experiments based on passive solar design and other renewable technologies have emerged. Second, alongside the search for alternative energy sources, the increasing cost of oil provides the necessary impetus to develop oil fields that previously were not economically viable, such as those below the ocean bed in seas that exceed a depth of 2,000 metres off the shores of Africa and Brazil and those that lie beneath the ice of the Arctic region.

Other solutions to inflationary problems are more oriented towards government economic strategy, and some of the policies followed in the last forty years or so are briefly reviewed in the next two sections.

CURES FOR INFLATION

Taking a broad chronological overview, three 'cures' for inflation have been deployed in turn since 1970:

- prices and incomes policies
- control of money supply
- interest rate manoeuvres.

Prices and Incomes Policies

During the 1970s, the favoured policy device for stopping, or at least slowing down, cost-push inflation was various versions of wage and price controls. In general terms, these involved employers, unions and governments getting together to agree an annual 'norm' for wage increases. The negotiations, however, proved difficult to manage and, at best, prices and incomes policies only effectively restrained cost-push inflation for a temporary period.

Control of Money Supply

The 1980s witnessed a shift in policy emphasis as some economists argued that inflation, especially demand-pull inflation, seemed to be inextricably related to the size of a nation's money supply. The glib explanation, popularised in the media, was that 'too many pounds were chasing too few goods'. It is difficult to fully appreciate the money supply argument until it is understood that 'money' in a modern society not only comprises notes and coins but also holdings in bank accounts, savings accounts and government bonds. We do not, however, need to be too concerned with the intricacies of this method of control: it has dropped from favour as governments throughout the world experienced too many difficulties in targeting and controlling money supply. In 1977, the OECD published a table showing how the 24 countries that were members of the organisation were using 23 different definitions of money supply. In fact, there were four different definitions of money supply used in the UK during the operation of the policy.

Interest Rate Manoeuvres

Since the 1990s, governments have tried to bring inflation under control using interest rates. This is done on the understanding that the prevailing rate of interest significantly influences spending decisions, and, in particular, affects the decisions of both businesses and households on whether to borrow (that is, to incur debts) to pay for consumption and investment goods. As interest rates become higher, and more volatile, businesses and consumers generally become less confident about making new investments and negotiating future contracts. That is, other things being equal, higher rates of interest should encourage saving and discourage investment and consumer spending. To follow one sequence, higher interest rates will tend to increase the cost of financing house purchases, and so reduce demand and lower (or slow the rate of increase of) house prices. In summary, therefore, changes to interest rates have a strong influence on the level of spending in an economy.

As detailed in Chapter 12, the process of deciding interest rates begins with the monetary policy committee (MPC). The MPC has full responsibility for determining the rate of interest used by the Bank of England when dealing with other financial institutions trying to raise funds in the money market. The official **base rate** is sometimes referred to in monetary circles as the **repo rate** (which represents the rate that a central bank is willing to lend funds to other banks). This phrase relates to repurchase agreements (and sales) of assets such as government bonds between the Bank of England and its counterparties in the money market. The eurozone equivalent set by the European Central Bank is called the **refinancing rate**, and the interest rate used by the Federal Reserve in America is called the **discount rate**. At the introductory level you don't need to worry too much about the difference between the American, British or European terminology: the important point is that in each case the central bank sets a carefully considered unique rate of interest that determines the rate at which it will lend short term to the banking sector. Subsequently this determines the short-term rate of interest that banks charge each other for loans. (This rate is sometimes called LIBOR, which is short for the London interbank offered rate. The precise sequence is neatly captured in Figure 14.3.)

In turn dealings in the interbank market affect the interest rates adopted in the wider economy by all other credit agencies and financial institutions, such as the high street banks. No bank would lend to customers at a lower rate than the one at which it borrows, as it is in business to make a profit not a loss. So, when banks find themselves facing more difficult economic times, they tend to increase the rates charged on their loans and they might even decide to ration the quantity of credit they are willing to extend – in effect, both these actions were captured in the phrase **credit crunch**. More importantly it sowed the seed that triggered the financial crisis that dominated the headlines from 2008 to 2012, and this is explained further in the next section.

Key Points 14.2

○ Demand-pull inflation occurs when the total demand for goods and services rises faster than the rate of growth of supply.

○ Cost-push inflation is due to one or more of the following: (a) wage rises, (b) widening profit margins, and/or (c) raw materials price increases.

○ Prices and incomes policies, control of money supply and managing interest rates have all been used as attempts to reduce inflationary pressures.

THE CREDIT CRUNCH

The so-called credit crunch emerged from instability in the US mortgage markets during 2007, where a combination of rising interest rates and falling house prices had exposed poor quality lending, which led to a sharp increase in mortgage defaults. Subsequently a period of panic struck financial markets across Europe and America; uncertainty spread, loan defaults increased, liquidity evaporated, and central banks were called upon to throw financial lifelines and introduce new instruments to keep many institutions afloat. The over-riding problem was that banks had insufficient capital to meet their obligations, and a significant number of American and European banks and mortgage lenders were forced into protective mergers, nationalisation and even bankruptcy. In fact, Countrywide, Fannie Mae, Freddie Mac, Bear Stearns, Lehmans, Merrill Lynch, Northern Rock, Alliance and Leicester, HBOS, Bradford and Bingley, KSF, Dexia and Bankia all fell victim to the credit crunch debacle.

The surviving banks and building societies tightened lending criteria and raised their rates of interest despite the fact that central banks such as the Bank of England cut their rates several times. This action was unprecedented as the usual practice is for bank interest rates to mirror adjustments and certainly to move in the same direction as changes made to the official (repo) rate – and usually on the same day. Between 2008 and 2012 however, institutions in the financial market did not react to repo rate changes in this way; they began to revise their margins and increase the costs of borrowing to move to a new market equilibrium. In short, the financial crisis represented a period of time when global financial markets sought correction.

Figure 14.3 The money market

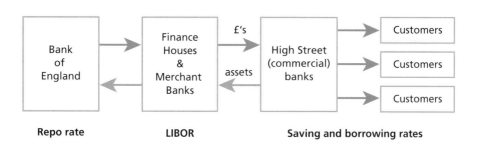

Part of the problem was the fact that central banks kept their rates close to zero for a significant period: a base rate of 0.5 per cent led the interbank market for more than three years from March 2009 onwards. However, a low base rate no longer provided a guarantee that other rates would stay low or that funds would actually be lent. The credit crunch had led to an abrupt change, financial stability was replaced by instability and markets went into 'crisis' mode. There was a distinct loss of trust between banks, and lending between banks dried up. They also became reluctant to lend to borrowers. This had severe consequences for the wider economy. Parkinson et al. (2009: 4) noted the downward cycle in the property and construction sector, where it quickly became evident that the financial turmoil meant that lenders would no longer lend, borrowers could not borrow, builders could not build and buyers could not buy. In the simplest of terms, the credit crunch had starved the economy of its productive capacity.

This raises a significant policy challenge, particularly if monetary policy focuses too much on inflation and not enough on financial stability. When official interest rates get close to zero the effect they have on regulating the economy becomes muted. Consequently, governments seek another way of affecting the price of money. One approach is to increase the supply of money circulating the economy. This is done through a process known as **quantitative easing**, and it has only been tried in Japan, the United States and the UK. Between March 2009 and July 2012, the Bank of England pumped an eye-popping £375 billion into the UK economy through this system. Note that the aim of the central bank here is still to bring down interest rates faced by companies and households and, most importantly, to create new money for use in an economy.

The monetary impact of quantitative easing spills out of asset trading in the repo market, and again Figure 14.3 is useful as an aid here. If a central bank such as the Bank of England wishes to inject money into the economy it can purchase assets (such as government bonds or high-quality debt issued by private companies) from insurance companies, pension funds, banks or non-financial firms. The outcome – regardless of the particular assets purchased – is that the seller's bank account is credited and the system finds itself with more funds. Moreover, when a government buys assets in such large amounts it tends to push their price up and the yield rate down, as the rate of return on these types of asset is usually fixed. In theory, therefore, quantitative easing increases money supply and lowers long-term interest rates. This approach should boost the economy in a variety of ways but unfortunately this has not happened, at least not to the extent that the Bank of England thought was necessary to revive consumer spending and economic growth.

The only consolation for governments pulling levers without getting the desired results is that it takes time for this type of economic instability to play out. The Bank of England macroeconomic model recognises that interest rate changes can take up to one year to affect demand and output, and nearly two years for all the effects to be reflected in the inflation rate (MPC 1999: 3). With the quantitative easing experiment, however, the evidence is not so clear. It has only been used as an emergency measure during the recession and the jury is still out as to whether it boosts or destroys confidence in an economy. It could easily prove to be counter productive. Banks may well find themselves holding more reserves but it is questionable whether this will encourage them to increase lending to consumers and businesses; they may prefer to hold the extra reserves without increasing lending, and this could ultimately lead to inflationary pressure.

Quantitative easing is an unorthodox policy and, to date, has only been used in a crisis. Nobody knows how much is required or how long it will take to have the desired effect. At the time of writing the Bank of England is working with the Treasury to launch a new emergency measure, the 'funding for lending scheme', designed to boost lending to businesses by allowing banks to borrow from the government. While this makes sense in theory, it remains to be seen if it promotes some growth in the economy. But it does act as some kind of acknowledgment that quantitative easing may not work.

Without doubt greater financial stability is required if economies are to recover. Businesses and individuals have had to adapt to new market rates of interest on their investments, savings and debts. Businesses have seen significant cut backs in the quantity of orders they have been received. Employment has fallen and earning levels have reduced and, through a multiplier effect, this has led to further reductions in domestic demand. Large firms have had to tighten up the contractual terms offered to suppliers and smaller firms often failed to survive. Many small and medium-sized firms became insolvent. Between 2009 and 2011 more than 50,000 companies registered as bankrupt in the UK and, of these, a significant 20 per cent were from the construction sector (Insolvency Service 2012).

Clearly the credit crunch has left a serious scar on the world's financial system. Greece, Portugal and Ireland have required formal assistance to alleviate problems of financing their national debt. Many banks in the United States and Europe have had to be rescued. Revelations about price rigging in the LIBOR market and money-laundering scandals have further damaged the reputations of some banks. Serious questions have been raised about the regulatory system in the financial sector. Recommendations to improve the regulatory regime and to regain confidence in the banking system across Europe are currently being considered to stave off further financial crisis. Going forward, there will be more extensive rules governing banks.

In broader policy terms, monetary policy should become equally concerned about its two areas of responsibility, namely maintaining financial and price stability. In the next two sections, we broaden out the theoretical economic arguments that explain the importance of managing expectations. This is presented, in its traditional context, as a way of maintaining price stability, but these theories could also help to inform the new emerging financial frameworks.

Key Points 14.3

○ The credit crunch represents a period of tighter lending conditions, both in terms of cost and availability. It raised questions about the effectiveness of lowering interest rates as a means of promoting economic growth.

○ Quantitative easing is a way of managing long-term interest rates, but whether it effectively promotes lending or assists in maintaining price stability is in debate.

○ The financial crisis of 2008 to 2012 highlighted the need to develop new and tighter ways to regulate the banking sector.

○ Monetary policy should be equally concerned about maintaining financial and price stability.

EXPECTATIONS

Today, economists and politicians believe that expectations influence all types of economic variables including inflation. In the past, these expectations had been considered as external influences that were beyond control. In formal terms, they represented an example of an exogenous variable. Modern economic theory, however, seeks to understand behaviour that is based on expectations. Initially, during the 1970s, the **adaptive-expectations hypothesis** was accepted.

Adaptive Expectations

The basic idea behind this theory is that people take what happened in the past as the best indicator of what will happen in the future. Look at Figure 14.4 (on page 258). Here we represent the actual rate of inflation by a black dashed line. Inflation is rising: in year one it is 6 per cent, during the second year it is 8 per cent, and in the third year it jumps to 12 per cent; thereafter it stays at 12 per cent.

According to the theory of adaptive expectations, the expected rate of inflation in any given year is whatever the rate of inflation was in the previous year. The coloured line in Figure 14.4 represents this expected inflation rate. During the second year, people believed inflation would be 6 per cent because that is what it was the year before. During the third year, they believed it would be 8 per cent (because that is what it was the year before). In this simplified model, people's expectations are always behind reality. When the rate of inflation is rising, they will always believe that inflation is going to be less than it actually turns out to be; when the rate of inflation is falling, they will always believe it is going to be more than it actually turns out to be. Only when the rate of inflation remains constant for a sustained period of time do expectations come into line with reality.

Figure 14.4 Adaptive expectations theory

According to this hypothesis, individuals formulate their expectation of inflation according to the rate in the previous year. In the figure, the actual rate of inflation is shown in black and the expected rate of inflation is shown in colour – and it clearly lags one year behind in pattern. For example, in year 2 people expect inflation to be 6% because that is what it actually was in year one.

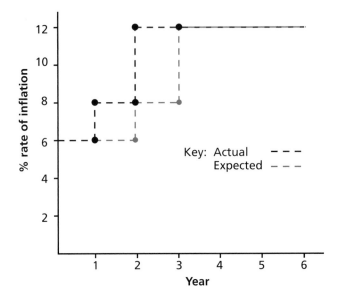

A more sophisticated model of expectations would not allow people to be consistently proved wrong. It would recognise that opinions are based on a range of data, which includes historical trends but not to the exclusion of everything else.

Rational Expectations

The **rational-expectations hypothesis** is the logical development of the adaptive-expectations model. It is more concerned with human behaviour and it allows for the sifting and weighing of all available information. In this theoretical framework, people do not simply look at what has happened in previous years. They consider what has happened in the past and combine it with their knowledge of current government policies and any anticipated policy changes. Rational expectations, therefore, require judgement about current and future policies combined with lessons from the past. For example, the shrewd forecaster considering house prices in 2006 would recognise that prices do not continue to rise forever – and that the rises in previous years could not necessarily be sustained.

Rational expectations take on more importance in some sectors of the economy than others. For example, city brokers, investment managers, property developers, fund managers, estate agents and those involved in other valuations depend upon their specialist skilled knowledge to make shrewd judgements. It is debatable, however, whether managers, workers and consumers in the general economy can be expected to have developed the same level of expertise. Consequently, the general public often adopts the expectations of the professionals. This may be better than the 'blind leading the blind', but it can lead to scenarios where power psychology seems to generate interesting outcomes. Two detailed examples are considered next.

THE WAGE-PRICE SPIRAL

Expectations can be regarded as the driving force that underlies the beginning and end of a wage-price spiral. If we consider Figure 14.5 it is evident that, regardless of the initial cause of inflation, workers eventually obtain a wage increase which subsequently leads to prices being marked up. The interesting point, however,

Figure 14.5 The wage-price spiral

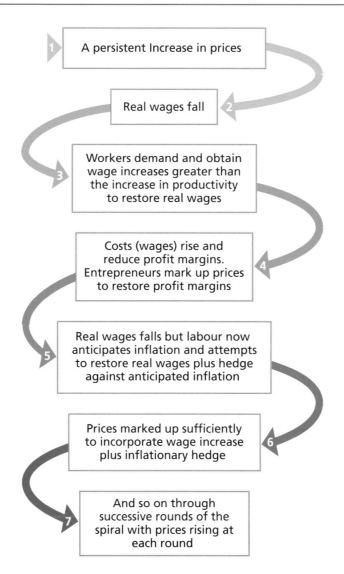

1. A persistent Increase in prices

2. Real wages fall

3. Workers demand and obtain wage increases greater than the increase in productivity to restore real wages

4. Costs (wages) rise and reduce profit margins. Entrepreneurs mark up prices to restore profit margins

5. Real wages falls but labour now anticipates inflation and attempts to restore real wages plus hedge against anticipated inflation

6. Prices marked up sufficiently to incorporate wage increase plus inflationary hedge

7. And so on through successive rounds of the spiral with prices rising at each round

is that what perpetuates and accelerates this spiral-type behaviour is expectations; most notably the expectations of those in charge of governments, trade unions and employer confederations. In turn these expectations are exacerbated further by the media coverage of inflation (Blanchflower 2008: 11).

Consider the position at box 5 in Figure 14.5. At this stage in the proceedings, it is most likely that labour correctly anticipates further inflation. Workers not only negotiate a wage increase to restore their purchasing power, but they also seek to incorporate a 'hedge' against future inflation. For example, if there has been a period of inflation at 15 per cent, workers will demand a greater than inflation pay increase of, say, 30 per cent. This would allow 15 per cent for the previous inflation, 10 per cent for anticipated inflation, and 5 per cent which can be given up during negotiations. Next, we can see – in box 6 – that entrepreneurs employ a similar rationale and increase prices sufficiently to cover the total increase in their wage bill plus a precautionary amount to avoid having to make further price increases too soon. In both cases, the wage claims and price increases are being made on the basis of judgements about the future. By box 7, therefore, it should be evident that judgements anticipating inflation become firmly entrenched and create a self-fulfilling prophecy; with workers seeking compensation for rising prices through higher wages and companies attempting to pass on cost increases to their customers. In short, it is rational expectations that drive the wage-price spiral.

Since the late 1980s, governments have tried to break the mentality associated with the wage-price spiral by convincing businesses, workers and the general public that their policies should slowly eradicate inflation, thereby encouraging people to lower their judgements about the future inflation rates. Indeed, this is why governments in countries such as New Zealand, Canada, Sweden and the UK introduced the idea of explicitly targeting inflation. As economic advisers to the UK government were keen to point out, an explicit target for the monthly consumer price index is desirable as 'it provides a clear anchor for inflation expectations, and makes it easier to attain price stability' (Balls and O'Donnell 2002: 72). Consequently, whenever the trend of low inflation shows signs of changing course, government ministers carefully account for these 'blips' in the monthly CPI figures and address people's worries as expectations must not be allowed to change. The problem with economic forecasting, however, is that unforeseen events occur, 'blips' can turn into 'blots', and prices can start to rise again. Once this occurs, rational expectations will see the wage-price spiral accelerate upwards with a vengeance. This underlines the fact that the psychology of expectations plays a crucial part in determining all economic destiny, especially aspects relating to financial and price stability.

TIME-LAG

A similar type of analysis can be applied to those market sectors in which the nature of the product means that a long time elapses before changes in output can be realised. The markets normally considered in this context are agriculture and construction. The problem is that the inflexible nature of these markets causes them to respond slowly to price changes. Furthermore, these markets are often structured in such a way that factors such as trade union practices, training requirements, information flows and legislation also restrict the dynamics of the market. Economists refer to these problems as structural rigidities.

Construction markets – as the cyclical booms and slumps demonstrate – struggle to arrive at stable market prices. The cobweb theorem can be used to explain the circumstances. As we described in Chapter 6, construction can be a lengthy and fragmented process. For example, it takes approximately 18 months to two years to complete a house from the initial planning stages and up to five years to deliver a modern office block. As a result, price instability within the construction sector and its associated markets is quite common. As an example, we show the percentage change in house prices according to four sets of indices in Figure 14.6. Though there is some divergence between them, the four indices each illustrate the key point that house prices tend to be unstable.

Figure 14.6 UK house price inflation

The graph shows the unstable nature of prices between 2000 and 2010. The cobweb theorem, illustrated in Figure 14.7, helps to explain these price fluctuations.

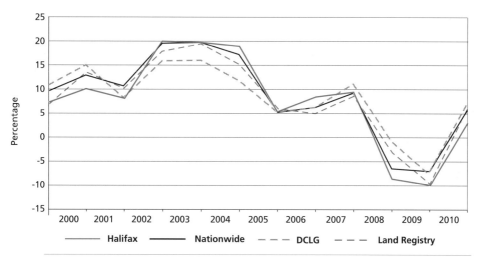

Source: DCLG (2011) *House Price Index*

To account for this instability, we need to accept two distinguishing features of the construction industry.

- The decision of builders to change their output is heavily influenced by current market prices. So, for example, house supplies in two years time are dependent on today's prices.
- There are many small building firms and they each take decisions to adjust their scale of production in isolation from each other.

If buildings happen to be scarce in one time period, then high market-clearing prices will prompt construction firms to begin new projects. This rise in production will in due course depress market prices, which sets off a major contraction in the scale of production. We show this process in Figure 14.7. The fixed amount of property available in the short run means that the market-clearing price may rise to P_1. The supply curve indicates that, at price P_1, construction firms would like to produce Q_2, although they cannot do so immediately. Building contractors begin to make plans to adjust output based on the current market price and try to produce Q_2. After a time-lag, a potentially unstable situation is likely. Indeed, assuming the pattern of demand remains unchanged, and output Q_2 is actually built, prices will have to drop to absorb the excess supply. This will ultimately result in a contraction of production, leading once more to higher prices. Note that in this example, market prices move above and below the equilibrium level and are not stable. This price instability could be greater or less depending on the elasticity of the supply and demand curves.

Figure 14.7 A cobweb diagram showing how property prices can fluctuate

Commencing from the equilibrium price, a change in the conditions will move the market from its equilibrium position. In the short term, the market is faced with a perfectly inelastic supply curve, as shown by the dashed line (P_{ES}). The longer-run supply curve S indicates that at price P_1, construction firms would like to build Q_2. These structures will take some time to construct. However, when Q_2 is eventually completed, the price will need to fall to P_2 to absorb the excess supply. This will result in a subsequent contraction of supply as shown by position 3. The shortage of supply will cause prices to be bid up and the process continues.

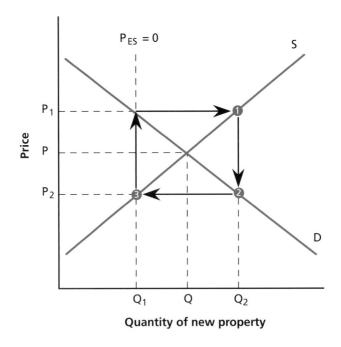

The pattern of oscillation around the equilibrium point is accounted for by the inherent time-lag in producing construction output, by imperfect information and, to a lesser extent, by the inflexible nature of the construction market. The oscillating pattern also provides the explanation for the name of this model – the cobweb theorem.

Obviously, there are limitations to this generalised theoretical model. First, builders are aware of the cyclical nature of their industry and attempt to make adjustments to relate starts to expected prices rather than current prices. Second, they can choose to manipulate their stocks by holding property off the market during the low price part of the cycle. Nonetheless, the economic analysis is still useful since it highlights the root of a problem, and amplifies the need for a stable economy to secure the necessary business confidence for construction firms to maintain a steady rate of production.

Key Points 14.4

○ The adaptive-expectations hypothesis argues that people make predictions about the current year's rate of inflation based on the previous year's rate of inflation.

○ The rational-expectations hypothesis assumes that individuals form judgements by examining all available information. That means that economic forecasters not only look at past data and trends but also at the impact of current and/or anticipated future government policy.

○ The wage-price spiral represents the inflationary process as one in which incomes and prices continually chase each other in an upward direction (see Figure 14.5).

○ Construction firms operate in a different business environment as significant time-lags exist before any planned changes in output can be realised. These production time-lags together with several structural rigidities cause construction markets to be unstable.

○ It is rational expectations that determine the speed and direction of price changes.

Reading 5

An important theme in Part C has been the idea that construction plays a leading role in the process of national development and macroeconomic management. The following two extracts analyse the shape of the Bon curve, which is derived by plotting GDP per capita against expenditure on construction to show the importance of this sector to economic growth. The first edited extract is based on the empirical evidence from 39 countries, at different stages of development, during 1994–2000. The authors adopt a clear and general approach to introduce and interrogate Bon's theory. The second extract is slightly more advanced, in the sense that it seeks to question the exact shape of the curve for smaller nation states, such as Hong Kong, Singapore, and Trinidad and Tobago, and provides a case study approach that focuses on data relating to Cyprus during 1998–2005. Some obvious questions are suggested by studying these extracts. For example, does the data reflect the current relationship between construction output and GDP in your country? In other words, use recent construction data to clarify your understanding of the sector's broader impacts on the current pace and direction of economic growth. What do you think of the closing idea posed in the second extract that micro-states with fragile ecosystems will inevitably experience sharper limits on construction growth than larger scale economies? Finally, what are the implications of the dynamic role of construction for macroeconomic management? In particular, consider the importance of government policies designed to secure growth, maintain stable prices and protect the environment.

Les Ruddock and Jorges Lopes (2006) 'The construction sector and economic development: the Bon curve' *Construction Management and Economics* 24: 717–23

Introduction

The relationship between a country's state of development and the level of activity in the construction sector is one, which has been the subject of study at the macroeconomic level for a number of years (Turin 1973; World Bank 1984; Wells 1987; Bon 1990). A major obstacle to such studies has been the lack of appropriate information on the sector, particularly in developing countries....

Bon (1992) analysed the changing role of the construction sector at various stages of economic development and presented a development pattern for the industry based on the stage of development of a country's economy. This notion is well-explained in basic terms by Tan (2002: 593): "In low income countries, construction output is low. As industrialization proceeds, factories, offices, infrastructure and houses are required, and construction as a percentage of gross domestic product reaches a peak in middle income countries. It then tapers off as the infrastructure becomes more developed and housing shortages are less severe or are eliminated." An important aspect of the proposition was that, in the early stages of development, the share of construction increases but ultimately declines, in relative terms, in industrially advanced countries – and even at some stage, the decline is not only relative but also in absolute terms, that is 'volume follows share'....

In Bon's 1992 paper, the link between economic development and construction is discussed; his basic premise is that the share of construction in total output first goes

Figure 1 Bon curve

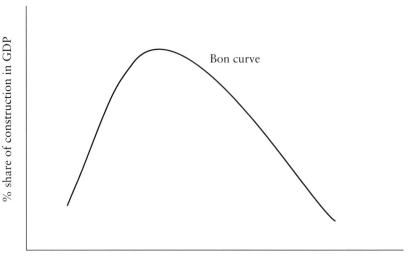

up and then comes down with economic development, and this is called the inverted U-shaped relationship plotted in Figure 1.... His paper concerned the entire path, from the least developed countries (LDC) through to those that are classified as advanced industrialized countries....

Data

The indicator of construction industry activity used for this analysis is gross value added (GVA) in construction.

GVA in construction is calculated the same way as in any other sector, but includes only the activities of the construction activity proper. For example, it excludes the building materials industry which is accounted in the manufacturing sector. The main indicator of general economic activity used in this study is GDP per capita. It adjusts the growth in the economy with the growth in population. It is a better indicator of a country's welfare particularly in developing nations, where the growth rate of population has been since the Second World War roughly twice as high as in developed economies....

Does the volume follow the share?

The analysis depicted clearly demonstrates that the inverse 'U'-shaped pattern holds for the share of construction in the national

economy. That is the share of construction in total output first goes up and then comes down with economic development. Table 1 shows Bon's grouping of countries, according to their status of development, categories 1 and 2 taken together represent the LDCs (and those that are emerging such as India and China), category 2 the newly industrialized countries (NICs) and category 3 represents advanced industrialized countries the AICs.

What about Bon's proposition that at some stage the decline of the construction industry activity switches from relative to absolute decline, that 'volume follows share'? Looking at Table 1, it can be seen that the share of construction in the advanced industrial countries increased by 6.5% in the period 1994–2000, which suggests that construction volume did not decrease in these countries over the period. This is consistent with Carassus' (2004) findings that the construction volume, measured in constant prices, in seven out of eight advanced industrial countries did not decline in the period 1990–2001. OECD (1998) also shed new light on the argument, as data for the European Union 15 shows that both GDP and gross fixed capital formation (GFCF) increased in the period

1960–1996, though remained generally stagnant in the period 1974–1982. The data also shows that the growth in GDP is higher than that of GFCF, thus the relative decline in capital formation is also evident for the entire period....

We do believe that Bon's (1990) analysis which concerned the period 1970–1985 was certainly influenced by the recessive period 1973–1982, characterized by two oil shocks, that followed the period 1960–1973 – the 'golden age' of the world economy...

Table 1 Relative growth rate in construction (1994–2000)

GDP per capita in 1994 ($US)	LDC		NIC	AIC
	<1000	1000 to 2499	2500 to 9999	10000 or over
Relative change in construction (%)	-5.0	11.1	10.7	6.5

Note: The figure shows the average for the countries in each category

References

Bon, R. (1990) The World Construction Market 1970–85, in *Building Economics & Construction Management: Proceedings of the CIB W65 Symposium*, Sydney.

Bon, R. (1992) The future of international construction: secular patterns of growth and decline. *Habitat International*, 16(3), 119–28.

Carassus, J. (2004) *The Construction Sector System Approach: An International Framework*, CIB: Publication 293, CIB, Rotterdam.

OECD (1998) *National Accounts: Main Aggregates 1960–1996, Vol. 1*, Organisation for Economic Co-operation and Development, Paris.

Tan, W. (2002) Construction and economic development in selected LDCs: past, present and future. *Construction Management & Economics*, 20, 593–9.

Turin, D.A. (1973) *The Construction Industry: Its Economic Significance and its Role in Development*, UCERG, London.

Wells, J. (1987) *The Construction Industry in Developing Countries: Alternative Strategies for Development*, Croom Helm Ltd, London.

World Bank (1984) *The Construction Industry: Issues and Strategies in Developing Countries*, The World Bank, Washington, DC.

....

Extract information: Edited and adapted from pages 717–18 and 721–2, plus relevant references from page 723.

Second part of Reading 5 continues overleaf

Ozay Mehmet and Vedat Yorucu (2008) 'Explosive construction in a micro-state: environmental limit and the Bon curve: evidence from North Cyprus', *Construction Management and Economics* 26: 79–88

The shape of the Bon curve

Although the existence of the Bon curve [see Figure 1, page 266] enjoys general acceptance, its shape is not yet settled. The exact nature of the construction–GDP relationship is a matter of empirical question and several questions of theory emerge over stages and transformations, over and above problems of statistics, sources and methodology. Thus, some question the slope of the Bon curve during the drive to, and beyond the maturity stage. Accordingly, as economies 'mature', construction activity slows down and the declining portion of the Bon curve becomes flatter, indicating a declining importance of the construction sector in the economy....

Structural transformation within the construction sector

Within the aggregate relationship revealed by the Bon curve, there are important transformations because the construction sector is a diverse sector, consisting of residential, institutional, commercial, industrial, infrastructure, maintenance and other subdivisions. During the course of economic development, the relative significance of these construction sub-sectors are altered in tandem with the Colin–Clark hypothesis (Mehmet 1999: 96) that with a rising GDP, the structure of the economy changes from being agrarian to industrial and then to service-based economic activity. Pietroforte and Tangerined (1999), using disaggregated Italian data for the 50 largest construction firms, have demonstrated that during the 1980s and 1990s these firms have responded to changing industrial policy to adapt a service orientation in construction activity in order to remain competitive. One finds Clark-type evidence in earlier construction activity research as well. Thus, Strassman (1970) argues that the construction sector, like agriculture or manufacturing, follows a pattern of change that reflects a country's level of development. After lagging in early development, construction accelerates in middle income countries and then falls off. The extent of construction activity in an economy is closely linked in particular to the extent of urbanization in the economy. The share of urban population increases initially and then decreases as GDP per capita rises.

Does size affect the Bon curve?

The importance of the size of an economy has traditionally been ignored. Generally smallness has been considered to be a negative, limiting factor on economic development....But this is misleading as some of the most prosperous countries in the world are micro-states, with less than 10,000 km sq, such as Singapore, Cyprus, Trinidad and Tobago, Hong Kong, etc. (Mehmet and Tahiroglu, 2002). At the same time, however, construction activity may be more volatile in micro-states. Thus one major project coming to an end can have a major impact on the aggregate data, whereas in larger economies any single project will only form a relatively small proportion of total construction output. This may exhibit greater short-term volatility than the large countries. In any event, construction activity in micro-states is an important topic in its own right given the fact that some of the wealthiest countries in the world are micro-states....Accordingly, it is not an idle question to inquire whether or not construction in small states contradicts or follows the shape of the Bon curve....

In Trinidad and Tobago Ramsaran and Hosein (2006) found that construction has become very much a leading sector of the economy, manifesting a powerful multiplier impact in both boom and bust periods, growing at twice the GDP rate and that the contribution of construction to economic development was positively correlated with the domestic resource content of construction output. This is the pattern observed in North Cyprus as well (Yorucu and Keles, 2007) where small size is encountering a critical ecological limit which larger states may be better able to postpone.

The Bon curve and the environmental limit — a hypothesis from North Cyprus

The other major consequence of smallness is the prospect of an earlier environmental limit than in larger states. Concern for the protection of a fragile ecosystem, watershed or natural heritage becomes evident as income per capita rises, and when explosive construction-led growth occurs, as in the case of North Cyprus after 2003, public demand for tougher environmental standards may lead to a sharper downturn in construction than in larger states. Theoretically, this implies that in micro-states the declining portion of the Bon curve may be steeper than in larger economies. In other words for micro-states like North Cyprus, an inverted V–shape may be more appropriate than an inverted U–shape and this is portrayed in Figure 2. In micro-states with land constraint and access to cheap labour supply, the expansion portion of the Bon curve may be steeper, the peak much more pronounced and the declining portion relatively flatter compared to the traditional Bon curve....

Figure 2 Bon curve

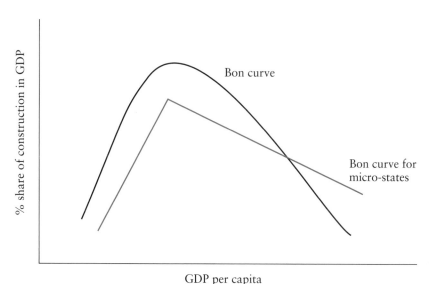

References

Mehmet, O. (1999) *Westernizing the Third World, Eurocentricity of Economic Development Theories*, Routledge, London and New York.

Mehmet, O. and Tahiroglu, M. (2002) Growth and equity in microstates: does size matter in development? *International Journal of Social Economics*, 29(1–2), 152–62.

Pietroforte, R. and Tangerined, P. (1999) From boom to bust: the case of Italian construction firms. *Construction Management and Economics*, 17, 419–25.

Ramsaran, R. and Hosein, R. (2006) Growth, employment and the construction industry in Trinidad and Tobago. *Construction Management and Economics*, 24, 465–74.

Strassmann, W.P. (1970) The construction sector in economic development. *Scottish Journal of Political Economy*, 17(3), 390–410.

Yorucu, V. and Keles, R. (2007) The construction boom and environmental protection in Northern Cyprus as a consequence of the Annan Plan. *Construction Management and Economics*, 25(1), 77–86.

....

Extract information: Adapted from pages 80–2 plus relevant references from page 87.

15

Sustainable Construction

Throughout this book we have tried to capture the main ideas that underpin the concept of sustainable construction. In this final chapter, we revisit some of the central issues that shape the sustainable construction agenda. At the outset, it is important to remove any lingering confusions caused by the term 'sustainability'. Sometimes, sustainability is used solely to refer to concerns surrounding the natural environment; at other times, it seems to have a broader connotation, including two other integral strands – sustainable communities and sustainable businesses. The narrow environmental focus is perhaps understandable – both for historical reasons and because, in discussing the construction industry, environmental impacts are significant. However, in this book, we have tried to stress that sustainability is formed of three constituent parts – the community, business and the environment – and this is the focus of the chapter.

SUSTAINABLE DEVELOPMENT

Sustainability is often misinterpreted as being synonymous with the terms 'green' or 'environmental'. However, as we explained in Chapter 2 (see, for example, Key Points 2.4) and reiterated above, this only represents a third of the agenda. It is impossible to talk of one strand of the sustainable development agenda without considering the others – especially if the agenda is to become more than a theoretical construct. Indeed, it is impossible to make an environmental decision without there being economic and social implications.

A useful starting point, therefore, comes from confirmation that sustainability embraces the three themes of environmental, social and economic accountability – often known as the triple bottom line. Sustainability can be achieved by minimising negative impacts and maximising benefits. The best way of doing this is to look for solutions that solve more than one problem at a time. These solutions are known as 'win-win-win', as they secure economic, social and environmental benefits simultaneously. Win-win solutions and triple bottom lines are an ideal way of thinking about sustainability as they emphasise the need to integrate social, environmental and economic issues.

A brief chronology of the development movement shows how the broader concept of sustainability has emerged. During the post-war period in the 1950s and 1960s, development was mainly discussed in relation to the less developed countries. Up until the 1970s, therefore, development thinking was concerned primarily with economic growth and its distribution. A common reference point from this period was Rostow's model of economic growth (see Figure 15.1 on page 272) which portrays development as a set of five linear stages. From this perspective, a key policy aim was to assist the less developed countries through aid and technology transfer to help them reach the critical 'take off' stage. In Chapter 13, we discussed some of the ideas which underpin this model of economic development. For example, we noted that investment in construction is especially important during the development phases of an economy (see, for example, Key Points 13.2).

Figure 15.1 Rostow's stages of economic growth

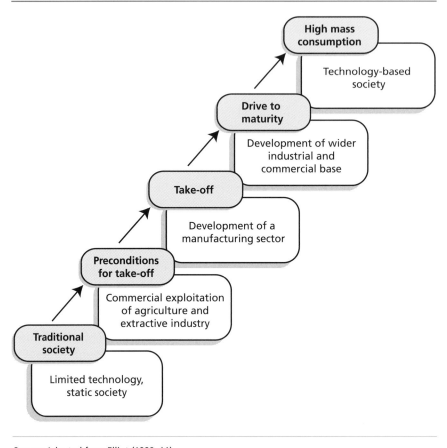

Source: Adapted from Elliot (1999: 11)

The origins of the sustainable development movement date from the 1970s. In this period, the environmental concerns of the world's poor – such as a lack of clean water or sanitation – began to take centre stage. In 1973, for example, the United Nation's Environment Programme (UNEP) was established to meet demands for an international environmental watchdog. In short, issues relating to the environment and development began to be viewed as interdependent.

The next milestone was the 1987 report of the World Commission on Environment and Development. Entitled *Our Common Future* (WCED 1987) – though more commonly referred to as the Brundtland report after the commission's chair, the former Prime Minister of Norway, Gro Harlem Brundtland – this report was translated into 24 languages and popularised the idea of sustainable development as being concerned with intergenerational equity and justice.

Perhaps the most well-known definition of **sustainable development** comes from the Brundtland report (see the first definition in Table 15.1). However, there are more than 70 definitions – Pezzey (1989) lists 60 – and Table 15.1, which lists just three interpretations, gives some idea of the breadth of the concept. For example, a comparison of the first and last definition in Table 15.1 shows how over a 25-year period the vision of sustainable development has been extended to include the mitigation of climate change and the protection of ecosystems. Clearly the definitions are sufficiently vague to become a breeding ground for disagreement.

Table 15.1 Three interpretations of sustainable development

1	Development that meets the needs of the present without compromising the ability of future generations to meet their own needs (WCED 1987: 43)
2	Development that delivers basic environmental, social and economic services to all residences of a community without threatening the viability of natural, built and social systems (ICLEI 1996)
3	Development that eradicates poverty, reduces inequality and makes growth inclusive, and production and consumption more sustainable, while combating climate change and respecting a range of other planetary boundaries (UN 2012: 6)

It is perhaps easier, therefore, to clarify the definition by starting at the other extreme and setting out features of 'unsustainability'. Unsustainable development is associated with ozone depletion, poor sanitation, extinction of species and habitat, social conflict, toxic pollution and resource depletion. The sustainability movement is a reaction to these problems and stems from concerns about the future capacity of the planet's life support systems. Indeed, since the 1970s academics have discussed the problems of prolonged economic growth pushing society beyond global limits – see, for example, any of the titles by Meadows et al. The movement gained some impetus from the ideas put forward by environmental economists, including work on the monetary value of externalities and the integration of economic and environmental systems. These ideas were outlined in Chapter 11, and it may be useful here to review Key Points 11.1 and 11.3. In Table 15.2 (on page 274) we contrast unsustainable development with modern ideas of sustainable development.

Most people fully support the concept of sustainable development, as the associated ideas are not contentious. The challenge lies in finding a way that society can develop sustainably. Throughout this book we have alluded to some signs of progress towards this goal – but we have also made it clear that there is much uncertainty about the most appropriate strategies. For example, in Chapter 11 we distinguished between the approach of neoclassical and environmental economists. The neoclassical school of thought promotes a free market approach, arguing that this should ensure that the depletion of natural resources will be countered by increases in man-made equivalents. According to this technocratic perspective, aggregate capital is kept intact over time. In contrast, environmental economists believe it is necessary to maintain a critical minimum level of natural capital.

Table 15.2 What makes sustainable development different?

Unsustainable Development	Sustainable Development
Aims to raise the standard of living, based solely on monetary measurements of gross domestic product.	Aims to improve the general quality of life, including non-monetary factors to do with the environment and community.
Treats the economy, society and the environment as three separate issues.	Sees economic, social and environmental issues as interlinked.
Focuses on improving things in the short term. Mainly leaves issues to do with the future up to those who will live in it.	Looks at the needs of future generations as well as people today, and seeks to avoid problems in the future by taking precautions today.
Treats the environment as a luxury to be protected if we can afford it.	Takes account of the environment and its capacity to support human activity.

In 1992, the United Nations Conference on Environment and Development – the first Earth Summit – took place in Rio de Janeiro, Brazil. At the time it was the largest international conference of any kind, with 118 heads of state in attendance. This was clearly a sign of the times, as societies felt threatened by the looming problems of resource depletion and environmental degradation. The main document signed at Rio contained 27 principles for sustainable development and an action plan known as Agenda 21. It was here that the three strands – environmental, economic and social – explicitly emerged.

Rather worryingly at Rio+20 – the Earth Summit of 2012 – the same old issues still needed to be addressed and some momentum appears to have been lost. No longer were all heads of state in attendance, notably Germany, the UK and the US chose to send representatives, claiming that the economic questions to be discussed at the same time at the G20 summit in Mexico needed to take precedence. However, 193 countries were represented and the principles of the Rio declaration were reaffirmed.

The 2012 conference reiterated the need to promote a civil society, a green economy and eradicate poverty (particularly in Africa). Interestingly, following the arguments outlined in Chapters 11 and 13 regarding the need to integrate natural capital into accounting systems, the 2012 conference also set ambitious aims for the UN to create a framework for GDP+ with 50 countries committing to including the value of natural resources in their national accounts.

The synergy between environmental, economic and social factors is summarised in Figure 15.2. However, it can also be argued that there are a number of conflicting tensions that limit the synergy between the three strands. Is it really possible to achieve continuous economic growth and manage emissions and pollution? Are opportunities to enjoy the countryside and its associated biodiversity equally available to all sectors of society? Do the fiscal burdens imposed by market-based instruments cause greater cost to the poor or do they apply fairly to everyone? The issues raised by these type of questions are symbolised by the trade-off zones in Figure 15.2.

Figure 15.2 The three strands of sustainability

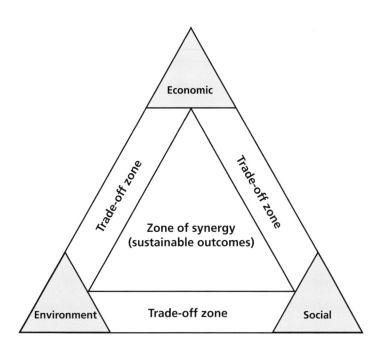

Over the years the Rio Earth Summits have prompted individual countries to adopt different approaches to reflect their own specific priorities. In the UK, a national *Strategy for Sustainable Development* was first introduced in 1994 and subsequently revised in 2005. This continues to inform the UK approach and, at the time of writing, the government's vision of a sustainable future is still based on the principles that were set out in the 2005 (Defra) policy, namely:

- prudent use of natural resources to secure sustainable consumption and production
- protection and enhancement of the natural environment
- strong and stable economy to provide prosperity and opportunities for all
- good governance based on sound scientific evidence.

These basic principles have a strong resonance with the three-part framework adopted for this book.

SUSTAINABLE CONSTRUCTION

In several countries, construction has been identified as the first sector to require specific attention in meeting the sustainable agenda. There are several reasons for this 'accolade'. First, in sustainable development terms, construction is consistently responsible for some of the most profound negative impacts. Examples of the industry's use of large amounts of materials and resources, and its reputation as a huge generator of waste were detailed in Chapters 9 and 11, where we reviewed each stage of the life cycle pertaining to the construction process and outlined the mass balance model (see for example Key Points 9.3 and 11.1). It was noted that the construction industry consumes more raw materials than any other industrial sector and is responsible for a significant proportion of the world's waste and carbon dioxide emissions. In the UK alone, it is estimated that more than 100 megatonnes of construction and demolition material ends up as waste every year, including a significant amount of material delivered to sites and then thrown away unused.

Second, the construction industry is a vitally important industry. This has already been reviewed at some length. In Chapter 13, we discussed how construction is commonly regarded as the engine of economic growth and illustrated the point with case studies from China, Japan and South Africa, and in Reading 5 we examined the idea further by exploring Bon's (1992) hypothesis that a country's level of construction activity is closely related to its stage of economic development. In purely quantitative terms, the contributions made by construction to employment and GDP are quite significant. Construction employs more than any other industrial sector – accounting for more than a 110 million jobs across the globe and creating an estimated 10 per cent of the world's GDP (UNEP 2012a). The industry is not just important economically, it is the key to the quality of life as it produces the built environment and puts in place the physical facilities and infrastructure that determine the degree of freedom and flexibility that society may enjoy. Its products also have a long lifetime, typically for anything up to one hundred years after construction. This puts it in a very different league to, say, producers of photocopier paper, washing machines and cars.

Finally, the most worrying reason that construction has been selected as warranting a special case in the sustainability agenda is because of its perceived lack of change. In nearly every other sector of the economy, technological developments have fuelled changes in business attitudes. For example, the manufacturing industry has become leaner, cleaner and quicker at all tasks. Yet, as we have pointed out at several points throughout this book, the construction industry is old fashioned. It suffers from inertia and there is a distinct lack of inclination for change. The problem was alluded to at some length in Chapter 6, where it was noted that similar sentiments are echoed throughout the world. To take a European example, a review of 30 market areas with high economic and societal value identified construction as requiring urgent and coordinated action. To paraphrase the report, the construction industry at national levels across the EU can be singled out as lacking innovation. It is a market of few large players and many small businesses. The owners of these businesses look for job opportunities in their local areas and are not inclined to invest time or money into research and development. Their main concern is to ensure order books for the next 6 to 12 months. This has an impact on the effectiveness of innovation, business strategy, design activities and training requirements. As a consequence, the industry of the twenty-first century is still made up of a high proportion of small contractors, who execute business plans across short time horizons, and persist with traditional methods of work (EU 2007: 10). Two years later a survey of the UK construction industry found a similar picture. It pointed out that the aim of genuinely embedding a spirit of change in the sector had not been achieved and there was no evidence of a united resolve across the diverse constituencies of UK construction to become a modern industry. In those exceptional cases where commitments to change had been made, they tended to be superficial and expedient, not tangible and sustainable (Wolstenholme 2009: 10).

Worldwide Definitions

The notion of sustainability therefore has a special relevance to construction and a specific agenda is evolving. The first international conference on sustainable construction was held in 1994. The conference defined sustainable construction as:

> the creation and operation of a healthy built environment based on ecological principles and resource efficiency.

These principles of sustainable construction have been taken to mean different things in different countries. Table 15.3 lists the countries that have developed a specific sustainable construction policy, and in most instances these policies refer to the narrowly defined construction industry. In some cases, the policies include economic, environmental and social considerations within their framework and these are referred to as a 'three strand policy'; others only identify an environmental strand. In practice, the main impacts of these policies are limited to the narrowly defined construction industry and generally tend to favour environmental aspects. All the countries listed in Table 15.3 are in the developed world.

Table 15.3 Countries following a sustainable construction agenda

Three strand policy	Environmental strand
Denmark	Australia
Finland	Canada
France	Germany
Ireland	USA
Japan	
Netherlands	
Norway	
Portugal	
Sweden	
United Kingdom	

Sources: Adapted from Manseau and Seaden (2001), EU (2007) and Kibert (2007)

Discussions, however, have also begun on a separate agenda for sustainable construction in developing countries, as the problems in these areas are greatly compounded by the huge populations that are poor and marginalised – often surviving in slums without the provision of clean water or sanitation. This means that there are significant differences in priorities, skill levels and capacity. Indeed much of the literature makes a distinction between a 'green agenda' that relates to issues of sustainability and a 'brown agenda' that relates to development. The green agenda is a response to the impact of ecologically detrimental developments, such as deforestation, climate change, pollution and the over-consumption of non-renewable resources on the earth's life-support systems, while the 'brown agenda' focuses on the problems of poverty and underdevelopment. However, as Du Plessis (2007) points out the green or brown agenda approach alone will not result in sustainable development; the brown agenda is the development part of the equation, the green agenda is what makes the development sustainable. The one without the other is not sustainable development. Thus, the challenge for the construction sector in developing countries is not just to respond to the development challenges of adequate housing, rapid urbanisation and lack of infrastructure, but also to do it in a way that is socially and ecologically responsible. As a consequence, in the developing world sustainable construction commences from a different point and needs to place more emphasis on the social aspects of sustainability. Du Plessis (2002: 8) defined sustainable construction in developing countries as:

> a holistic process aiming to restore and maintain harmony between the natural and the built environments, and create settlements that affirm human dignity and encourage economic equity.

This definition and its related agenda was put together by 22 experts from 12 countries, and was launched at the World Summit on Sustainable Development in Johannesburg in 2002, however, Chrisna du Plessis is attributed as the main author and she continues to coordinate this theme for the International Council for Research and Innovation in Building and Construction (often abbreviated as CIB). A quick scan of the agenda proposed by Du Plessis et al. (2002) shows that while the concept of sustainable construction may have become mainstream, the debate around how sustainability should be interpreted has changed to accommodate the realities of adapting to climate change and its uncertainties, and the realisation that doing a bad thing less badly is not good enough. This has led to a far greater emphasis on energy efficiency and a whole new generation of technologies, design practices and international targets on emissions, and a focus on the elimination of corrupt practice. Yet, despite the rapid rate of urbanisation in most developing countries and the increasing rate of infrastructure improvement in the last decade the idea of sustainable construction still eludes this part of the world.

The sustainable construction agenda in Europe and America is slightly more advanced, but it is still a slow uphill struggle. Stakeholders in building and construction appear to have become more aware of the need to embrace sustainability, as evidenced by the interest shown in membership of Green Building Councils across the world following the establishment of its World Council in 2002. Interestingly the 26 fully operational members are mostly from the rich developed world, the exceptions being India and South Africa (WGBC 2012). However, the general complexity of the construction sector, and the even greater complexities of sustainability – given the economic-social-ecological system within which it operates – impose limits on the effectiveness of current policies, regulations, assessment schemes, and preferential financing mechanisms put forward as incentives to change. As Du Plessis and Cole (2011: 443) point out, by offering simple linear solutions to complex problems, current mainstream sustainable construction practices and incentives often lead to good intentions having unintended consequences and driving perverse behaviour, which Wolstenholme (2009) referred to above as making 'expedient and superficial commitments'.

Three interpretations of what sustainable construction has involved in practice in Europe are shown in Table 15.4. The key ideas that recur in these definitions are to minimise the amount of energy and resources used in the construction process, reduce the amount of waste and pollution, and to respect the various stakeholders – particularly the users – both now and in the future. This last aspect seems to be a particular problem in construction where the client and contractor are so often separate and indifferent to one another's needs. It seems obvious that an efficient building is one that supports the client's needs, yet in Chapter 6 we noted the problematic nature of the relationship between clients and contractors. It is not surprising that research has shown that well-designed and well-constructed hospitals aid the recovery of patients, that good schools are those that are designed with

Table 15.4 Three interpretations of sustainable construction

In Finland, since 1998, it means

✓ Intensified energy efficiency & extensive utilisation of renewable energy resources
✓ Increasing the sense of wellbeing over a prolonged service life
✓ Saving of natural resources and promotion of the use of by–products
✓ Reducing waste and emissions
✓ Recycling building materials
✓ Supporting the use of local resources
✓ Implementation of quality assurance and environmental management systems

In The Netherlands, since 1999, it means

✓ Consume a minimum amount of energy and water over the life span
✓ Make efficient use of raw materials
✓ Generate a minimum amount of pollution and waste
✓ Use a minimum amount of land and integrate well with the natural environment
✓ Meet user needs now and in the future
✓ Create a healthy indoor environment

In the United Kingdom, since 2000, it means

✓ Minimising the consumption of carbon-based energy
✓ Improving whole life value by supporting best practice construction
✓ Delivering buildings and structures that provide greater satisfaction, wellbeing and value to customers and users
✓ Respecting and treating stakeholders more fairly
✓ Enhancing and better protecting the natural environment
✓ Being more profitable and using resources more efficiently

Source: Adapted from EU (2001) and HM Government (2008)

learning in mind. As an architect would describe it, in any building 'form' and 'function' should be complementary. But there are too many instances where form and function are not effectively thought through.

In practical terms, sustainable construction can be reduced to three important messages for the way the industry should work.

* Buildings and infrastructure projects should become more cost effective to produce and run, because they have been constructed with less and yield more.
* Construction projects should contribute positively to their environment, using materials and systems that are easily replenished over their full life cycle.
* Contractors and clients should, wherever possible, create higher standards of respect for people and communities involved with the project, from the site workers through to the final community of users.

A central purpose of this text has been to encourage students of construction to engage with modern economic analysis – and, in particular, analysis that includes, at the very least, a study of economic efficiency, environmental externalities and social equity. As Professor Charles Kibert (1999: 3–4) one of the first economists to address the question of sustainable construction perceptively pointed out: 'Economics is of crucial importance in dealing with the subject of sustainability, both to demonstrate the ultimate advantages of creating a sustainable built environment as well as to demonstrate the greatly undervalued role that natural systems play in our economic system'.

Table 15.5 confirms the key ideas that have been raised in this text to assist the construction industry to move towards achieving sustainable outcomes. From the list it is evident that the discipline of economics has a lot to offer anyone interested in understanding ways to overcome the barriers to sustainability.

Key Points 15.1

○ Sustainability embraces three broad themes: (a) the environment, (b) the community and (c) the economy.

○ To achieve sustainable outcomes, a more holistic approach needs to be developed.

○ In several countries pursuing sustainability, construction has been singled out to require special attention. This is partly due to its significant contributions to the economy, partly due to the negative impacts that it makes to any sustainable development agenda, but mainly because construction is so far behind other industrial sectors.

○ There are many interpretations of sustainable construction due to the fact that in its most modern guises it extends to adapting to climate change and achieving a low carbon economy.

○ Understanding the language and concepts of economics is important to anyone who is serious about dealing with the subject of sustainability.

Table 15.5 Factors that contribute to sustainable construction

Part of text	Key ideas
Part A Effective use of resources	✓ Competition and efficient pricing ✓ Achieving productive efficiency ✓ Targeting equity, between present and future generations ✓ Partnering and building long-term relationships ✓ Reducing asymmetric information ✓ Taking advantage of economies of scale ✓ Maximising profit (minimising risk)
Part B Effective protection of the environment	✓ Understanding market failure ✓ Using life cycle analysis ✓ Increasing resource efficiency ✓ Waste minimisation ✓ Environmental valuation of externalities ✓ Distinguishing between private costs and social costs ✓ Taking advantage of opportunities to gain competitive advantages by differentiating products
Part C Economic growth that meets the needs of everyone	✓ Understanding interest rates and interpreting expectations ✓ Forecasting and reading economic signals ✓ Encouraging efficient investment into construction for economic growth ✓ Avoiding supply constraints ✓ Understanding macroeconomic management ✓ Recognising the tension between increasing GDP and the ecological footprint ✓ Delivering buildings and structures that provide greater satisfaction, wellbeing and value to clients and users

THE ROLE OF MARKETS

Governments across Europe currently demonstrate a preference to let markets allocate resources rather than incur the expense of government intervention. (We introduced this trend in Chapter 2 and reviewed some of the associated policies in Parts B and C.) This means that difficult questions of resource allocation are frequently left to the private actions of individuals and companies as they seek to maximise respectively their satisfaction and profit. This is certainly the case in the UK today, where a significant amount of trust is placed in the virtue of the market to allocate resources efficiently. There is a strong argument to suggest that construction markets are inefficient as the industry tends to be short term in outlook, and slow to adopt innovation and adapt practice through experience. Sustainable construction, on the other hand, is a long-term goal that requires change.

Loose and Tight Couplings

In the specialised literature of construction economics and organisational theory, the industry is frequently defined as a **loosely coupled system**.

> Loosely coupled activities have few variables in common as the units involved in the production process function relatively independently of each other.

In other words, in loosely coupled systems the left hand frequently does not know what the right hand is doing. A good example of loosely coupled activities is the contrast between the manufacture of building materials and the way these materials are actually used when on site. In fact, one of the hurdles to overcome if sustainable construction is to become a reality involves closing the gap between traditional, project-based approaches to more integrated approaches to building involving on-site installation of purpose-made components. The concept of loosely coupled systems builds on the problems of imperfect information that were cited as forms of market failure in Chapters 6 and 10.

A construction project can be thought of as a network of firms that are drawn together to complete a specific set of operations on site. As the construction process is mainly about co-ordinating specialised tasks, loose couplings can occur across a number of relationships: among individuals, between subcontractors, at the organisational level, and between developers and users. The problems generated by these loose couplings are compounded in the construction industry as the activity is inevitably carried out at a specific and localised site where different teams have been brought together to complete a unique project. This means that in most cases the working relationships on site are based largely on trust in each other's judgement; and if a project fails, the trust relationship can quickly sour and turn into an adversarial relationship based on blame. In other words, each project involves a temporary network of firms coming together for the sole purposes of the specific project. At the same time the firms are possibly involved in other projects in which they are expected to co-ordinate their activities and resources with a completely different set of other firms. For example, in Figure 15.3 firm C needs to consider at least three couplings.

1   It needs to co-ordinate with other firms involved in firm C's individual projects – represented by the projects employing resources C1, C2, C3 and C4.
2   It needs to co-ordinate with the firms involved in its supply chain, for example firms D and E.
3   Individual projects exist within a broader, more loosely connected, permanent network and firm C needs to co-ordinate with associated firms A and B on activities and resources that lie beyond the scope of each individual project.

In contrast to the construction sector, manufacturing industry tends to be typified by **tightly coupled systems**. Manufacturing usually involves fewer independent elements and a far greater level of co-ordination from a centralised management team. This in turn encourages the use of standardised procedures, uniform product quality, improved management and economies of scale. Long-term relationships,

Figure 15.3 A network of construction projects

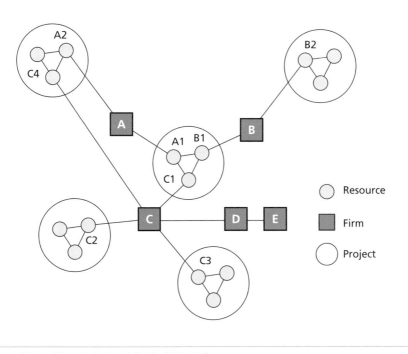

Source: Adapted from Dubois and Gadde (2002: 625)

formed in doing the same task with the same team, also encourage product development and learning. David Gann (2000) argues that it is the difficulties of creating the tight couplings enjoyed by manufacturing that lie at the heart of the problems faced by construction, where a lack of integration and innovation are serious constraints.

To achieve a shared vision, clear goals and an industry-wide strategy requires a good level of communication between the participating organisations. This explains why such a great deal of promise is held in the different forms of partnership that were discussed in Chapter 6 (see pages 88–92). Partnering, whether formalised through arrangements such as PFI or simply achieved by informal relationships, can potentially improve the integrated view. For example, the Dutch, Swedes and Danes have bridged many of the gaps between organisations in the house building sector. It is becoming the common practice in these countries to form multidisciplinary teams early in a project to brainstorm how they may contribute to the sustainable construction process. The strength of this type of network is the team building that emerges across disciplines, building links between developers, architects, contractors and, even, tenants. This tightening of the relationship between firms has enabled the Scandinavian and Dutch countries to progress furthest in the transition towards sustainable construction. Indeed they claim to be some of the few countries in the world to achieve economic growth over the past 20 years without increasing the consumption of fossil energy (Jensen and Gram-Hanssen 2008).

In the UK and elsewhere there is still a mind-set which tends to fragment the responsibility for sustainable construction. Construction firms argue that they can only adopt holistic approaches if clients ask for them, developers imply that there is no demand for sustainability and investors are hesitant to fund risky new ventures. There is a so-called 'circle of blame' that perpetuates the existing traditional approach to construction. The basic idea is captured in Figure 15.4, which characterises four typical views from the industry.

Figure 15.4 The circle of blame

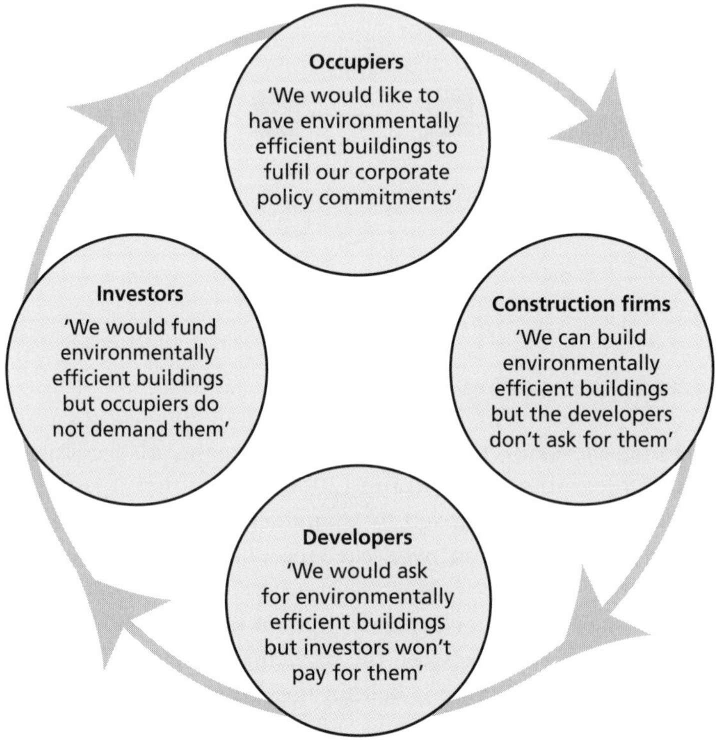

Occupiers
'We would like to have environmentally efficient buildings to fulfil our corporate policy commitments'

Construction firms
'We can build environmentally efficient buildings but the developers don't ask for them'

Developers
'We would ask for environmentally efficient buildings but investors won't pay for them'

Investors
'We would fund environmentally efficient buildings but occupiers do not demand them'

Source: Adapted from Cadman (2000)

Figure 15.4 highlights the conservative nature of the construction industry, and serves as a good reminder of the argument introduced in Chapter 9 where it was suggested that the firms who choose to adopt the product differentiation message will be the first to break out of this vicious circle. Although first presented more than 12 years ago the concept still has relevance, as there are still few examples of change (and there has been a four-year period of recession). One possible sign of change lies in the financial sector where property development finance is increasingly predicated on the basis that money will only be forthcoming once the proposed scheme has achieved a certain percentage of pre-lets. Differentiation in the conditions that financial markets agree to provide loans may slowly change the nature of the blame culture and promote a more integrated approach.

Key Points 15.2

○ Across Europe there is an increasing reliance on market forces to resolve questions of resource allocation in most industrial sectors.

○ The construction industry is characterised by a low level of inter-firm connection, which means that the same team seldom works together on more than one project.

○ In the UK and elsewhere there is a tendency to fragment the responsibility for sustainable construction.

○ The tightening of the relationships between firms is essential if new areas of competence are to emerge in the field of sustainable construction.

THE ROLE OF GOVERNMENTS

The construction industry has a significant impact on the nation's livelihood on all fronts – economic, social and environmental – and so it is entirely appropriate that governments show an interest in the sector's development and performance. A detailed review of public policies across Europe, the United States, Australia, Japan, Brazil, Russia, India and China suggest that governments adopt very different approaches to managing the industry. In the transitional economies, the level of government intervention can be very low key to the point of being neglectful; in the richer economies, government can be innovative and proactive in their support for construction. The contrast in approach is most evident in policies encouraging sustainable construction. The official view from the panel dealing with sustainable construction at the International Council for Research and Innovation in Building and Construction (2012) is that some of the 'low-hanging fruit' may have been plucked but the challenge still remains to extend and continue the road map towards sustainable construction across the globe.

As a consequence, the government of many developed countries have commissioned and published a steady stream of reports on the construction industry. These tend to promote research, integration of the supply chain, innovation in methods and process, improved economic performance, adaption to climate change and greater sustainability. As an example of the strategies that have recently been encouraged we review the construction reports commissioned and published by the government of the United Kingdom since 1998. The respective titles are shown in Table 15.6 and their content is *briefly* outlined below, as we have already referred to these reports several times. They all address the question of a 'new approach' – sustainable construction. For a detailed review of the main construction reports published from 1944 to 1998 see Murray and Langford (2003), or for a concise summary see Wolstenholme (2009: 29–30).

Table 15.6 Government reports 1998 to 2011

Year	Report
1998	The Egan Report *Rethinking Construction*
2008	HM Government *Strategy for Sustainable Construction*
2010	Innovation & Growth Team *Low Carbon Construction*
2011	Cabinet Office *Government Construction Strategy*

The Egan Report (1998) noted the difficulties posed by the fragmented nature of the construction industry. It argued that this inhibits the opportunities for efficient working, leads the industry to underachieve and limits the possibility of change. The report stressed the advantages of partnering, and the need to eliminate waste and to investigate the ideas of 'lean thinking'. In many ways, Egan paved the way for the first sustainable construction agenda (*Building a Better Quality of Life*) that was launched two years later by the Department of the Environment, Transport and the Regions (DETR 2000). As the government observed in its introduction to this agenda, 'the benefits which can flow from a more efficient and sustainable construction industry are potentially immense' (DETR 2000: 7). Unfortunately, the industry did not engage wholeheartedly with the ethos or step change involved.

In 2008 the *Strategy for Sustainable Construction* was launched with a far higher profile. This report was endorsed by six government departments and supported by several public sector clients including the Highways Agency, Defence Estates and the Housing Corporation. The strategy set specific actions and targets to be met by the industry during the next decade. It covers both the 'ends' and 'means' of sustainable construction. The 'ends' relate directly to sustainability targets, such as climate change, waste reduction and biodiversity; the 'means' describe processes such as innovation, procurement, regulation and design (HM Government 2008).

In 2010, the Innovation and Growth Team (IGT), a team drawn from right across the construction industry, reported on how the sector could become more sustainable, specifically investigating how it might support the transition to a low carbon economy. IGT was keen to stress that its plans should complement the *Strategy for Sustainable Construction*, and many of its 65 recommendations covered similar territory, although the approach was more comprehensive and covered issues relating to housing, infrastructure and non-domestic buildings. The 230-page report commences by highlighting the complex and challenging nature of the industry. Therefore it is not surprising that it found that the degree of awareness, engagement and readiness to deliver products and services that would enable the transition to a low carbon world appeared to vary considerably across the sector (IGT 2010: 1).

The *Government Construction Strategy* (2011) reiterated many of the themes that had emerged during the last decade in so far as there was recognition that construction underperforms in terms of its capacity to deliver value, and that there is a lack of investment in construction efficiency, low levels of standardisation and a fragmented public sector client base. But, most importantly, this report made a further call for the industry to reduce its costs by 20 per cent and to improve its long-term sustainability (Cabinet Office 2011: 5–6).

In all these reports the government's role is acknowledged:

- as a regulator with regard to aspects such as building and planning regulations
- as a policy maker for issues that directly affect but go wider than the industry, such as energy efficiency, waste management and climate change
- and, as a major client.

It is in its role as a client that a government can make the greatest progress in implementing the sustainability agenda. As pointed out at several points throughout this text, the government has responsibility for a large variety of public buildings and infrastructure, extending to an estate of around 7,500 buildings that occupy more than 10 million square metres of space (HM Government 2012). This is one reason why public sector work accounts for nearly 40 per cent of the industry's workload. So the government is in a position where it can literally lead by example, as well as by setting policy targets and making and enforcing regulations.

The last two reports cited in Table 15.6 both express the hope that the industry will adopt **building information modelling (BIM)**. This is a tool used to generate and manage multidimensional data with the aim of increasing productivity in building design, construction and facility management. It is far more than a 3D model or a simple visualisation. The key word in BIM is 'information' as it integrates data relating to costs, time, energy and sustainability. In effect, it is more of a 6D model, providing a complete picture that allows for complex analyses from concept to completion and beyond. The government's goal is that all construction projects procured in the public sector should be using BIM by 2016 (Cabinet Office: 2011: 14). BIM is seen as a way to integrate the construction team and get it to work collaboratively, to reduce errors and unnecessary changes, to cut out layers of waste and reduce transaction costs in the transmission of information across the supply chain (IGT 2010: 66). In short, it streamlines and integrates the processes of design, procurement, construction and asset management. But whether it will succeed in being the game-changer that it is set up to be and unlock opportunities to support sustainable design, construction and operation remains to be seen. This challenge is further complicated by the fact that buildings and infrastructure – the products of the industry – have such lengthy life spans. Developments that are under construction today will still be in use in 50 – or even 100 – year's time.

To close this section, the government clearly has a role to play. It takes this role seriously, evidenced to some extent by the 2009 appointment of a senior civil servant to advise the government (as head of the Construction Sector Unit), and, in fact, the IGT report was his first responsibility. However, as pointed out in Chapters 9, 10 and 11, government policies do not always achieve the desired results. Critics have claimed that over several generations the construction industry has suffered from initiative overload, too much bureaucracy and too many targets.

Complexity

If you have studied the previous chapters – in particular Chapters 1, 6, 7, 8 and 9 – you should recognise that construction is a complex industry compared to other industrial sectors. In these chapters, we demonstrated how construction involves the assembly of a range of materials from a broadly spread range of sources. In turn, each of the material inputs is subjected to different types of processing according to their particular use on a specific project. Assembling these materials are teams of subcontractors gathered for the sole purpose of completing the project. In Chapter 6 we compared the construction process to film production to clarify why each product is regarded as 'unique' and to stress the short-term relationship of the project team. We represented some of the problems associated with this method of production in Figure 15.3. To illustrate the complexity of the construction industry – and the way it shies from the use of standardised components – Egan (1998: 30) referred to the rather startling fact that while the average car contains about 3,000 components, the standard house has about 40,000 parts.

The debate about the practical problems inherent in securing any changes to the construction process has a long history. The industry has become notoriously defensive about the difficulties that emerge from new initiatives and in reviews of the progress of sustainable construction it would not be surprising to see defensive remarks couched in terms of the 'complexities of sustainability'. Research carried out over the last 50 years confirms the complex nature of the construction process – see, for example, Cox and Goodman (1956), Turin (1966, 1975), Winch (1987), Gidado (1996) and Pearce (2003). This research simply underlines the fact that the challenge of engaging the industry in a common vision of any nature is truly difficult. So achieving a sustainable construction industry could prove to be a long and difficult journey

As emphasised at several points in the text, it is the diverse and fragmented nature of the industry that constrains the opportunities to improve performance, learn from experience and introduce innovation. The construction industry is typified by a compartmentalised approach, with each of the specialist trades and numerous gangs of subcontractors tending to be primarily concerned with their own domains. This approach leads to an emphasis on the costs of each component part or process without due consideration to the quality, efficiency and general value for money across the project as a whole. To a large extent this explains why so much hope is currently pinned on the successful adoption of BIM as it offers the potential for a truly revolutionary change in the way in which construction-related activities are executed across the entire supply chain. Computer, automobile, and aircraft manufacturers have already benefited from the integration of design and manufacturing, harnessing automation technology and using electronic standards throughout the supply chain. But the construction industry has not yet followed this path. There is still a widespread dependency on paper as a means of capturing and exchanging information, as the industry is noted for its incapacity to operate harmoniously across two or more hardware devices or software routines – this problem is referred to in the literature as 'inadequate inter-operability'. The cost of inadequate inter-operability can be quantified by comparing traditional business activities to BIM. In the latter electronic data exchanges are made across common computer-based platforms, which means that project data needs to only be entered once, and it is then available to all stakeholders instantaneously through networks. There is a seamless database of all processes, from the initial designs through to procurement, construction and engineering specifications, to details for operation and management. The difference between the current and BIM scenarios represents not only the total economic loss associated with inadequate inter-operability but also the missed opportunities that could create significant benefits for the construction industry and the public at large. The lack of integrated computer technology for construction and facilities management has been estimated to impose a cost burden in the region of $16 billion per year. Of these costs, roughly 68 per cent ($10.6 billion) were incurred by building owners and operators (Gallaher et al. 2004). Obviously these estimates are speculative, but they highlight the important point that adoption of a comprehensive digital model throughout the life cycle of a building would be a step in the right direction to generating cost savings. And that is why the adoption of BIM is so central to the government ambitions of cutting construction costs by 20 per cent over a five-year period (Cabinet Office 2011).

As pointed out in Chapter 1 when we reviewed the ongoing debate between the broad and narrow definitions of construction, it is difficult to know where precisely the boundaries of the industry actually lie. This is why government initiatives to enhance productivity, performance integration and sustainability are so important, as the sector is not only complex, but at present it also lacks cohesion and vision. One of the extracts in the final reading (on page 295) reviews the problematic nature of this approach and raises questions about the effectiveness of government policy designed to secure sustainability.

Key Points 15.3

○ There are three roles that government plays in supporting construction, as a regulator, policy maker, and major client.

○ In its role of client the government can lead by example, and much hope is pinned on the industry adopting BIM for use on public sector projects.

○ The construction process is complex, and this is exacerbated by the diverse nature of the construction industry. Note the number of government departments that have taken an interest in the work of the industry over the last decade.

○ The construction industry is characterised as having a conservative nature and this problem is also compounded by its fragmented structure. Again it might be interesting to note the number of parties identified in the text as being concerned about change across the sector.

CONCLUDING REMARKS

As an industry, construction clearly needs a cultural change and this has been recognised for many years. Indeed the European Union in its first broad agenda for sustainable construction recommended that 'employers and professional bodies should phase out recognition of training and educational courses which do not take adequate account of the topic of sustainable construction' (EU 2001: 23). Certainly training, research, collaboration and working as a team could be major drivers of change. Arriving at a consensus on what should be delivered under the heading of construction economics at universities should also help to close the gap between government rhetoric and construction industry practice.

The new approach outlined in this book has contributed towards meeting this requirement. It has introduced some of the strategies that could be adopted by professionals entering traditional construction firms. We suggest four options following *The Greening of Industry for a Sustainable Future*, a framework produced by Schot et al. (1997). These are:

- the defensive option
- the offensive option
- the eco-efficiency option
- the sustainability option.

These four options offer a useful framework for analysing whether sustainable construction can become a reality.

THE DEFENSIVE OPTION

This option refers to output being changed to comply with regulations, on the basis that a primary motivation to make improvements is to avoid costs that might be imposed by non-compliance. In this context, Stern (2007: 434) referred to an OECD study that confirmed that 19 out of 20 countries surveyed had mandatory building regulations (codes) designed to increase energy efficiency. This has been an important strategy in reducing the environmental impacts of business activity. At the very least, it has caused most construction firms to acknowledge some of the environmental aspects of the agenda, and it has had some impact on the use of landfill and virgin aggregates that are both subject to tax. From a client's perspective, it is the prospect of potentially more onerous legislation in the future that is leading some developers and investors to adopt a 'beyond compliance' culture, either to achieve higher returns on projects or to reduce risks. In short, as Sayce et al. (2007: 630) have argued legislation and regulations will be used as important drivers to break into the circle of blame and slowly move us towards greener construction.

THE OFFENSIVE OPTION

This option was introduced in Chapter 9. A firm adopting this strategy would seek to break with convention in order to gain a competitive advantage by differentiating its products from those of its competitors. A few of the major construction firms in Europe involved with the passive house movement and similar low energy projects have taken advantage of this option and shown that major improvements in sustainability are technically feasible. Similarly the World Green Building Council was proud to announce that eight of the most outstanding green office buildings around the world consumed 85 per cent less energy and sent 70 per cent less waste to landfill than their traditional counterparts (WGBC 2009). The exceptional firms involved in these developments have made substantial contributions to their own financial performance while demonstrating the ability to incorporate zero carbon systems into all building types, as well as to creating sustainable approaches to the management of waste and infrastructure. Unfortunately, however, the work of most construction firms falls far short of these standards, and large-scale developments to secure sustainable outcomes continue to be exceptions to the rule almost everywhere.

THE ECO-EFFICIENCY OPTION

This strategy is about identifying win-win solutions by reducing environmental impacts and costs. It builds on the ideas presented in Parts A and B of the text and, in terms of construction, it forms the most promising option. It is largely about getting more from less, since it emphasises the importance of achieving value for money as a holistic and long-term objective. These ideas were introduced in the section on value management in Chapter 6 and extended further under the heading of resource efficiency in Chapter 9, where we gave a range of examples of how to increase efficiency in buildings and infrastructure. For instance, Italy and the Czech Republic both continue with notoriously low recycling rates, which are due, in part, to a strong consumer preference for virgin aggregate materials and the lack of any significant price difference between virgin and recycled materials. In contrast, construction firms in the UK and Sweden face higher landfill taxes that encourage far better rates of recycling (EEA 2008: 8). The UK and Sweden could therefore claim to be relatively more eco-efficient. In other words, enlightened companies in the right cultural setting understand this option, but the vast majority of small firms remain indifferent and uninvolved.

THE SUSTAINABILITY OPTION

This is the most difficult option as it seeks to achieve triple bottom line benefits – addressing economic viability, environmental soundness and social responsibility. Any business in any sector would find this difficult, as most economic activity still takes place with very little concern for the wider environmental and social effects. As the government minister responsible for launching the 2011 Construction Strategy reminded the audience, the government seeks to enable more to be constructed within the funds available. So the point to remember is that most contractual decisions are ultimately signed by individuals or departments who often have a low price as their priority. This option, however, demands adopting new values – values that are more holistic, that show concern for future generations and which delegate responsibility to the lowest level. Perhaps surprisingly some developments have taken place within construction to suggest that some of the players in the industry are prepared to meet these challenging criteria. For example, many PFI projects, and their equivalents across the globe have demonstrated that environmental and social benefits can be explicitly considered as part of a whole life cost appraisal. Reading 4 has also highlighted some strategic developments in corporate social reporting that suggest that some large construction companies are beginning to engage with widening the set of criteria on which they promote and support their business. However, given current government cuts in expenditure and the aftermath of the late 2000s financial crisis, the sustainability option is difficult to progress effectively in the short term.

The Future

The sustainable construction agenda has made, and is continuing to make, the industry think about a broader range of strategies. It is possible to see players within the construction industry take up opportunities within each of the four options listed above. Clearly far more needs to be done to integrate inputs and outputs in a holistic and schematic manner. It is important, therefore, that at some point in the near future, all construction firms, and not just the brave few seeking to gain a competitive advantage, heed the government messages. At least students of this new approach should recognise the value of breaking the mould and adopting values that could carry this forward. As the literature on sustainability makes clear, it is those working at the present time who determine the quality of life of future generations.

Reading 6

As we have emphasised throughout, the scope and meaning of both 'construction' and 'sustainability' are wide ranging and complex. And, for the reasons we have rehearsed in this book, sustainable construction is a challenging goal. Some might find it difficult to imagine sustainable construction as a future for the industry. However, the following extracts encourage you to finish the study of this *new approach* on a positive note. The first extract effectively picks up on the distinction between the broad and narrow definitions of construction first introduced in Chapter 1, and addresses the current inability of statistical analyses to capture the full scope and impact of construction on sustainable development. The second extract identifies several barriers that explain why sustainable construction is not yet standard industry practice (although other issues exist that the authors overlook). The closing aim, therefore, is to encourage you to think broadly and rigorously about the nature of the problems that need to be resolved before the journey towards sustainable construction can progress.

Mariagrazia Squicciarini and Anna-Leena Asikainen (2011): 'A value chain statistical definition of construction and the performance of the sector', *Construction Management and Economics*, 29:7, 671–93

Economic downturns, like the one that started in 2008, emphasize the need to address structural and sectoral problems, and to identify ways to increase productivity and competitiveness. They also under-line the necessity to monitor the implementation and to assess the effectiveness of the policies put in place to meet these challenges. This in turn calls for a precise and systemic definition and measurement of the industries to be targeted in order to determine the most suitable policy tools to be chosen. Whether employment, innovation or sustainability related, to maximize impact policies often need to be cross-cutting and to encompass the entire value chain of the industry in question. This is especially true for sectors like construction, with its multiplicity of heterogeneous actors, specialities and trades (Kokkala, 2010).

Despite the strategic importance of the construction sector for economies worldwide, attempts to capture its true scope have been relatively incomplete, and have changed across countries and over time (Francis, 1997). As Ruddock (2000, 2003, 2009) underlines, this has caused dissatisfaction about the state and quality of construction statistics among researchers and practitioners (Briscoe, 2006; Runeson and de Valence, 2009; Lewis, 2009), and has raised concerns about the incompleteness and narrowness of the statistical definition of the sector. Such a shortcoming is addressed by proposing a definition of the sector that builds on Pearce (2003) and encompasses the most important activities performed within the construction value chain. To achieve this, the codes defined within the Statistical Classification of Economic Activities used in the European Community (NACE1) are analysed to identify those activities outside the official definition of the construction sector that are fundamentally linked to construction. The 'wide' definition of construction proposed is, to the best of our knowledge, the first to explicitly formalize in terms of NACE classes the extensive range of activities traditionally considered as part of the construction industry value chain. As it is NACE-based, the implementation of this approach does not require changes in the way statistical data are gathered or aggregated.

Equivalents of the definition are also developed for other established industrial classifications notably the North American Industry Classification System (NAICS) of business establishments as used primarily in the US, Canada and Mexico; and the International Standard Industrial Classification (ISIC) of all Economic Activities developed by the United Nations. Although the taxonomies considered may sometimes differ in the breadth and depth of their definitions, the NACE–SIC–NAICS correspondence table should facilitate the use of the broad definition proposed, enable a better quantification of the construction sector's value chain, and allow for comparisons across countries and over time.

[T]he extent to which derived statistics and performance indicators of the construction sector can differ when accounting for the full value chain of the industry is discussed. The different roles played by core and non-core activities and the way they shape the performance of the broad construction sector are highlighted. This in turn warns about designing policies that target only the 'strict' NACE-defined construction industry (Section F), which covers the construction of buildings, civil engineering, and construction specialized activities, while overlooking components like the manufacturing of construction products and architectural and engineering activities. These activities and components are fundamental for the functioning and advancement of the construction sector and may cause interventions to fail, if left outside the scope of too narrowly defined vertical policies.

The first section [of this paper] characterizes the construction sector, underlines its relevance for economies worldwide. The second section...introduces an alternative value chain statistical definition of construction, and supplies NAICS- and ISIC-based equivalents of the taxonomy proposed...The third section offers some statistical evidence about how much of the sector is overlooked when only NACE Section F activities are considered and describes the performance of the wide construction sector in terms of firm size, turnover, and innovative input and output.

References

Briscoe, G. (2006) How useful and reliable are construction statistics? *Building Research and Information*, 34(3), 220–9.

Francis, M. (1997) *Developments in the Construction Sector of Trinidad & Tobago: Outlook for the 1990s*, Research Report, 63-72, Central Bank of Trinidad and Tobago, Port of Spain.

Kokkala, M. (2010) *Seven Development Proposals: Public Research on Housing, Construction and Use of Land in Finland* [in Finnish], Report of the Ministry of the Environment No. 10/2010.

Lewis, T.M. (2009) Quantifying the GDP–construction relationship, in Ruddock, L. (ed.) *Economics for the Modern Built Environment*, Taylor & Francis, London.

Pearce, D. (2003) *The Social and Economic Value of Construction: The Construction Industry's Contribution to Sustainable Development*, nCRISP, Davis Langdon Consultancy, London.

Ruddock, L. (2000) An international survey of macroeconomic and market information on the construction sector: issues of availability and reliability. *RICS Research Paper Series*, 2(11).

Ruddock, L. (2003) Measuring the global construction industry: improving the quality of data. *Construction Management and Economics*, 20, 553–6.

Ruddock, L., (ed.) (2009) *Economics for the Modern Built Environment*. London: Taylor & Francis, pp. 79–94.

Ruddock, L. and Ruddock, S. (2009) The scope of the construction sector: determining its value, in L. Ruddock (ed.) *Economics for the Modern Built Environment*, Taylor & Francis, London.

Runeson, K.G. and de Valence, G. (2009) The new construction industry, in L. Ruddock (ed.) *Economics for the Modern Built Environment*, London: Taylor & Francis, pp. 199–211.

....

Extract information: Adapted and edited from pages 671–2 plus relevant references from pages 691–2.

Mohamed Matar, Maged Georgy and Moheeb Ibrahim (2008) 'Sustainable construction management: introduction of the operational context space', *Construction Management and Economics* 26: 261–75

Introduction

As a discipline, sustainable construction has been evolving since the late 1980s. It continuously gains momentum as increasing evidence about depletion of the environment and environmental loadings becomes obvious. However, regardless of its importance and the expanding foundation of knowledge in the field, sustainable construction is by no means standard industry practice in many world countries (Landman, 1999). Although environmental protective measures, environmental management systems (EMS) and frameworks have become very common at manufacturing and industrial facilities, only a few construction companies have considered the use of a full EMS system in their projects (Christini et al 2004). Several barriers contribute to hindering sustainable construction being the dominant trend of the industry. Principally, these barriers can be summarized into two main categories: general barriers and technical barriers (Matar et al 2004; Matar 2007).

General barriers to sustainable construction

(1) The lack of expressed interest from different project parties. A survey made in the US by the Reed Research Group revealed that almost 60% of industry professionals do not even try to attempt green projects. Only 32% of construction clients have shown interest in pursuing green construction (RRG 2003).
(2) The lack of training/education in sustainable design/construction (Du Plessis 2002).
(3) The slow recovery of investment in technology to promote sustainable construction practices, such as off-site manufacture (Du Plessis 2002; RRG 2003).
(4) The higher initial cost of sustainable building (Landman,1999).

Technical barriers to sustainable construction

(1) The lack of a well-defined set of sustainable construction practices that can be practically engineered into projects (Pearce and Vanegas 2002). Although several best practice guidelines exist covering different environmental performance aspects, there is little formally structured information about procedures associated with the inclusion of environmental issues in the construction process (Cole 2000).
(2) The need for a mature and integrated framework of application for sustainable practices in construction. While certain efforts have been ongoing to control and enhance individual aspects of the environmental qualities of built facilities, detailed comprehensive approaches have generally been lacking. Each of the currently available systems has its own set of assumptions and limitations, and is designed for utilization by specific participants in the building process (Scheuer and Keoleian 2002). This hinders the capacity of different industry stakeholders to cooperate in a uniform and positive way.
(3) The disagreement about an optimum project delivery structure...Very few studies have considered the effect of project planning on the capacity to adopt sustainability practices (Chen et al, 2005). As sustainability is concerned with the complete life cycle, a building needs to be considered from its design and construction, through the operational stage, to its deconstruction...
(4) The need for effective drivers for change for different parties in the construction industry (Vanegas and Pearce, 2000; Du Plessis, 2002).

References

Chen, Z., Li, H. and Wong, C.T.C. (2005) Environmental Planning: analytic network process model for environmentally conscious construction planning. *Construction Engineering and Management*, 131: 92–101.

Christini, G., Fetsko, M. and Hendrickson, C. (2004) Environmental management systems and ISO 14001 certification for construction firms. *Construction Engineering and Management*, 130: 330–6.

Cole, R.J. (2000) Building environmental assessment methods: assessing construction practices. *Construction Management and Economics*, 18: 949–57.

Du Plessis, C. (2002) *Agenda for Sustainable Construction in Developing Countries: A discussion document*. Report for CIB and UNEP

Landman, M. (1999) Breaking through the barriers to sustainable building: insights from professionals on government initiatives to promote environmentally sound practices, MA thesis, Tufts University, USA.

Matar, M.M. (2007) An integration framework for sustainable construction, MSc thesis, Cairo University, Egypt.

Matar, M.M., Georgy, M.E. and Ibrahim, M.E. (2004) Towards a more applicable set of sustainable construction practices, in *International Conference: Future Vision and Challenges for Urban Development*, HBRC, Cairo, Egypt, 20–22 December pp. CE16: 1–12.

Pearce, A.R. and Vanegas, J.A. (2002) Defining sustainability for built environment systems: an operational framework. *International Journal of Environmental Technology and Management*, 2(1–3), 94–113.

RRG (Reed Research Group) (2003) Where our readers stand on sustainability. *Building Design and Construction*, 11: 14–17.

Scheuer, C.W. and Keoleian, G.A. (2002) *Evaluation of LEED Using Life Cycle Assessment Methods*, National Institute of Standards and Technology (NIST), Gaithersburg, MD, USA.

Vanegas, J.A. and Pearce, A.R. (2000) Drivers for change: an organizational perspective on sustainable construction, in Walsh, K.D. (ed) *ASCE Construction Congress VI*, Orlando, Florida, USA, 20–22 February, pp. 406–15.

Extract information: Adapted and edited from pages 261–2 plus relevant references from pages 274–5.

Glossary

adaptive-expectations hypothesis A theory of economic behaviour which states that people's expectations of the future rate of inflation are informed primarily by the rate of inflation in the immediate past. This process is also known as extrapolative expectations, and can be applied to many economic variables.

aggregate demand (AD) All expenditures for the entire economy summed together.

aggregate supply (AS) All production for the entire economy summed together.

allocative efficiency The use of resources that generate the highest possible value of output as determined in the market economy by consumers. Also referred to as economic efficiency.

asymmetric information A situation where two parties to a transaction involving a good or service have unequal knowledge of the properties or risks involved in making that transaction.

average fixed costs (AFC) Total fixed costs divided by the number of units produced.

average total costs (ATC) Total costs divided by the number of units produced.

average variable costs (AVC) Total variable costs divided by the number of units produced.

balance of payments A summary of monetary transactions with overseas nations. It is compiled as an account of inflows and outflows recording visible and invisible trade, investment earnings, transfers and financial assets.

barriers to entry Conditions in the marketplace that make it either impossible or very difficult for firms to enter an industry and offer competition to existing producers or suppliers. Factors which can make it very difficult for a firm to enter a market include the high start-up costs facing companies new to an industry, the legal requirements imposed by government and technological constraints.

base rate The rate of interest that UK financial intermediaries use as a reference for all other interest rates for lending and receiving deposits. For example, large financial intermediaries will borrow from one another at interest rates close to the base rate.

base year The year which is chosen as the point of reference for comparison to other years.

boom A period of time during which overall business activity is rising at a more rapid rate than its long-term trend.

Building Cost and Information Service (BCIS) Set up by the Royal Institution of Chartered Surveyors in 1962, this service provides detailed information on construction costs. The BCIS publishes a building cost index and a tender price index. The data is revised on a quarterly basis. Copies are available from most academic libraries.

building cycles Refers to the fluctuations in construction output – from boom to bust and back again. This sequence has been studied by many economists. The cycles have been measured as varying in duration from 2–25 years according to the sector and country. For example, Ball and Wood (1994) identified housing cycles of 13 years in Germany and 25 years in Sweden.

building information modelling (BIM) Software used to generate and manage data on building and construction projects on a multidimensional basis. The tools and processes in a BIM system provide digital documentation about a building, its performance, its construction, its cost and its operation.

building regulations A code of practice which specifies the type and minimum quality of materials to be used in a building. These regulations are legally enforced by district councils through, for example, building controls officers.

Building Research Establishment (BRE) A set of government laboratories based in Garston, Watford. The BRE researches general developments, such as those concerning the control of fire, energy and the environment, that relate to all types of construction.

business fluctuations A type of cycle found in overall economic activity, as evidenced by changes in national income, employment and prices. Sometimes referred to as a business cycle.

capital All manufactured resources, including buildings, equipment, machines and improvements to land.

capital goods These are goods that are used in the production of other goods. Examples include cranes, factories and foundries. Consumers do not directly consume capital goods.

capital value The monetary worth of an asset; for example, the price it could be purchased for.

cartel Any arrangement made by a number of independent producers to co-ordinate their buying or selling decisions. The members of a cartel agree in effect to operate as if they were a monopoly.

central bank The official institution of a country which guarantees the liquidity of that nation's banking system. In some cases, the central bank monitors and supervises commercial banks on the government's behalf. It normally acts as banker to the banks and other nationally important institutions. It is usually owned by the government and manages the national debt, exchange rates and the issuing of currency.

centrally planned model A theoretical system in which the government controls the factors of production and makes all decisions about their use and about the distribution of income. This system is often associated with communist nations. A centrally planned economy is also known as a pure command system.

ceteris paribus The assumption that all other things are held equal, or constant, except those under study.

chaos theory This suggests that the natural order of things is stormy and erratic rather than efficient and predictable.

circular flow model A model of the flows of resources, goods and services, as well as money, receipts and payments for goods and services in the economy.

claimant unemployment This is a record of the number of people claiming unemployment related benefits on one particular day each month.

cobweb theorem A dynamic model which tries to explain why cyclical fluctuations in output and prices – such as those in the property sector – can occur.

coincident indicators Economic statistics that are used by economic forecasters to track movements in the economy. For example, changes in output and the stock levels of raw material confirm that an economy is changing.

collusion An agreement, written or unwritten, between producers to determine prices, share out markets and/or set production levels, to avoid the danger of price wars and excessive competition.

commercial banks These are privately owned profit-seeking institutions, sometimes referred to as joint-stock banks to highlight the fact that they have shareholders. Most high street banks, such as HSBC, NatWest and Barclays, are commercial banks.

community infrastructure levy A planning charge imposed by local authorities on developers. In effect, a type of development tax, the tariff contributes towards infrastructure requirements. It came into force in April 2010.

competition policy This is a collection of government measures to monitor and control firms attempting to operate in an anti-competitive manner. For example, the government attempts to prevent mergers between companies that are not in the public's interest.

competitive tendering The process of inviting contractors to bid for work on a particular project. The opportunity to tender may be open to all companies or restricted to a set of selected (preferred) contractors.

complementary goods Two goods are considered complementary if both are used together. The more you buy of one, the more you buy of the other – and vice versa. For example, bricks and cement are complementary goods.

Confederation of British Industry (CBI) Founded in 1965, the CBI represents the interests of British firms. Membership consists of thousands of companies, plus hundreds of trade associations and employers' federations. The CBI's main aim is to express business views to government.

concentration ratio A measure of the degree to which an industry is dominated by its largest firms.

constant capital approach This approach to sustainable development argues that man-made capital can act as a substitute for natural resources (natural capital) and, therefore, that as long as aggregate capital stock remains constant, there will be a future.

constant prices Monetary value expressed in terms of real purchasing power, using a particular year as the base, or standard of comparison, to allow for price changes. For example, by expressing GDP at constant prices, comparisons can be made over a number of years.

constant returns to scale A situation in which the long-run average cost curve of a firm remains horizontal as output increases.

consumer goods Goods that are used directly by households to generate satisfaction. Contrast with capital goods.

consumer price index (CPI) A weighted average of prices of a representative set of goods and services purchased by the typical household. It has been the official measure used by the UK government to monitor inflation since January 2004. The CPI is similar to the retail prices index (RPI), but there are differences of coverage and methodology.

consumer sovereignty The concept that the consumer is king. In other words, the idea that consumers ultimately determine which goods and services are produced in the economy. This may not apply in markets dominated by very large firms.

contestable markets This refers to markets in which there is strong potential (or actual) competition because there are no barriers to entry or exit. In a contestable market, firms have to price products competitively and profits are constrained.

contingent valuation method This is a technique used to identify the price of an externality. It involves a survey of the interested parties in an attempt to attribute a hypothetical (monetary) value to an environmental gain or loss.

contracting out Used in the context of privatisation, contracting out refers to the transfer of publicly provided activities to private contractors. For example, a county council may contract out the management of its property portfolio to a private sector surveying firm.

cost-benefit analysis This is a way of appraising an investment proposal. It involves taking into account the external costs and benefits of a proposed development as well as the conventional private costs and benefits. This is done by estimating monetary values for aspects such as health, time, leisure and pollution.

cost-push inflation A rise in price level associated with a rise in production costs, such as the price of raw materials.

cover pricing This occurs when a company submits a bid price during a tender process that is not necessarily designed to win the contract but is intended to give the appearance of competition. The client is left with a false impression of the market price and may end up paying more than is necessary.

credit crunch A termed coined to describe the financial markets of 2007–2008 in which lenders began to raise the cost of borrowing and restrict the supply of loans.

current prices Monetary values expressed in terms of today's prices. In other words, what you have to pay for a good or service today. Also called absolute or nominal prices. Contrast with constant prices.

cyclical indicators Economic statistics that are used by economic forecasters to analyse the state of the economy. See entries for leading, lagging and coincident indicators.

deflation This is a sustained persistent fall in the general price level. It is the opposite of inflation.

demand function A symbolised representation of the relationship between the quantity demanded of a good and its various determinants. It looks like an algebraic equation but it is actually just shorthand notation.

demand management Government policies designed to control the level of total demand in an economy. Demand management policies are closely associated with Keynesian economics.

demand-pull inflation An increase in price level caused by total demand exceeding the current level of supply. Demand-pull inflation is often an unwanted outcome of Keynesian policies introduced to achieve full employment.

demand schedule A set of pairs of numbers showing various possible prices and the quantities demanded at each price. The demand schedule shows the rate of planned purchase per time period at different prices of the good.

demand-side economics This term generally refers to government policies which attempt to alter the level of aggregate demand.

depression A period of prolonged and deep recession in the economy, characterised by significant falls in output and soaring unemployment.

deregulation Describes the process by which state monopolies are opened up to competition – a strategy adopted in the UK since 1979. Now used to describe a situation in any industry in which statutory barriers to competition are liberalised or removed.

derived demand Demands created in a market to help meet other demands. For example, the demand for factory buildings is derived from the demand for manufactured goods. The term highlights the distinction between investment goods and consumer goods.

design and build An all-embracing agreement in which a contractor agrees to undertake building, engineering work, design and cost estimating as part of a package for a client.

direct policy This phrase is used to distinguish direct government intervention from broader macroeconomic policies. Direct policy tends to be of a legislative nature.

direct relationship A relationship between two variables that is positive, such that an increase in one is associated with an increase in the other, and a decrease in one is associated with a decrease in the other.

discounting A mathematical procedure by which the value of a sum or a stream of sums due to be received at specific dates in the future is expressed in terms of its current value.

discount rate The interest rate at which the Federal Reserve lends on a short-term basis to the US banking sector.

diseconomies of scale When increases in output lead to increases in long-run average costs.

ecological footprint An accounting tool used to measure how much productive land and water an individual, city or country requires to produce all the resources it consumes and to absorb all the waste it generates.

economic good Any product or service that is scarce and allocated by the price mechanism. Contrast with free good.

economic growth An increase in an economy's real level of output over time. Usually measured by the rate of change of national income from one year to the next.

economic system The institutional means through which resources are used to satisfy human wants.

economics A social science studying human behaviour and, in particular, the way in which individuals and societies choose among the alternative uses of scarce resources to satisfy wants.

economies of scale When increases in output lead to decreases in long-run average costs.

ecosystem services Refers to biological processes that contribute to human wellbeing, such as clean air and water, biodiversity, pollination, climate control and waste assimilation. These services tend to be public goods with no price and no market and they are not effectively factored into the cost of development. This raises questions about their value. Several examples are listed in Table 11.2.

effective demand Demand that involves desire and ability to pay. In other words, it is the demand that can be measured by actual spending.

efficiency See glossary entries for allocative efficiency and productive efficiency.

elasticity A measure of the level of response to a specific change in market conditions such as a change in price or income.

embodied energy The energy used in the process of extracting raw materials, manufacturing components and transporting them to site.

endogenous variables These are economic factors which affect other aspects of a theory or model from within. For example, the level of unemployment will affect the amount of income tax collected. Contrast with exogenous variables.

entrepreneur The fourth factor of production involving human resources that perform the functions of raising capital, organising, managing, assembling the other factors of production (land, capital and labour), and making basic business policy decisions. The entrepreneur is a risk-taker.

environment In modern economic analysis, environmental assets such as clean air, species diversity and tropical rainforests are considered alongside the allocation of resources of traditional economic goods and services. Also see the entry for environmental policy.

environmental economics The origins of environmental economics date back to the 1960s when green thinking became a popular concern. A central tenet of the environmental economics school is that as the economic system cannot operate without the support of the environment, financial value should be placed on environmental services.

environmental policy Governments across the world are accepting responsibility for the global and local problems that arise as a result of market failure. In the UK, for example, the government has set out a strategy for sustainable development in the general economy (DETR 1999) as well as a sustainable strategy for the construction industry (DETR 2000).

equilibrium A situation in which the plans of buyers and sellers exactly coincide, so that there is neither excess supply nor excess demand.

equilibrium price The price that clears the market, at which there is no excess quantity demanded or supplied. In other words, the price at which the demand curve intersects the supply curve. Also known as the market clearing price.

equity See glossary entries for horizontal equity and vertical equity.

exogenous variables These are economic factors which impinge upon a theory or model from the outside, such as the weather. They are sometimes referred to as autonomous variables, and they contrast with endogenous variables.

expenditure approach A way of computing national income (GDP) by adding up the values of all spending on final goods and services.

external economies of scale These are the savings a firm can achieve in long-run average costs due to changes that benefit the whole industry. For example, a big increase in state-funded training could help all firms in a particular sector become more efficient.

external costs The costs experienced by third parties due to the production or consumption of a good or service; not usually expressed in monetary terms.

externalities The benefits or costs that are experienced by parties other than the immediate seller and buyer in a transaction. These third-party effects are also known as external costs or benefits.

factor markets In factor markets, households sell resources such as labour, land, capital and entrepreneurial ability. Businesses are the buyers of these resources to generate output (see Figure 3.1).

factors of production Often grouped under four headings – land, capital, labour and entrepreneurial ability – these are the resources and inputs required to produce any good or service.

financial intermediaries Institutions such as commercial banks and building societies that link savers (people and organisations that have extra funds) with borrowers (people and organisations that need funds).

Financial Services Authority The organisation responsible with the Bank of England for regulating and supervising all financial intermediaries in the United Kingdom.

fine tune A term used in economics to suggest that an economy has an engine like a car, and this can be set up to run at different levels of efficiency by adjusting various flows.

fiscal policy A combination of government spending and taxation used to achieve macroeconomic management.

fixed costs The costs that do not vary with output. Examples of fixed costs include the rent on a building and the price of machinery. These costs are fixed for a particular period of time, although in the long run they are variable.

foreign exchange market The market for buying and selling foreign currencies.

free enterprise A system in which private businesses are able to obtain resources, to organise those resources and to sell the finished product in any way they choose.

free good Exceptional goods, such as watercourses, sunshine and air, that have a zero price because they do not require the use of scarce factors of production.

free market model A theoretical economic system in which individuals privately own productive resources and can use these resources in whatever manner they choose. Other terms for this system are a pure market or pure capitalist economy.

free riders Individuals who do not pay for the goods and services that they consume.

full employment A situation where job vacancies exceed the number of registered unemployed. The concept recognises that even when the economy is doing well there will be some people out of work searching for jobs that suit their skills.

game theory The study of how decisions are made, particularly in the context in which decision-makers have to take other people's responses into account. It is used to analyse oligopolistic forms of competition.

golden rule A fiscal objective that constrains a government only to borrow to finance investment not to fund current spending.

government failure The concept that government policy intervention may not necessarily improve economic efficiency.

government intervention Measures undertaken by the state to achieve goals not guaranteed by the market system.

great depression The worldwide economic downturn that started in the United States in 1929 and led to the most severe depression in the world to date.

gross domestic fixed capital formation A national income accounting category representing expenditure on fixed assets (such as buildings, vehicles, plant and machinery). It is more common to refer to this expenditure as investment. Spending on maintenance and repairs is excluded from this measure.

gross domestic product (GDP) The most common measurement of a nation's wealth, based on income generated from resources within its own boundaries; the monetary value of its output of goods and services.

gross national income (GNI) A measurement of a country's wealth. It represents the total output of goods and services produced by the country in a year, plus the net value of overseas assets. In the UK, there is little difference between GDP and GNI. In 2010, UK GDP was £1,458 billion and GNI was £1,479 billion.

harmonised index of consumer prices (HICP) A standard measure of consumer price inflation that allows comparison between European Union countries. The series commenced in January 1996 and became the official target of government policy in 2004. It is usually referred to as the consumer price index.

headline inflation rate The change in the retail price index that is announced by the UK government in a monthly press release. It contrasts with the official rate of inflation, which is based on the consumer price index.

hedonic pricing method A technique used to identify the price of an externality. It provides an estimate of the implicit price of an environmental attribute by comparing the value of two identical goods, one with the environmental element and the other without.

horizontal equity The concept that all people should be treated identically. The idea that underpins equal opportunities policies.

human capital Investment in education and training that enhances the productivity of individuals.

ILO unemployment rate A measure of unemployment produced by the International Labour Organisation. It defines unemployment as people who are without work yet actively seeking employment. Data is gathered through labour force surveys.

imperfectly competitive markets A broadly used term to refer to all markets that do not have the characteristics of perfect competition. Commonly referred to as imperfect competition.

income approach A way of measuring gross domestic product by adding up all factor rewards – that is, the sum of wages, interest, rent and profits.

income elasticity of demand A measurement of the responsiveness of the quantity demanded to a change in income.

index numbers A way of expressing the relative change of a variable between one period of time and some other period of time selected as the 'base year'. For example, the base year index number is set at 100 and the value of the variable in subsequent years is expressed above or below 100 according to its percentage deviation from the base.

inferior goods Products or services for which demand decreases as consumer income increases. Goods of this nature are exceptions. Contrast with normal goods.

inflation A sustained rise in prices, formally measured by the consumer price index.

injections Supplementary inputs into the circular flow of income in an economy. Typical injections are investment, government spending and export income.

interest rates These determine the cost of obtaining credit and the rewards paid to owners of capital.

internal economies Formal term for economies of scale. Contrast with external economies of scale.

inverse relationship A relationship between two variables in which an increase in one variable is associated with a decrease in the other, and a decrease in one variable is associated with an increase in the other.

investment Spending by businesses on things like machines and buildings which can be used to produce goods and services in the future.

joint production A term coined to denote the phenomenon that several outputs inevitably emerge from economic activity.

joint profit maximisation A hypothesis put forward by Fellner (1949) to suggest that firms in the same market will make pricing and output decisions that are designed to maximise the overall profits of the group of firms. In essence, the member firms of the market collude to act as an oligopoly.

labour The human resource involved in production. In other words, the contributions (both thinking and doing) of people who work.

lagging indicators Economic statistics (such as unemployment and investment) that change approximately twelve months after a change in overall activity (gross domestic product).

land One of the factors of production. In economic terms, land consists of both the physical space (in which economic activity can be located) but also natural resources such as coal, oil, water, natural vegetation and climate.

law of demand Quantity demanded and price are inversely related. In other words, the law of demand states that more is bought at a lower price, less at a higher price (other things being equal). Also known as the theory of demand.

law of diminishing (marginal) returns States that after some point successive increases in a variable factor of production, such as labour, added to fixed factors of production will result in less than a proportional increase in output.

law of increasing opportunity costs An economic principle that states that in order to get additional units of a good, society must sacrifice ever-increasing amounts of other goods. It is also referred to as the law of increasing relative costs.

law of supply The relationship between price and quantity supplied (other things remaining equal) is a direct one. That is, as price increases so does the quantity supplied and vice versa.

leading indicators Economic statistics (such as retail sales and consumer credit) that change approximately six months in advance of gross domestic product and are used to predict changes in the economic cycle.

leakages A withdrawal from the circular flow of income. Examples are net taxes, savings and imports.

liabilities All the legal claims for payment that can be made on an institution or company. In short, the amount owing to others.

life cycle analysis This approach aims to take into account the whole life costs incurred during a project. The analysis covers the total cost of ownership, including the initial investment, and production, operation and maintenance costs.

liquidity This describes the ease with which an asset can be used to meet liabilities. Cash is the most liquid asset.

long run That time period in which all factors of production can be varied.

long-run average cost curve (LAC) This represents the cheapest way to produce various levels of output given existing technology and current prices. It is derived from a compilation of short-run positions (see Figure 7.4b, page 114).

loosely coupled systems Every industrial activity is to some extent interdependent with a number of other activities; that is, the activities are coupled in some way. Some of these couplings are 'tight' while others are 'loose'. Construction is typified as a loosely coupled system as the firms involved in the industry operate fairly independently of another.

low inflation A trend for annual price increases to be below 5 per cent.

macroeconomics The study of economy-wide phenomena, such as total consumer expenditure.

macroeconomic objectives Targets relating to the whole economy, such as price stability, employment levels and the balance of payments.

marginal cost (MC) The change in total costs due to a one-unit increase in the variable input. The cost of using more of a factor of production.

marginal physical product (MPP) The output that the addition of one more worker produces. The marginal physical product of the worker is equal to the change in total output that can be accounted for by hiring one more worker.

marginal propensity to leak (MPL) The proportion of an increase in national income that is withdrawn from the circular flow. For example, a 0.2 MPL indicates that out of an additional £100 earned, £20 will 'leak' in imports, savings or tax.

marginal revenue (MR) The change in total revenues resulting from the sale of an additional unit of a product.

market An abstract concept concerning all the arrangements that individuals have for exchanging goods and services with one another. Economists often study the market for particular goods and service, such as the labour market, the car market, the commercial property market, the housing market, the building materials market, the credit market, and so on.

market-based instruments These involve various incentive systems designed to operate through the price mechanism to encourage environmentally friendly behaviour. Examples include the climate change levy and the landfill tax.

market-clearing price Another term for equilibrium price.

market economy An economy in which prices are used to signal the value of individual resources to firms and households.

market failure A situation in which the free forces of supply and demand lead to either an under- or over-allocation of resources to a specific economic activity.

market mechanism See market economy and price mechanism.

market structures The characteristics of a market which determine the behaviour of participating firms, such as the number of buyers and sellers and the ease of entry into (and exit from) a market.

market supply schedule A set of numbers showing the quantity supplied at various prices by the firms comprising the industry. The horizontal summation at each price suggests the market supply.

mass balance model This model provides a framework to describe the relationships between the economy and the environment by focusing on natural resource depletion and pollution.

maximum price legislation A price ceiling set by a government agency specifying a level in a specific market beyond which prices must not rise.

menu costs The resources used up revising contracts due to inflation.

merit good A product or service that is deemed socially desirable by politicians. If left to the private market, merit goods may be under produced.

mesoeconomics A study of economic activities at the level of sectors or industries.

microeconomics The study of economic behaviour of individual households and firms and how prices of goods and services are determined.

midpoint method A mathematical technique used for estimating elasticity. It involves calculating percentage changes in price and quantity compared with the average of the initial and final values.

minimum efficient scale The lowest rate of output at which long-run average costs reach their minimum point.

mixed economy An economic system in which decisions about how resources are used are made partly by the private sector and partly by the government.

mobility of labour The ease with which labour can be transferred from one type of employment to another. Mobility of labour is considered in terms of geographical and occupational mobility. The converse concept, the immobility of labour, is often employed by economists.

models Simplified representations of the real world used to make predictions and to provide greater clarity to economic explanations.

modern methods of construction The phrase covers a wide range of products and solutions, encompassing everything from new innovative building components to entire factory-built structures and building modules.

monetarists Economists who believe that changes in money supply are important in the determination of the full employment level of national income.

monetary policy A policy, often implemented by a central bank, to influence the growth of aggregate demand and money supply, to control price inflation and to maintain financial stability.

monetary policy committee (MPC) A Bank of England committee established in 1997 that sets interest rates independently of HM Treasury in order to achieve the UK government's predetermined target rate of inflation.

money Anything that is generally accepted as a means (medium) of payment for goods and services, or the settlement of debts (that is, deferred payments). Ideally, money should act as a store of value and unit of account, although during periods of high inflation it may become deficient in these respects.

money supply A generic term used to denote the amount of 'money' in circulation. There are many definitions of money supply, with individual variants including different types of bank deposits in the overall money supply measure.

monopolistic competition A market situation in which a large number of firms produce similar but not identical products. There is relatively easy entry into the industry.

monopoly A market structure in which a single supplier dominates the market.

monopsony A market in which there is only one buyer.

multiplier The number by which an initial injection into an economy must be multiplied to find the eventual change in national income. Mathematically, it is the reciprocal of the marginal propensity to leak (MPL).

national accounts An annual record of an economy's aggregate performance.

National Audit Office (NAO) A government agency responsible for auditing the tax revenues spent by central government. It investigates all aspects of value and publishes around 60 reports each year on its findings. Electronic versions of these reports are freely available online for download (see Part C web reviews on page 202 for details).

national income A generic term for all that is produced, earned and spent in a country during one year. Strictly speaking, it is defined as gross national product (GNP) minus capital depreciation.

nationalisation Taking into public ownership some part, or all, of an economic activity previously located in the private sector.

natural capital approach An approach to sustainable development that regards the environment as critically important to future economic wellbeing. Its basic premise stands in stark contrast to the constant capital approach.

natural monopoly An industry best suited to production through a single firm. Such situations arise when production requires extremely large capital investments.

neoclassical economics Economists in the neoclassical school follow the traditions of classical economists such as Adam Smith and believe that free markets are best suited to steer economies.

neutral equilibrium A theoretical concept closely associated with a two sector economy where the established levels of activity persist forever, since there are no pressures for change.

new economy Coined in the late 1990s, this term describes the use of information technology in business, both within and between firms and between firms and consumers. More generally, it describes a way of restructuring economies through the use of information technology.

nominal values The values of variables such as gross domestic product and investment expressed in current prices. Also called money values. In other words, the actual market prices at which goods and services are sold.

normal goods Goods for which demand increases as income increases. Most products and services that we deal with are normal goods.

normal profit The minimum rate of profit necessary to ensure that a sufficient number of people will be prepared to undertake risks and organise production. In more formal terms, it is the normal rate of return on investment – which differs from industry to industry.

normative economics Analysis involving value judgements about economic policies; relates to whether things are good or bad. A statement of 'what ought to be'.

notes and coins The currency of a nation, normally referred to as cash.

novation A term coined during the early 1990s to describe a client-led contractual bonding between an architect and building contractor. See design and build.

Office for Budget Responsibility (OBR) This is the UK's official fiscal watchdog. It was created in 2010 to provide an independent analysis of the UK's public finances. There are several equivalent bodies around the world, such as the Danish Economic Council and the German Council of Economic Experts, both of which have existed for more than 50 years.

Office for National Statistics (ONS) The UK government agency responsible for compiling and distributing economic, demographic and social statistics. It publishes an array of economic data on the Internet (see page 30 for details).

official rate of inflation In most industrialised countries, the target, or, typically, the midpoint of a target range, for consumer price inflation is between 1 per cent and 2.5 per cent. In the UK, the current official target rate of inflation is 2 per cent as measured by the consumer price index.

off-site production (OSP) A term used to describe any construction process that is carried out away from the building site, such as in a factory or in a specially created temporary production facility close to the construction site.

oligopoly A situation in which a large part of the market is supplied by a small number of firms. The firms may behave as if they are interdependent.

OPEC see Organisation of Petroleum Exporting Countries.

opportunity cost The highest-valued alternative that has to be sacrificed for the option that has been chosen.

opportunity cost of capital The normal rate of return or the amount that must be paid to an investor to induce him or her to invest in a business. Economists consider the opportunity cost of capital as a cost of production.

optimism bias A measure of the extent to which actual project costs (and project duration) might exceed the original estimates.

Organisation for Economic Cooperation and Development (OECD) The OECD publishes many economic commentaries each year. It produces data allowing international comparison of economic and environmental indicators. Based in Paris, the OECD's members comprise all the major capitalist countries, including Australia, Japan, the United States, Canada and western European nations.

Organisation of Petroleum Exporting Countries (OPEC) This is a consortium of 13 petroleum producing nations. The organisation was formed in 1960 to control oil supply and prices. It is an example of a cartel.

output approach A way of measuring national income (see gross national product) by adding up the value of the output produced by each sector of the economy.

partnering A broadly defined term to describe a situation in which two or more organisations work openly together to improve performance by agreeing mutual objectives and ways for resolving any disputes. See also public-private partnership and strategic partnering.

per capita A Latin phrase meaning per head of the population.

perfect competition A market structure in which the decisions of buyers and sellers have no effect on market price.

perfectly competitive firm A firm that is such a small part of the total industry picture that it cannot affect the price of the product it sells.

perfectly elastic A supply or demand curve characterised by a reduction in quantity to zero when there is the slightest increase or decrease in price. Producers and consumers are completely responsive to any change of price.

perfectly inelastic The characteristic of a supply or demand curve for which quantity supplied remains constant, no matter what happens to price. Producers and consumers are completely unresponsive to price changes.

planning curve Another name for the long-run average cost curve.

planning gain A trade-off agreement between a local authority and developer that permits the developer to build in return for some compensatory benefits to the community. A common example involves the provision of social housing as an integral part of a proposal for a commercial housing development.

planning horizon Another name for long-run cost curves. All inputs are variable during the planning period.

planning regulations Each local authority has plans on how its area will develop and a body responsible for deciding what is allowed to be built. The terms of reference are determined by centrally produced planning policy guidance notes.

polluter pays principle A strategy based on market incentives to assure that those who pollute are encouraged to reduce the costs that fall on society. The principle was succinctly set out in 1987 by the Secretary of State for the Environment: 'The polluter must bear the cost of pollution.'

positive economics Analysis that is strictly limited to making purely descriptive statements and scientific predictions, such as 'if A, then B'. Positive statements can be checked against the evidence – they are statements of what is. Contrast with normative economics.

post occupancy evaluation A scheme that offers users (occupiers) of a building the opportunity to review its engineering and design performance after completion. The process has been in use since the 1970s but it is only just becoming accepted as a means of keeping designers and constructors involved beyond the completion stage.

precautionary principle A recommendation that the costs of avoiding unacceptable environmental consequences should be met even if the precautions may turn out in the long run to have been greater than necessary.

present value A future monetary value expressed in today's prices; the most that someone would pay today to receive a certain sum at some point in the future.

price-elastic demand When a price change of a product or service results in a more than proportionate change in demand, then demand is said to be price elastic.

price elasticity A measurement of the responsiveness of the quantity demanded or supplied to a change in unit price.

price elasticity of supply A measurement of the responsiveness of the quantity produced for the market due to a change in price.

price index The cost of today's basket of goods expressed as a percentage of the cost of the same basket during a base year.

price-inelastic demand When a change in price results in a less than proportionate change in demand, then demand is said to be price inelastic.

price mechanism Prices are a signalling system between firms and households concerning the use of resources. If a price mechanism operates, there is a market economy. The terms 'price' and 'market' are interchangeable.

price system An economic system in which (relative) prices are constantly changing to reflect changes in supply and demand for different commodities. The prices of commodities are signals to everyone within the system about which are relatively expensive and which are relatively cheap.

price-taker A characteristic of a perfectly competitive firm. A price-taker is a firm that must take the price of its product from the prices of its competitors.

principal-agent A concept used to highlight one possible cause of market failure – inexperienced clients (the principal) cannot be confident that their best interests are properly represented by contractors (the agent) acting on their behalf.

principle of exclusion This simply means that anyone who does not pay for a product or service will not be allowed to benefit from consuming that particular good – they will be left out or excluded from the product's or service's benefits.

principle of rivalry The principle that private goods cannot be shared. If person A uses a private good, than that prevents the possibility of person B using that good. Persons A and B cannot eat the same apple simultaneously.

private commercial A sector of the construction industry concerned with privately funded commercial developments such as shops, offices and leisure facilities (see Tables 1.3 and 1.4).

private costs The specific monetary payments made by firms and individuals in the production and consumption of goods and services.

private finance initiative (PFI) A form of procurement to encourage private investment in public sector projects. PFI suppliers are typically contracted not only to construct a facility but also to finance, manage and maintain the infrastructure for a period of several years following construction. Introduced in the UK in 1992, the initiative is designed to reduce the pressure on public sector borrowing.

private goods Goods that can only be consumed by one individual at a time. Private goods are subject to the principles of exclusion.

private industrial A sector of the construction industry concerned with privately funded industrial developments like factories and warehouses. Since the privatisation of the public utilities – gas, water, and electricity supply and generation – the significance of this sector has increased (see Tables 1.3 and 1.4).

privatisation In very general terms, this involves the transfer of assets from the public sector to the private sector.

procurement A generic term used by professionals within the built environment to describe the general process of obtaining, acquiring and securing some property or land.

product differentiation This occurs in imperfect markets when individual producers introduce differences in the characteristics of products and services that are broadly alternatives to each other in order to gain a competitive advantage. In construction markets, product differentiation may be achieved by craftsmanship, foreign language ability, associated financial packages or some technological superiority. A differentiated product gives producers greater freedom to determine their prices.

product markets A market in which businesses sell consumer goods to households (see Figure 13.1).

production function The relationship between inputs and output. A production function is a technological, not an economic, relationship.

production possibility curve A curve representing the maximum combinations of two goods that can be produced assuming (a) a fixed amount of productive resources and (b) efficient use of those resources.

productive efficiency The utilisation of the cheapest production technique for any given output rate; no inputs are wilfully wasted. Also known as 'technical' efficiency.

profit The income generated by selling something for a higher price than was paid for it. In production, profit is the difference between total revenues received from consumers who purchase goods and the total cost of producing those goods.

profit maximisation model A model based on the assumption that the central aim of the firm is maximising profit. This can be achieved by setting prices so that marginal revenue equals marginal cost. If marginal revenue is greater than marginal cost, total profit can be increased; conversely, when marginal cost is greater than marginal revenue, total profits decline.

progressive income tax A tax system in which a higher percentage of income is taxed the more a taxpayer earns. Put formally, the marginal tax rate exceeds the average tax rate as income rises.

project partnering An arrangement between the main contractor and the client working together on a single project. It embraces a range of possibilities, but it usually comes into effect after the contract for the project has been awarded. It has been devised to overcome the adversarial relationships that typify traditional construction contracts. It is designed to prevent contractors regarding a project as a sequence of separate operations and making no long-term commitment to its success.

public choice theory A relatively new concept to mainstream economics that seeks to explain how governments decide policy.

public goods Goods for which the principles of exclusion and rivalry do not apply. Public goods can be jointly consumed by many individuals simultaneously, at no additional cost and with no reduction in the quality or quantity of the provision concerned.

public (non residential) A sector of the construction industry concerned with the construction of roads, prisons, schools, etc. In short, public sector works other than housing (see Tables 1.3 and 1.4).

public private partnership This is a particular type of contractual arrangement between the public and private sector. Similar to PFI, the private sector is allowed a role in financing, building and maintaining public sector facilities, although the government retains a stake in the PPP company. In contrast to the private finance initiative, under public private partnership arrangements the government is not liable for a fixed stream of annual payments.

public sector The simplest definition is all forms of ownership by central and local government.

public sector net borrowing requirement (PSNBR) This is a measure of the public sector's annual financing requirement, covering the combined funding requirement of both central and local government.

quantitative easing (QE) The practice of a central bank buying financial assets with the aim of increasing the reserves of commercial banks. In the simplest of terms, it is electronically created money pumped into a financial system.

quasi-public goods Goods or services which by their nature could be made available for purchase by individuals, but which the state finds administratively more convenient to provide for all the nation (such as roads).

rational-expectations hypothesis A theory that suggests that individuals combine all available information to form judgements about the future.

real values Measurement of economic values after adjustments have been made for inflation.

recession A period of time during which the rate of growth of business activity is consistently less than its long-term trend. Most commonly identified as two consecutive negative quarters of GDP.

refinancing rate The interest rate at which the European Central Bank lends on a short-term basis to the euro banking sector.

registered social landlords A group of private organisations that manage nearly two million homes in the UK for tenants on lower incomes with support from the government. Examples of organisations on the register include charitable companies, housing associations and co-ownership societies.

rent controls A price ceiling on rents charged for private rented accommodation. First introduced in 1915, rent controls were designed to protect tenants from unscrupulous landlords. However, many politicians argued that rent controls distorted the housing market and they are no longer popular.

repo rate The interest rate at which the Bank of England lends on a short-term basis to the UK banking sector.

research and development policy Research and development (R&D) is undertaken by a wide range of institutions and organisations, although about 50 per cent of UK R&D is government funded. Historically, the UK has been strong in research but less effective in development.

resource allocation The assignment of resources to various uses. More specifically, it means determining what will be produced, how it will be produced, who will produce it, and for whom it will be produced.

resources Inputs used in the production of goods and services. Commonly separated into four categories: land, labour, capital and entrepreneur (see separate glossary entries for details). Also called factors of production.

retail price index (RPI) A statistical measure of the change in prices of a representative set of goods and services purchased by the average household in the UK.

right to buy The Housing Act 1980 gave council tenants in the UK the right to acquire their houses or flats at a discount of the market value and effectively allowed local authorities to transfer their housing stock to the private sector.

Royal Institute of British Architects (RIBA) The principal professional body in the United Kingdom concerned with architecture. Established in 1834, it currently has 27,000 members, with 6,000 registered overseas.

Royal Institution of Chartered Surveyors (RICS) The main UK professional body concerned with surveying in its various guises. It was founded in 1868 and now has over 85,000 members across its seven different divisions.

scarcity A reference to the fact that at any point in time there is a finite amount of resources, while people have an infinite amount of 'wants' for goods and services.

services Goods that do not have physical characteristics. Examples of services include the 'goods' provided by doctors, lawyers, dentists, educators, retailers, surveyors, wholesalers and welfare staff.

short run The time period in which a firm cannot alter its current size of plant.

sick building syndrome Defined by the World Health Organisation as a syndrome of complaints covering general feelings of malaise, the onset of which is associated with the occupancy of certain modern buildings. The problem particularly affects the health of some office workers.

social costs The full cost that society bears when a resource-using action occurs. For example, the social cost of driving a car is equal to all the private costs plus any additional cost that society bears, such as air pollution and traffic congestion. (Some authors use this term to simply refer to external costs.)

social price The total price when all costs and benefits have been considered – that is, when private costs and benefits are added to the external costs and benefits.

stable equilibrium A situation in which if there is a shock that disturbs the existing relationship between the forces of supply and demand, there are self-corrective forces that automatically cause the disequilibrium to be remedied.

stagflation A term coined in the 1970s to represent the combined problem of high unemployment and rising prices.

strategic partnering An agreement between a contractor and client to work together on a series of construction projects in order to promote continuous improvement. Contrast with project partnering.

structural rigidities These are obstacles within markets that prevent a swift response to changing forces of supply and demand. They are more prevalent in some markets than others. One example of a structural rigidity is the way that commercial leases affect the dynamics of the property market.

subnormal profits A rate of return that is below the rates being earned elsewhere. More commonly these would be referred to as a loss.

substitute goods Two goods are considered substitutes if one can be used in place of the other. A change in the price of one, therefore, causes a shift in demand for the other. For example, if the price of butter goes up, the demand for margarine will rise; if the price of butter goes down, the demand for margarine will decrease.

suicide bid A term coined by journalists to describe the desperate measures that a construction firm may resort to during a recession to win business. In other words, the practice of submitting a tender that contains no profit element.

supernormal profits A rate of return that is greater than the rates being earned elsewhere. Also known as abnormal profits.

supply constraints These occur when it is either not possible to increase output – or, it is only possible at high costs.

supply curve The graphical representation of the supply schedule; a line showing the supply schedule, which slopes upwards (has a positive slope).

supply schedule A set of numbers showing prices and the quantity supplied at various prices for a specified period of time. In effect, therefore, the schedule shows the rate of planned production at each price.

supply-side economics Government policies designed to create incentives for firms and individuals to increase productivity. Supply-side economics is concerned with the level of aggregate supply.

sustainable construction An international agenda set for the construction industry designed to deliver products that perform better over their full life cycle and work in a more economically efficient and socially responsible way. For other definitions, see Table 15.4 (page 279).

sustainable development In general terms, developments that balance social, environmental and economic concerns. The sustainable development agenda recognises that decisions made today can have serious implications for future generations. For other definitions see Table 15.1 (page 273), which includes the most widely quoted definition from the World Commission on Environment and Development.

sustainable investment rule A fiscal rule stating that public sector net debt as a proportion of GDP should be held over the economic cycle at a stable level – that is, below 40 per cent of GDP.

tax bracket A band of income to which a specific and unique marginal tax rate is applied.

tax burden The incidence of tax within society as a whole – and the contribution in tax of different sections of society.

theory of the firm A theory of how suppliers of any product or service make choices in the face of changing constraints.

third party Persons external to negotiations and activities between buyers and sellers. If person A buys a car with no brakes and then runs person B over, person B is a third party to the deal struck between person A and the seller of the car, and person B's suffering is a negative externality.

tightly coupled systems An industry typified by close relationships and high levels of co-ordination between producers. For example, manufacturing displays many tightly coupled systems. Contrast with loosely coupled systems.

total costs All the costs of a firm combined, such as rent, payments to workers, interest on borrowed money, rates, material costs, etc.

total expenditure The total monetary value of all the final goods and services bought in an economy during a year.

total income The total amount earned by the nation's resources (factors). National income, therefore, includes wages, rent, interest payments and profits that are received, respectively, by workers, landowners, capital owners and entrepreneurs.

total output The total value of all the final goods and services produced in the economy during the year.

total revenue The price per unit times the total quantity sold.

trade-off A term relating to opportunity cost. In order to get a desired economic good, it is necessary to trade off some other desired economic good whenever we are in a world of scarcity. A trade-off involves the sacrifice that must be made in order to obtain something.

transaction costs All the costs associated with exchanging, such as the informational costs of finding out price and quality, service record and durability, and the cost of enforcing the contract.

transfer payments Money payments made by governments to individuals for which no services or goods are concurrently rendered. Examples are social security payments and student grants.

transition economies Refers to the economies of the 25 countries in Eastern Europe and the former Soviet Union that are moving from a centrally planned to a market system. The term, however, can be used less strictly to refer to any economy changing its system of resource allocation.

travel cost method A technique used to identify the price of an externality. The central premise involves estimating how much people are willing to pay to travel in order to experience an environmental asset.

unit elastic demand If a percentage change in price leads to an identical percentage change in demand, then the product has an elasticity of unity – that is, an elasticity of one.

unit elastic supply A property of the supply curve where the quantity supplied changes exactly in proportion to changes in price. In this hypothetical situation, revenue is invariant to price changes.

u-value A traditional measurement of heat loss. As the u-value coefficient moves nearer to zero the insulation quality of the material being measured improves. In other words, as the u-value of a wall lowers there is less heat lost through the fabric of the building.

value management A term used to embrace all activities and techniques used to establish where costs are being incurred and where value could be added. In effect, value management is a process that should achieve better value for money for the client. It is commonly abbreviated to VM.

variable costs Costs that vary with the rate of production. Examples include wages paid to workers, the costs of materials, and so on.

vendor A seller – especially one who sells land and property.

vertical equity The concept behind measures to achieve social justice or fairness by providing benefits targeted at people with specific needs. The idea that underpins policies such as means-tested benefits and taxing the rich more heavily than the poor.

voluntary instruments This term has emerged as the green agenda has developed, and it refers to the various assessment schemes for buildings and management. Ironically, the common thread is that these schemes always have to be 'paid for' as legislation has not been introduced to make them compulsory.

wage-price spiral An inflationary process in which incomes and retail prices follow each other in an upward direction (see Figure 14.5).

wages councils Bodies set up by the government to determine the pay of those in occupations that are traditionally poorly rewarded. At their peak these bodies regulated wages for more than one million employees in trades such as retailing and agriculture. Wages councils were abolished in the UK in 1993.

x-inefficiency Describes the organisational slack that results in higher unit costs than would occur within a more competitive marketplace.

References

Ball, M., Farshchi, M. and Grilli, M. (2000) 'Competition and the Persistence of Profits in the UK Construction Industry', *Construction Management and Economics*, 18: 733–45

Balls, E. and O'Donnell, G. (eds) (2002) *Reforming Britain's Economic and Financial Policy: Towards greater economic stability*, Palgrave: Basingstoke

Banwell, H. (1964) *The Placing and Management of Contracts for Building and Civil Engineering Works*, HMSO: London

Barker, K. (2003) *Review of Housing Supply: securing our future needs – Interim report*, HM Treasury: London

Baumol, W.J. (1982) 'Contestable markets: an uprising in the theory of industrial structure', *American Economic Review*, 72: 1–15

Bell, M., Wingfield, J., Miles-Shenton, D. and Seavers, J. (2010) *Low Carbon Housing: Lessons from Elm Tree Mews*, Joseph Rowntree Foundation: York

Bennett, J. and Jayes, S. (1995) *Trusting the Team*, University of Reading: Centre for Strategic Studies in Construction

Blanchflower, D. (2008) Inflation, expectations and monetary policy, speech delivered at the Royal Society, 29 April. Available from Bank of England: London

BIS (2012) *Skanska boss appointed co-chair of Green Construction Board*, Department for Business Innovation and Skills, press release, 2 July

Blismas, N., Pasquire, C. and Gibb, A. (2006) 'Benefit evaluation for off-site production in construction', *Construction Management and Economics* 24: 121–30

Bon, R. (1992) 'The future of international construction: secular patterns of growth and decline', *Habitat International*, 16(3): 119–28

Bon, R. and Crosthwaite, D. (2000) *The Future of International Construction*, Thomas Telford: London

Boulding, K.E. (1966) 'The economics of the coming spaceship earth', in H. Jarrett (ed.) *Environmental Quality in a Growing Economy*, Johns Hopkins University Press: Baltimore

Brochner, J. (2011) 'Developing construction economics as industry economics', in G. de Valence (ed.) *Modern Construction Economics: Theory and Application*, Spon Press: London and New York

Brockmann, C. (2011) 'Collusion and corruption in the construction sector', in G. de Valence (ed.) *Modern Construction Economics: Theory and Application*, Spon Press: London and New York

Broughton, T. (2001) 'In Person: Brian Wilson Interview', *Building*, 14 December

Cabinet Office (2011) *Government Construction Strategy*. Cabinet Office: London

Cadman, D. (2000) The vicious circle of blame, cited in Keeping, M., *What about demand? Do investors want 'sustainable buildings'?* The RICS Research Foundation: London

Carassus, J. (ed.) (2004) *The construction sector system approach: an international framework*, CIB paper 293, International Council for Research and Innovation in Building and Construction: Rotterdam

Carbon Trust (2011) *Driving Green Growth: Annual Report 2010/11*. The Carbon Trust: London

CEEQUAL (2012) *Awards – Case Studies*. Available from www.ceequal.com

Chiang, Y., Tang, B. and Wang, F.K.W. (2008) 'Volume building as competitive strategy', *Construction Management and Economics*, 26: 161–76

Choy, C.F. (2011) Revisiting the 'Bon Curve', *Construction Management and Economics*, 29: 695-712

CIDB (2012) *Strategic Plan of the Construction Industry Development Board 2011/12 to 2015/16*, Construction Industry Development Board, Department of Public Works: Johannesburg

Constructing Excellence (2010) *Lean study tour of the Japanese Construction industry*, promotional leaflet, Constructing Excellence: London

Cooke, A.J. (1996) *Economics and Construction*, Macmillan: London

Costanza, R. et al. (1997) 'The value of the world's ecosystem services and natural capital', *Nature*, 387: 253–60

Cox, R. and Goodman, C. (1956) 'Marketing of house building materials', *Journal of Marketing*, 21: 36–61

Daly, H.E. (1999) 'Uneconomic growth and the built environment: In theory and in fact', in C.K. Kibert (ed.) *Reshaping the Built Environment: Ecology, Ethics and Economics*, Island Press: Washington

DCLG (2007) *The Future for Building Control*, Department for Communities and Local Government: London

DCLG (2008) *Greener Homes for the Future*, Department for Communities and Local Government: London

DCLG (2012) *Code for sustainable homes and energy performance of buildings*, statistical release, February, Department for Communities and Local Government: London

De Valence, G. (2006) Guest Editorial, *Construction Management and Economics*, 24: 661–8

De Valence, G (ed.) (2011) *Modern Construction Economics: Theory and Application*. Spon Press: London and New York

Defra (2005) *UK Strategy for Sustainable Development*, Department for Environment, Food and Rural Affairs: London

Defra (2007) *An Introductory Guide to Valuing Ecosystem Services*, Department for Environment, Food and Rural Affairs: London

Defra (2011) *Environmental Statistics – Key Facts*. Available from www.defra.gov.uk

Desai, P. and Riddlestone, S. (2002) *Bioregional Solutions: For Living on one Planet*, Green Books: Devon

DETR (2000) *Building a Better Quality of Life: A Strategy for More Sustainable Construction*, Department of the Environment, Transport and the Regions: London

Dey-Chowdhury, S. (2007) 'House price indices of the UK', *Economic and Labour Market Review*, 1: 54–8

DfT (2007) *Adding Capacity At Heathrow Airport consultation document*, Department for Transport: London

Drew, D. and Skitmore, P. (1997) 'The effect of contract type and size on competitiveness in bidding', *Construction Management and Economics*, 15: 469–89

Du Plessis, C. (2002) *Agenda 21 for Sustainable Construction in Developing Countries: A discussion document*, report for CSIR, CIB and UNEP-IETC, Pretoria. Available at www.cibworld.nl

Du Plessis, C. (2007) 'A strategic framework for sustainable construction in developing countries', *Construction Management and Economics*, 25(1): 67–76

Du Plessis, C. and Cole, R.J. (2011) 'Motivating change: shifting the paradigm', *Building Research and Information*, 39(5): 436–49

Dubios, A. and Gadde, L. (2002), 'The construction industry as a loosely coupled system: implications for productivity and innovation', *Construction Management and Economics*, 20: 621–31

Dulaimi, M.S., Ling, F.Y.Y. and Ofori, G. (2001) *Building a World Class Construction Industry: Motivators and Enablers*, Singapore University Press: Singapore

EBRD (2011) *Transition Report – Crisis and Transition: the people's perspective*, European Bank for Reconstruction and Development: London

Economist, The (2002) 'Gordon Brown's Manifesto', *The Economist*, 20 April

EEA (2008) *Effectiveness of Environmental Taxes and Charges for Managing Sand, Gravel and Rock Extraction in Selected EU Countries*, European Environment Agency: Copenhagen

Egan, J. (1998) *Rethinking Construction: The Report of the Construction Task Force*, DETR: London

Egan, J. (2002) *Rethinking Construction: Accelerating Change*, Strategic Forum Secretariat: London

Elkins, P. (2008) 'Path of least resistance', *The Guardian*, 13 February

Elliott, J. (1999) *An Introduction to Sustainable Development*, Routledge: London and New York

Ellis, R.C.T., Wood, G.D. and Keel, D.A. (2005) 'Value management practices of leading UK cost consultants', *Construction Management and Economics*, 23: 483–93

Ends report (2012) *Green Deal delayed for commercial buildings*. Available from endsreport.com, 25 April 2012

EU (2001) *Competitiveness of the Construction Industry: An Agenda for Sustainable Construction in Europe*, European Commission: Brussels

EU (2007) *Accelerating the Development of the Sustainable Construction Market in Europe*, European Commission: Brussels

EU (2011) *Roadmap to a Resource Efficient Europe*, European Commission: Brussels

Evans, R., Haryott, R., Haste, N. and Jones, A. (1998) *The Long-term Costs of Owning and Using Buildings*, Royal Academy of Engineering: London

Fairclough, J. (2002) *Rethinking Construction Innovation and Research: A review of government R&D policies and practices*, DTLR: London

Fellner, W. (1949) *Competition Among the Few: Oligopoly and Similar Market Structures*, Alfred A. Knopf: New York

FIEC (2012) *Construction in Europe Key Figures*, Fédération de l'Industrie Européenne de la Construction: Brussels

Flyvbjerg, B. (2003) *Megaprojects and Risk: An Anatomy of Ambition*, Cambridge University Press: Cambridge

Forum for the future (2006) *The Mass Balance Movement: The Definitive Reference for Resource Flows within the UK Environmental Economy*. Biffaward: Nottinghamshire

Foster, V. and Briceno-Garmendia, C. (2009) *Africa's Infrastructure: A Time for Transformation*, World Bank: Washington DC

Friedman, M. (1962) *Capitalism and Freedom*, (with new preface in the 40th anniversary edition in 2002), University of Chicago Press: Chicago and London

Gann, D. (2000) *Building Innovation: Complex Constructs in a Changing World*, Thomas Telford: London

Gardiner, J. (2011) 'Over half of QSs have seen clients accept suicide bids', *Building*, 10 June

Gardiner, J. (2012) 'Where are the homes?', *Building*, 25 May

GFN (Global Footprint Network) (2012) *National Footprint Accounts' underlying methodology*, data and methods available from www.footprintnetwork.org

GHA (Good Homes Alliance) (2011) *Good Homes Alliance Monitoring Programme 2009–11: technical report. Results from Phase 1: post-construction performance of a sample of highly sustainable new homes*. GHA: London

Gidado, K.I. (1996) 'Project complexity: the focal point of construction production planning', *Construction Management and Economics*, 14: 213–25

Goodland, R., Daly, H. and El Serafy, S. (1992) *Population, Technology and Lifestyle*, Island Press: Washington DC

Gruneberg, S. and Ive, G. (2000) *The Economics of the Modern Construction Firm*, Macmillan: London

Hardoon, D. and Heinrich, F. (2011) *Bribery Pays*, Transparency International: Berlin

Harris, F. and McCaffer, R. (1989) *Modern Construction Management*, BSP Professional Books: Oxford

Hedges, A. and Wilson, S. (1987) *The Office Environment Survey: A Study of Building Sickness*, Building Use Studies: London

Heerwagen, J. (2000) 'Green buildings, organizational success and occupant productivity', *Building Research and Information*, 28: 353–67

Hellowell, M. and Pollock, A.M. (2007) *Private Finance, Public Deficits: A report on the cost of PFI and its impact on health services in England*, Centre for International Public Health Policy, University of Edinburgh: Scotland

Hillebrandt, P. (1974) *Economic Theory and the Construction Industry*, Macmillan: London

Hillebrandt, P. (1984) *Analysis of the British Construction Industry*, Macmillan: London

Hillebrandt, P. (1985) *Economic Theory and the Construction Industry*, second edition, Macmillan: London

Hillebrandt, P. (2000) *Economic Theory and the Construction Industry*, third edition, Macmillan: London

HM Government (2008) *Strategy for Sustainable Construction*, BERR: London

HM Government (2012) *The State of the Estate*, Cabinet Office: London

HM Treasury (1997) *Statement of Intent on Environmental Taxation*, HMSO: London

HM Treasury (2011) *PFI signed project list – November 2011*, Excel spreadsheet available from www.hm-treasury.gov.uk

HM Treasury (2012) *Forecasts for the UK economy: A comparison of independent forecasts*, May, No. 301. Available from www.hm-treasury.gov.uk/forecasts

ICLEI (1996) *The Local Agenda 21 Planning Guide*, International Council for Local Environmental Initiatives

IFS (2012) *Fuel for Thought: The What, Why and How of Motoring Taxation*. Institute for Fiscal Studies, RAC Foundation: London

IGT (2010) *Low Carbon Construction*. Innovation and Growth Team, HM Government: London

Insolvency Service (2012) *Insolvency Statistics Archive*, available from www.insolvency.gov.uk

International Council for Research and Innovation in Building and Construction (CIB) (2012) *Appointment of Chrisna du Plessis as sustainable construction priority theme coordinator*, CIB press release, 10 May

International Monetary Fund (2007) *World Economic Outlook: Globalization and Inequality*, IMF: Washington DC

IPCC (Intergovernmental Panel on Climate Change) (2007) *Climate Change 2007: The Fourth IPCC Assessment Report*, Cambridge University Press: New York and Cambridge

Ive, G. and Gruneberg, S. (2000) *The Economics of the Modern Construction Sector*, Macmillan: London

Jensen, J.O. and Gram-Hansenn, K. (2008) 'Ecological modernization of sustainable buildings: a Danish perspective', *Building Research and Information*, 36: 146–58

Jessop, S. (2002) 'Environmental Supply Chain Management: A Contractors Perspective', in CIEF meeting notes *Environment and the Supply Chain – the weakest link?*, 16 April, London

Jewell, C.A., Flanagan, R. and Cattell, K. (2005) The effects of the informal sector on construction. In *Broadening Perspectives – Proceedings of the Construction Research Congress*, San Diego, USA

Kibert, C. (1994) 'Principles of sustainable construction', in *Proceedings of the First International Conference on Sustainable Construction*, Tampa, Florida, US, 6–9 November: 1–9

Kibert, C. (ed.) (1999) *Reshaping the Built Environment: Ecology, Ethics and Economics*, Island Press: Washington DC

Kibert, C. (2007) 'The next generation of sustainable construction', *Building Research and Information*, 35: 595–601

Latham, M. (1994) *Constructing the Team*, HMSO: London

Layard, R. (2011) *Happiness: Lessons from a New Science*, second edition, Penguin Books: London and New York

Leibenstein, H. (1973) 'Competition and X-Inefficiency: Reply', *Journal of Political Economy*, 81: 765–77

Lipsey, R. and Crystal, A. (1995) *An Introduction to Positive Economics*, Oxford University Press: Oxford

Lovins, A., Lovins, H. and Hawken, P. (1999) 'A road map for natural capitalism', *Harvard Business Review*, May-June: 145–58

Mactavish, A., Quartermaine, R. and Woollam, C. (2012) 'Energy Ratings Rented Property', *Building*, 30 March, 50–53

Manseau, A. and Seaden, G. (eds) (2001) *Innovation in Construction: An International Review of Public Policies*, Spon Press: London and New York. A paper presenting the framework and conclusions preceded the comprehensive detailed report in *Building Research and Information*, 2001, 29: 182–96

Marshall, A. (1920) *Principles of Economics*, eighth edition, Macmillan: London

Marx, K. (1844) *Economic and Philosophical Manuscripts*, Pelican Marx Library: London

Matar, M.M., Georgy, M.E. and Ibrahim, M.E. (2008) 'Sustainable construction management: introduction of the operational context space (OCS)', *Construction Management and Economics*, 26: 261–75

McMeeken, R. (2008) 'The Squeeze', *Building*, 15 February: 41–3

Meadows, D.H., Meadows, D.L., and Randers, J. (1992) *Beyond the Limits – Global Collapse or a Sustainable Future*, Earthscan Publications: London

Meadows, D.H., Meadows, D.L., Randers, J., and Behrens, W. (1972) *The Limits to Growth – A Report to the Club of Rome*, Universe Books: New York`

Meadows, D.H., Richardson, J., and Bruckmann, G. (1982) *Groping in the Dark – The First Decade of Global Modelling*, John Wiley: New York

Mehmet, O. and Yorucu, V. (2008) 'Explosive construction in a micro-state: environmental limit and the Bon curve: evidence from North Cyprus', *Construction Management and Economics*, 26: 79–88

Meikle, J. (2011) *Note on: Informal Construction*, International Comparison Program (Fifth Technical Advisory Group Meeting April 18–19) World Bank: Washington D.C.

Milford, R. (2009) *Greenhouse Gas Emission Baselines and Reduction Potentials from Buildings in South Africa*. United Nations Environment Programme: Paris

Miller, J.D. (2011) *Infrastructure 2011: A Strategic Priority*. Urban Land Institute: Washington D.C.

MPC (Monetary Policy Committee) (1999) 'The transmission mechanism of monetary policy', *Bank of England Quarterly Bulletin*, Spring, Bank of England: London

Murray, M. and Langford, D. (eds.) (2003) *Construction Reports 1944–1998*, Blackwell Science: Oxford

NAO (2001) *Modernising Construction*, National Audit Office, TSO: London

NAO (2005) *Using Modern Methods of Construction to Build Homes More Quickly and Efficiently*, National Audit Office, TSO: London

NAO (2007) *Homebuilding: Measuring Construction Performance*, National Audit Office, TSO: London

NAO (2007a) *The Carbon Trust: Accelerating the Move to a Low Carbon Economy*, National Audit Office, TSO: London

NAO (2008) *Making Changes in Operational PFI Projects*, National Audit Office, TSO: London

NBER (2010) *Business Cycle Dating Committee*, 20 September, National Bureau of Economic Research, available from www.nber.org/cycles.

NHBC (2010) *Zero Carbon Compendium: Who's Doing What in Housing Worldwide*, National House Building Council: Amersham

OECD (2008) *African Economic Outlook*, OECD: Paris

Ofori, G. (1994) 'Establishing Construction Economics as an Academic Discipline', *Construction Management and Economics*, 12: 295–306

OFT (2009) *Bid Rigging in the Construction Industry in England*, Office of Fair Trading: London

ONS (2006) *UK Input-output analyses*, Office for National Statistics: London

ONS (2011a) *Construction Statistics Annual*, Office for National Statistics: London

ONS (2011b) *Construction Output Time Series Data*, available in Excel format from www.ons.gov.uk

ONS (2011c) *UK National Accounts: The Blue Book*, Office for National Statistics: London

Ozorhorn, B., Abbott, C., Aouad, G. and Powell, P. (2010) *Innovation in Construction: A Project Life Cycle Approach*, Salford Centre for Research and Innovation in the built environment (SCRI): University of Salford

Pan, W., Gibb, A.G.F. and Dainty, A.R.J. (2008) 'Leading UK housebuilders' utilization of offsite construction methods', *Building Research and Information*, 36(1): 56–67

Parkinson, M., Ball, M., Blake, N. and Key, T. (2009) *The Credit Crunch and Regeneration: Impact and Implications*, Department for Communities and Local Government: London

Pearce, D. (2003) *The Social and Economic Value of Construction: The Construction Industry's Contribution to Sustainable Development*, nCRISP: London

Peterson, G. (2010) 'Growth of ecosystem services concept', *Resilience Science* 21 January, available from http://rs.resalliance.org

Pezzey, J. (1989) *Definitions of Sustainability*, UK CEED: Peterborough

Power, A. (2008) 'Does demolition or refurbishment of old and inefficient homes help to increase our environmental, social and economic viability?', *Energy Policy*, 36: 4487–4501

Pratten, C.F. (1971) *Economies of Scale in Manufacturing Industry*, Cambridge University Press: Cambridge

Price, R. (2010) *Government Economic Service Review of the Economics of Sustainability*. DEFRA: London

Raynsford, N. (2012) 'All talk and no trousers', *Building*, 13 April: 22–3

Reeves, K. (2002) 'Construction business systems in Japan: general contractors and subcontractors', *Building Research and Information*, 30: 413–24

Ricardo, D. (1817) *On the Principles of Political Economy and Taxation*, John Murray: London

Richardson, S. (2012) 'Once upon a time in the east', *Building*, 23 March: 36–40

Ruddock, L. and Lopes, J. (2006) 'The construction sector and economic development: the Bon curve', *Construction Management and Economics* 24: 717–23

SafetyNet (2009) *Cost-benefit Analysis*. European Commission: Brussels

Samuelson, P.A. and Nordhaus, W.D. (2010) *Economics International Edition*, 19th edition, McGraw Hill: New York

Sayce, S., Ellison, L. and Parnell, P. (2007) 'Understanding investment drivers for UK sustainable property', *Building Research and Information*, 35: 629–43

Schot, J., Brand, E. and Fischer, K. (1997) *The Greening of Industry for a Sustainable Future: Building an International Research Agenda*, The Greening of Industry Network: The Netherlands

Shash, A.A. (1993) 'Factors considered in tendering decisions by top UK contractors', *Construction Management and Economics*, 11: 111–18

Shiers, D. (2000) 'Environmentally responsible buildings in the UK commercial property sector', *Property Management*, 18: 352–65

Smith, R.A., Kersey, J.R. and Griffiths, P.J. (2003) *The Construction Industry Mass Balance: resource Use, Wastes and Emissions*, Viridis Report VR4, Viridis and CIRIA: London

Solow, R. (1974) 'The economics of resources or the resources of economics', *American Economic Review*, 64: 1–14

Stern, N. (2007) *The Economics of Climate Change: The Stern Review*, Cambridge University Press: Cambridge. A web edition was released in December 2006, it is available as an independent review from www.hm-treasury.gov.uk

Sweet, R. (2002) 'The Long March', *Construction Manager*, April: 17–19

Taipale, K. (2010) *Buildings and construction as tools for promoting more sustainable patterns of consumption and production*. Sustainable Development Innovation Briefs, Issue 9. United Nations Department of economic and social affairs: New York

Taipale, K. (2012) 'From light green to sustainable buildings', in *State of the World 2012: Moving Toward Sustainable Prosperity*, Worldwatch Institute: Washington, DC

The City UK (2012) *Response to HM Treasury's Call for Evidence on PFI Reform*, consultation letter, 10 February, available from www.thecityuk.com

Transparency International (2010) *Transparency International Annual Report*, available from www.transparency.org

Turin, D.A. (1966) Building as a process, proceedings of the Bartlett Society: London. *Reprinted in Building Research and Information*, 2003, 31: 180–7

Turin, D.A. (ed.) (1975) *Aspects of the Economics of Construction*, George Godwin: London

UKGC (2009) *Construction in the UK Economy: The Benefits of Investment*, UK Contractors Group, LEK Consulting: London

UN (2012) *Resilient People, Resilient Planet: A Future Worth Choosing*. Secretary-General's High-level Panel on Global Sustainability, United Nations: New York.

UNEP (2012) *Inclusive Wealth Report 2012: Measuring Progress toward Sustainability*, United Nations Environment Programme, Cambridge University Press: Cambridge

UNEP (2012a) *Sustainable Buildings and Climate Initiative*, UNEP-SBCI brochure, United Nations Environment Programme: Paris

Urge-Vorsatz, L.D., Harvey, D., Sevastianos, M. and Levine, M.D. (2007) 'Mitigating CO_2 emissions from energy use in the world's buildings', *Building Research and Information*, 35: 379–98

Vickers, J. (2011) *Independent Commission on Banking: final report*, 12 September, available from http://bankingcommission.independent.gov.uk

WCED (1987) *Our Common Future*, World Commission on Environment and Development, Oxford University Press: Oxford

Weizsäcker, E. von, Lovins, A. and Lovins, H. (1998) *Factor Four: Doubling Wealth, Halving Resource Use*, Earthscan: London

Weizsäcker, E. von, Hargroves, K., Smith, M.H. and Stasinopoulos, P. (2009) *Factor Five: Transforming the Global Economy through 80% Improvements in Resource Productivity*, Earthscan: London

WGBC (2009) *Six Continents One Mission: How Green Building is Shaping the Global Shift to a Low Carbon Economy*. World Green Building Council: Toronto Secretariat, available at www.worldgbc.org

WGBC (2012) *World Green Building Council Directory*. World Green Building Council: Toronto Secretariat, available at http:// www.worldgbc.org

Whittaker, M. and Savage, L. (2011) *Missing Out: Why Ordinary Workers are Experiencing Growth without Gain*. The Resolution Foundation Commission on Living Standards: London

Winch, G. (1987) 'The construction firm and the construction process: the allocation of resources to the construction project', in P. Lansley and P. Harlow (eds), *Managing Construction Worldwide, Volume 2, Productivity and Human Factors*, E & FN Spon: London

Wolstenholme, A. (2009) *Never Waste a Good Crisis: A Review of Progress since Rethinking Construction and Thoughts for our Future*. Construction Excellence: London

World Bank (2012) *World Development Indicators*, World Bank: Washington DC

Wright, E. (2010) 'Fast build nation', *Building*, 26 March: 32–3

WWF (2012) *Living Planet Report*, World Wide Fund for Nature International: Switzerland.

Yang, W. and Kohler, N. (2008) 'Simulation of the evolution of the Chinese building and infrastructure stock', *Building Research and Information*, 36: 1–19

Index

accounting profits *see* profits

adaptive-expectations hypothesis 257–8, 297

aggregate demand 228–9, 297

aggregate supply 230–2, 297
    aggregate supply curve 230
    perfectly elastic 231–2
    perfectly inelastic 231

aggregates levy 169

allocative efficiency *see* efficiency

asymmetric information 166, 297
    government intervention 171–2

balance of payments 205, 297

Bank of England
    base rate 253, 297
    Inflation Report 22, 202, 208
    monetary policy committee 208, 253, 308
    role in monetary policy 208, 253
    website 202

banks
    role of central bank 208, 298
    see also Bank of England

barriers to entry 134, 136, 297

bidding for work
    bidding for contracts 140–1
    contractors bid 111
    cover pricing 132, 300
    suicide bid 135, 315
    OFT enquiry into bid rigging 132–3

BIS *see* Department for Business Innovation and Skills (BIS)

Blue Book *see* UK National Accounts

Bon curve 222, 265–9

booms 205, 211–13, 268, 297

Brundtland report 272–3

Building Cost and Information Service (BCIS) 248, 250, 297

building information modelling (BIM) 121, 287, 288–9, 298

building industry *see* construction industry

building regulations 172, 176, 290, 298

Building Research Establishment (BRE) 144, 298

Building Research Establishment Environmental Assessment Method (BREEAM) 147

business fluctuations 211–13, 214, 298
    booms 205, 211–13, 268, 297
    recessions 203, 211–13, 220, 313

capital 9–10, 298
    see also factors of production

CarbonBuzz 144

central bank 208, 253, 298

centrally planned (model) *see* command economy

ceteris paribus 51, 66, 298

China
    construction industry 233–4
    ecological footprint 240
    macroeconomic statistics 220–1
    transition towards market system 37

circle of blame 284

circular flow model 15, 224–5, 299
    leakages 224–5, 306
    injections 224–5, 305

Civil Engineering Environmental Quality Assessment and Award Scheme (CEEQUAL) 151

climate change levy 169

cobweb theorem 261–2, 299

code for sustainable homes 66, 172

collusion 130–3

command economy 34–6